Beam Weapons

Roots of Reagan's 'Star Wars'

Jeff Hecht

Laser Light Press
Auburndale, MA, USA

Library of Congress Cataloging in Publication Data

Hecht, Jeff
Beam weapons.

Bibliography: p. Includes Index.
1. Directed energy weapons 1. Title
UG486.5.H43 1984 355.8'2595 83-24713

Laser Light Press
525 Auburn St., Auburndale, MA 02466 USA

Previously published as *Beam Weapons: The Next Arms Race* in 1984 by Plenum Press. 2000, 2014 by iUniverse

Includes Epilogue written in 2015

Cover Photo courtesy of Boeing

Preface to Laser Light Press edition

Originally published in 1984 by Plenum Press, *Beam Weapons* was the first book to examine the development of directed-energy weapons and plans for their use in missile defense. After three decades, it offers an important historical perspective on the crucial role that beam weapons played in the birth of the Reagan Administration's massive effort to develop advanced technology for nuclear defense. For that reason the only changes I have made in material from the original edition are to correct typos and to adjust the book for electronic publishing. Odd formatting, including blank spaces at the bottom of pages before illustrations, are artifacts of the republishing process and my lack of page-layout skills.

Reagan had not named his program at the time this book went to press. Critics dubbed it "Star Wars" because they thought it was more science fiction than real technology. Later, the Reagan Administration officially named it the "Strategic Defense Initiative," but most people except defense companies bidding for contracts still called it "Star Wars." SDI was not limited to beam weapons, but directed energy was a very high-profile part of the program, along with space technology.

I have added a brief epilogue to bring readers up to date on both strategic defense and directed energy. Lasers and optical technology have come a long way, but we are still far from a fleet of orbiting battle stations able to stop a massive nuclear missile attack. Military priorities have changed since the breakup of the Soviet Union, and the technological challenges of deploying megawatt-class directed-energy weapons in space proved overwhelming. Instead, a new class of solid-state lasers have emerged for possible short-distance use on land, at sea and in the air against weapons launched by insurgents. It's a modern technology for use against modern threats, and you can watch it on YouTube and read about it on the Internet. The epilogue also includes references to some books and articles on recent developments.

For an overview of laser weapon development from 1959 to 2009, see my article "A Half Century of Laser Weapons," in the February 2009 *Optics & Photonics News*. I hope to someday write a longer history.

Jeff Hecht Auburndale, MA
June 2015

Preface

I first wrote about laser weapons in 1976 in an article that was published in the October 1977 issue of *Analog Science Fiction.* The article was fact, but a science fiction magazine seemed an appropriate place for it then. Now that the President of the United States can talk about beam weapons in deadly serious tones on national television, it seems time for a more serious look.

This book is based on my analysis of information I've gathered in years of writing about laser technology. I have never had a security clearance, and though that limits the information I have been able to obtain, it means that I can say things that otherwise would be replaced by [deleted]. I have been watching the field long enough to play "fill in the blanks" with censored government documents and to make reasonable guesses of what many of the [deleted]s originally said.

There are no simple answers here, just complex technical and defense issues that deserve careful study and that are important to the future of our nation and the world. The defense issues raised in the latter half of the book may seem more immediate to the general reader, but they are inevitably related to the technological background laid in Chapters 4-9. Throughout the book, I have documented my sources as much as possible, while respecting the confidentiality of information given in private. I feel that documentation will add to the value of this book, particularly because a disturbing number of articles on beam weapons have contained serious inaccuracies.

I have tried to make this book as current as possible, but no book can keep up with the daily newspapers. As I write this, for example, Congress is in the midst of its haggling over the fiscal 1984 defense budget, and Pentagon analysts are working on a major study on prospects for missile defense technology. By the time this book is in print, the results of the budget battle and the Pentagon study will be known, but there is no way they can be included here. Thus, I have focused more on central defense and technology issues than on politics because the issues are not going to go away.

In correcting proofs, I have tried to update details, but was unable to make extensive changes. The emphasis of the beam weapons program has shifted some since I finished the manuscript, with battlefield applications fading while pop-up X-ray lasers missile defense have gained attention. The X-ray laser, in particular, deserves a few extra words of explanation.

My impression is that the X-ray laser weapon is a concept which has "just grown" around an apparent breakthrough in physics and the "wish lists" of military planners. The notion of a pop-up system, which would be launched into space only when needed, seems to have evolved from concerns of people like Edward Teller, who told me in a brief telephone interview that he had never advocated basing systems in orbit because they would be too vulnerable there. Having a system that would be stationed on the ground or in submarines for launch when needed evidently would get around those objections. Unfortunately, this concept has received very little critical examination in public. Although the X-ray laser program is shrouded in exceptionally heavy layers of government secrecy, a general idea of the concept has leaked out, and is discussed in Chapters 6 and 11. It may be possible to build a compact array of X-ray lasers that can be launched into space by a single missile, but I don't see how the equipment needed to track the targets and aim the lasers can be put into the same package. It might be possible to put some target identification and tracking gear on a separate satellite, but that would not avoid the fundamental problem, and would introduce the added complication of having to align the satellites precisely with respect to each other.

If someone does find a way around the formidable technical difficulties, a pop-up X-ray laser weapon system could be the realization of the nightmares of the harshest critics of beam weapons. Building one would almost certainly mean a repudiation of two decades of arms control, dating back to the 1963 Limited Nuclear Test Ban Treaty. It is hard to imagine any way the weapons could pop up from the ground or submarines to do their job in the 15 minutes it takes submarine-launched ballistic missiles to reach their targets-or even in the half-hour transit times of ground-launched intercontinental ballistic missiles. Yet putting them in place beforehand–but during a crisis–would be severely destabilizing, to say the least. Given those assumptions, you don't have to be a radical critic of the beam-weapon program to conclude that the most credible role for a pop-up X-ray laser system is as part of a first-strike system, which would be put in place before the attack to defend against any enemy missiles which survived a massive nuclear first strike.

I am not convinced that is the explicit intent of the X-ray laser program, though it would explain the prevalent secrecy. I think it more likely that the Department of Defense is only studying the possibilities of the X-ray laser. But as mentioned in Chapter 14, there are "first strikers" within the Pentagon, and they may be actively pushing for such capabilities.

Ultimately, this book is my own, particularly the mistakes my uneasy nerves tell me must be lurking somewhere. Yet it would not have been possible without help from many people. Linda Greenspan Regan of Plenum believed in it back before beam weapons hit the headlines, and kept me working on it. Barry Smernoff and Paul Nahin provided many stimulating phone conversations and a wealth of ideas. Ervin Nalos, Ray Schaefer, Alan Stein, Ron Waynant, and Gerry Yonas made helpful comments on portions of earlier drafts and shared their ideas with me. Col. Fred Holmes and Louis C. Marquet of the Department of Defense helped me get the facts straight on the department's unclassified activities. Others too numerous to mention have also taken the time to talk with me. Several people have given me copies of unpublished or hard-to-find reports; among them are Michael Callaham, Richard Garwin, Sen. Howell Heflin, Kosta Tsipis, and Sen. Malcolm Wallop. Col. Holmes and Jack Powers of the Department of Defense have been most helpful in getting me unclassified information and illustrations. Art Giordani turned my crude pencil sketches into clear and publishable drawings. And my wife Lois not only put up with the whole time-consuming business of writing a book but also offered her own helpful comments and did much of the legwork needed to get the pieces together.

Auburndale, Mass.

Dedication:
For Leah, who asked why there were bombs, and for my parents, who taught me to seek a better world where there wouldn't be any.

Contents

1.

An Introduction To Beam Weaponry

President Ronald Reagan thrust beam weapons into the spotlight on March 23, 1983, when he urged development of a system that "could intercept and destroy strategic ballistic missiles before they reached our own soil or that of our allies."[1] Although he did not specify the technology, White House aides explained after the speech that possibilities included lasers, high-energy particle beams, and microwaves.[2] Those three technologies are all beam weapons, known as "directed-energy" weapons in the halls of the Pentagon.

What has been called Reagan's *Star Wars* speech was not something new but was rather the first presidential-level recognition of an ongoing program. His speech did not stimulate this book–the first draft was sitting in the publisher's offices at the time. Through the end of fiscal 1982, the United States had spent over $2 billion trying to develop beam weapons. The fiscal 1983 budget–proposed by the Reagan Administration in January 1982 and finally approved by Congress in modified form at the end of that year–included over $400 million for beam weapons. Not all that money was earmarked for ballistic missile defense; the Pentagon is looking at many other potential uses for beam weapons. Nonetheless, efforts to develop beam weapons had escaped widespread attention until Reagan's speech. Controversy surrounded the Reagan proposal in the days after his speech, much as it has enveloped the whole idea of beam weapons wherever it has surfaced. Congressional Democrats and civilian critics of the Administration's defense policy described the proposal for ballistic missile defense as dangerous and unworkable. Soviet leader Yuri Andropov labeled it "irresponsible" and "insane."[3] Some of the names attached to the debate were new, but most of the ideas were familiar to those of us who have been watching the scene for years.

What's going on? A new technology is being drawn into the arms race. Military planners in the United States–and evidently in the Soviet Union as well–hope that beam weapons can provide dramatic new military capabilities. Defense against nuclear-armed ballistic missiles would have the most far-reaching impact, but there are other uses that could also be important–and probably easier to realize. Beam weapons have been suggested for use against satellites, aircraft, or missiles, in outer space, in the air, at sea, or on the ground. To military planners they offer the promise of a new generation of weaponry more capable than existing arms. To other observers they represent a boondoggle in the making–a particularly dangerous one that wouldn't work but could destabilize the international balance of power.

Despite Ronald Reagan's hawkish reputation, the wording of his *Star Wars* speech put him on a middle ground: the technology is promising, but by no means either a clearly achievable goal or one that is just around the corner. That's also the Pentagon's official position, and it is far from totally unreasonable. The massive investment in what is now the Pentagon's biggest ongoing technology development program[4] has paid some dividends in new technology but has also left many questions unanswered. Some answers may be hidden under government security wraps, but it is clear that many major issues remain unresolved. Pencil-and-paper studies have shown that beam weapons can work, but other studies have shown just the opposite. The Pentagon's lead research organization, the Defense Advanced Research Projects Agency, has even set up opposing teams of researchers, one dedicated to showing that beam weapons are practical, the other to shooting them down.[5] The Pentagon's hesitation is not without reason, but it has drawn sharp criticism from both sides: those who call for a crash development program and those who say it won't work.

Artist's conception of an orbiting laser weapon attacking another satellite. In real life the beam would be invisible in space--regardless of its wavelength-because there would be nothing to scatter light from the beam back to the observer. (Courtesy of Department of Defense.)

The issues go beyond beam weapon technology to the larger questions of the arms race and military posture. Here, too, there is vigorous debate. Political rhetoric notwithstanding, it is clear that beam weaponry is caught up in the overwhelming tide of the arms race between the United States and the Soviet Union. Certain uses of beam weapons, most notably for missile defense, would represent a major shift in defense posture. Major policy decisions must be made that will affect our lives and those of our children. We need to look at beam weapons in the context of military technology and policy and to try to grasp the whole complex picture well enough to make the right choices. To lay the ground for looking at that broad picture, the rest of this chapter will provide a brief overview of beam weaponry and its possible impacts, and how they are viewed by the military mind.

The Search for an Ideal Weapon System

Many military planners see beam weapons as an important step toward an "ideal" weapon system. The specifications for an ideal system are easy to write: it should hit its target with enough energy to disable it. That simple sentence carries many implications. The ideal weapon should *always* hit its target and should deliver enough energy to make *sure* that the target is disabled. It should disable any target in sight, including those that are large, hard to kill, and far away. It should deliver its energy almost instantaneously so the target can't escape. And, of course, it should be possible to build in a reasonably inexpensive form that the military can easily use.

Projectile weapons–guns, rockets, and missiles–don't exactly fill the bill. They have limited speed, range, and lethality; they can miss their targets, or even hit them. and fail to disable them. All projectile weapons face a fundamental problem in hitting their targets: the target can move while the weapon is on its way, and forces such as gravity and air resistance can deflect the weapon from the proper course.

Directed-energy technology could permit a quantum jump closer to the ideal weapon. A high-energy laser would shoot a beam of light, moving at 300,000 km (180,000 miles) per second. A particle-beam generator would shoot subatomic particles or atoms, accelerated almost to the speed of light and carrying large amounts of energy. In both cases the lethal bolt of energy would not hit the target instantaneously, and forces could conspire to deflect it from the target; however, in theory at least it seems more potent than guns, artillery, or missiles. Major questions remain to be answered before this promise can be realized. For particle beams the most immediate issue is whether they can be made to travel far enough and accurately enough through the air or through space to be militarily useful. This is the central question now being addressed by the Pentagon's particle-beam program. If demonstrations work out well, a large development effort will follow, but success is far from certain.

More time and more money have gone into laser technology than particle beams, but there are still doubts about its feasibility for weapons. Experimental laser weapons have shot down helicopters and small missiles in controlled tests. A laser able to emit around 2 million watts of power in a continuous beam has been built, and optimists argue that a 10-million-watt laser should soon be possible. Progress has been made in working out ways to track targets and to point and focus lasers at them.

Yet much remains to be done. Sticky problems remain in translating pointing, tracking, and focusing concepts from laboratory demonstrations into functional hardware. Potential ways to foil laser weaponry must be evaluated, to see if they might work well enough to make laser weapons useless. The lasers themselves, now bulky and temperamental devices, must be "militarized" so that the small army of PhDs now needed to operate them can be replaced by ordinary soldiers. And Pentagon brass has to be convinced that all the pieces can be put together into a lethal and cost-effective weapon system.

Elements of a Weapon System

The notion of a weapon *system* is a vital one in understanding beam weaponry. Neither a laser nor a particle-beam generator is a weapon in itself. True, anyone foolish enough to stand in the way of a powerful enough beam could be injured or killed. However, military targets are rarely that cooperative. The target must be found, the beam must be aimed at it, the energy must make its way to the target, some target-damaging interaction must take place, and the damage must be verified. The hardware that could do these tasks would turn a laser or particle-beam generator into a weapon system.

A soldier firing a rifle performs those tasks almost without thinking. He locates the target and aims the gun. Once he fires the gun, the explosion of gunpowder in the cartridge propels the bullet through the air toward the target. The soldier checks to see if he has knocked out the target; if he hasn't, he fires again.

Science fiction heroes generally use their futuristic weapons in about the same way. One example is Luke Skywalker in the film *Star Wars.* Even when piloting a spaceship equipped with a sophisticated computer system, he decides to trust "The Force" rather than the computer when firing the weapon that will demolish the evil empire's *Death Star.*

In real directed-energy weapons the job would be done by what's called a "fire-control" system, which gets its name because it controls the firing of the weapon. With minimal or perhaps no aid from a human operator, the fire- control system must identify a target, point the weapon at it, fire the weapon so the beam hits the target, track the target long enough to deliver a lethal dose of energy, and verify that the target has been "killed." The fire-control system must compensate for any atmospheric or magnetic field effects that might bend the beam away from the target. For missions such as missile defense, all this may have to be done in a fraction of a second, faster than a human operator could react.

Fire control is not enough, however. A weapon system must be "operational" in the strict military sense of the word-ready to work reliably whenever needed with little or no special attention, even after sitting unused for months or even years. This means a beam weapon should fire and hit its target on command, without the need for continual adjustments by highly trained technicians. What's more, the system must operate reliably, not in the benign environment of the laboratory, but in the much less friendly (and typically downright hostile) environment of combat, whether under enemy fire on the battlefield or in orbit in outer space.

It all adds up to a formidable problem in systems engineering, far more difficult than simply building the required components. In fact, contrary to what intuition might indicate, building a large laser or particle-beam generator is one of the easier parts of the job. This fact has sometimes led to considerable misunderstanding on the part of the general press and the public.

As far back as 1973, the Air Force shot down a winged drone in flight with a high-energy laser in what were then highly classified experiments conducted at Kirtland Air Force Base near Albuquerque, New Mexico. I saw a freshly declassified film of those experiments in April 1982, during the annual Conference on Lasers and Electro-Optics. Even though it was shown at 3:30 in the afternoon on Friday, the last day of the meeting-normally a horrible time for a technical presentation–it played to a full meeting room, including one Soviet visitor, V. A. Danilychev of the Lebedev Physics Institute in Moscow. The film was quite convincing; a bright spot appeared on the drone as the invisible infrared laser beam[6] heated it, and the drone caught fire and fell to the ground in well under a minute. A laser can bag its target on the military equivalent of a skeet shoot.

One demonstration does not make an effective, operational weapon system, however, as the Air Force found out to its embarrassment in mid-1981. After completing tests on the ground, overconfident officials announced at a January 1981 press conference that they planned airborne tests of the Airborne Laser Laboratory, a 400,000-W laser mounted in a KC-135 aircraft (the military version of a Boeing 707).[7] The first two airborne tests that June failed to destroy the targets, air-to-air missiles. Air Force officials insisted that they had learned valuable information from the tests and refused to classify them as failures. The press didn't buy that argument, however, and produced news stories with headlines like "Laser weapon flunks a test."[8] Public embarrassment convinced the Air Force to conduct further tests in secrecy.

The critical distinction between a large laser and an operational laser weapon system is sometimes missed in press reports, particularly those based on incomplete leaks of information. For example, published reports of Soviet plans to launch a high-energy laser in a spacecraft[9] or to install one on board a battleship[10] –if accurate–may well refer to large lasers rather than to weapon systems with militarily useful capabilities.

Weapon System Design

Weapon system design is a complex, multidimensional problem. The detailed design depends on the mission, operating environment, and the type of weapon. For beam weapons, there are several major design considerations:

- Ability to spot and track a target
- Beam power and quality
- Beam-direction capability (a term which covers pointing and focusing of beams of particles as well as laser beams
- Beam propagation through the air or space
- Beam-target interactions, including countermeasures taken by the enemy and any "counter-countermeasures" that may be taken to defeat the countermeasures
- Verification that the target has been destroyed or disabled
- Reliable operation of the weapon system

• Operating environment

Each of these factors has to be considered in making sure that a beam weapon system can do its job. Some, such as beam power and quality, depend largely on the design and nature of the beam source. Others, such as beam propagation and target interactions, depend largely on physical phenomena that affect how the beam interacts with the air and other objects. Designers must also consider mission requirements, such as the type of targets and operating environment.

The hardware that goes into the weapon system falls into some logical groupings. For its planning purposes the Department of Defense considers a laser weapon system to be made up of four subsystems:

• The laser itself (or laser subsystem)

• The target-acquisition system, which identifies and tracks the target

• The beam-control subsystem, which steers the beam, focuses it onto a vulnerable area on the target, and compensates for any propagation effects that might disperse or misdirect the beam

• The fire-control system, which controls system operation

Details of a particle-beam weapon system would differ, but the overall approach would be similar.

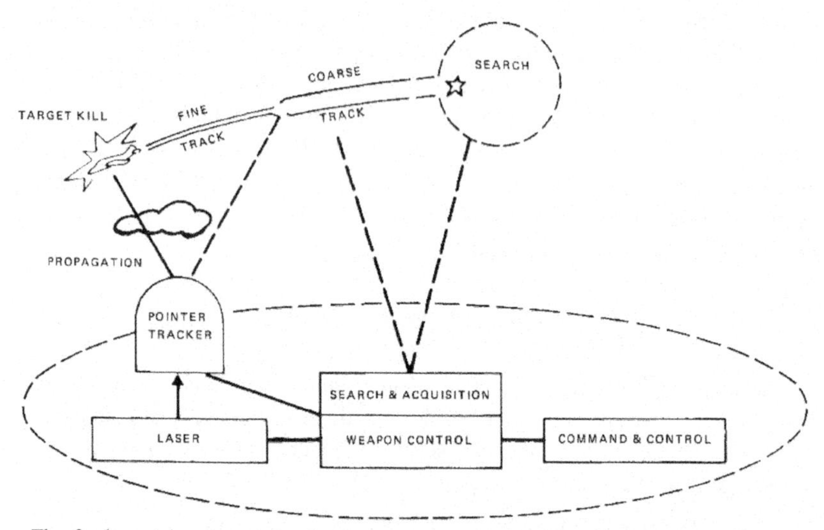

The fundamental components of a laser weapon system, according to Pentagon planners. This conceptual diagram comes from the Department of Defense's fact sheet on laser weapons. (Courtesy of Department of Defense.)

Roles That Weapons Play

Modem weapons are designed for specific functions. No single projectile weapon is effective against the full range of potential targets-from individual soldiers to heavily armored tanks, and from fast-moving missiles and fighter planes to more ponderous aircraft carriers. Nor is any single projectile weapon ideal for the full range of military environments. The same holds for beam weapons. A laser system able to blind the delicate sensors of a spy satellite would have no discernible effect on a nuclear warhead "hardened" to withstand much higher powers. Similarly, a laser designed for use on aircraft where light weight is critical would not be likely to meet the needs of a battleship, where resistance to high humidity is a must.

Missions for beam weapons fall under both of the traditional military categories, tactical and strategic. Tactical weapons are those intended for use in battles between armed forces on the ground, at sea, or in the air. They generally operate over short to moderate ranges measured in terms of kilometers or less. Strategic weapons are intended for use against other targets, such as arms factories or civilian populations of cities, or for defense of such strategic targets against enemy attack. Spy satellites, intercontinental ballistic missiles, and long-range bombers are considered strategic armaments; rifles, helicopters, short-range missiles, and most fighter aircraft are considered tactical. Nuclear weapons may be either strategic or tactical; generally high- yield bombs are intended for strategic use, while low-yield nuclear explosives such as the "neutron bomb" are intended for tactical use.

Three major tactical roles are envisioned for beam weapons. So far most effort has been on lasers because particle-beam and microwave technology are less mature and are expected to require much bulkier hardware. Bulk can be a serious problem because portability is essential for almost all tactical missions except those on large ships. The basic categories of beam weapons are:

Ground-Based Weapons. The goal being pursued by the Army is a moderate-or high-power laser that would be housed in a tank or other heavily armored vehicle. Pentagon spokesmen talk about using such weapons to defend "high-value targets," which in these days of tanks costing well over $1 million each can include many things, particularly sophisticated weapon systems and control centers. Ground-based lasers would operate over ranges of a few kilometers under extremely hostile conditions, including being subjected to dust, dirt, smoke, and enemy attack. They could be used against anything that moved on or flew above the battlefield, with the exception of heavily armored tanks. These lasers could destroy targets by causing mechanical damage, triggering explosions of fuel or munitions, or knocking out their sensors. They might be used to blind soldiers, temporarily or permanently, or to keep them "pinned down" under cover; but it will be a long time before there's enough laser firepower to incinerate individual enemy soldiers. Besides, bullets are cheaper.

Sea-Based Lasers. The Navy has considered putting laser weapons on ships to destroy attacking missiles-hopefully much faster and more effectively than conventional weapons. The PHALANX Gatling gun system now in use can fire 6000 shots/min, but the Navy is worried that it might not be able to blunt a cruise-missile attack. Laser weapons would have to be able to survive in a humid environment, but compactness would not be critical (aircraft carriers are big), so the Navy has also considered beams of charged particles. Operating distances would be several kilometers or less.

Air-Based Lasers. The Air Force is thinking of putting laser weapons in planes to defend against missile attack and against other aircraft. The biggest problem is bulk and weight; a laser weapon can't defend a plane unless it can fit inside one. The Air Force would like to put lasers in fast and maneuverable fighters, but it may have to settle for bombers, which are much larger and slower. Operating range would probably have to be over 10 km (6 miles) because aircraft can see targets farther away than forces on the ground or at sea.

Because each armed service has a separate mission for laser weapons, each has its own tactical laser program; overall coordination is through a central office at the Pentagon headed by Air Force Brigadier General Robert R. Rankine, Jr. This is the Department of Defense's Assistant for Directed Energy Weapons, which coordinates work on lasers, particle beams, and microwave weapon technology as well.

The strategic weapon program falls into somewhat different categories; it is mostly the responsibility of the Directed Energy Office headed by Louis C. Marquet at the Defense Advanced Research Projects Agency (DARPA).[11] There are two basic categories: comparatively near-term research in antisatellite weapons and long-term efforts to develop missile defense.

Antisatellite Lasers. The military role of satellites, particularly in surveillance (a polite word for electronic spying), arms-control verification, and communications, is making them potential military targets. The sensitive electronic eyes of optical and infrared spy satellites are vulnerable to an overload of light and hence comparatively easy targets for laser attack, but other satellites are also vulnerable. The Air Force is heavily involved and, as this book was being finished, was considering adding antisatellite lasers to its shopping list of new weapon systems.[12] The lasers might be on the ground, in aircraft, or in orbit, with each approach offering its own advantages and disadvantages. Beam weapons are also being considered to defend satellites against enemy attack.

Nuclear Defense Lasers. The most spectacular, and probably most significant, role proposed for lasers is to destroy enemy missiles and bombers before they can deliver their nuclear weapons. This could be done by orbiting laser battle stations, or by using mirror satellites to direct beams from powerful lasers on the ground. Certain types of particle beams have also been suggested for orbiting battle stations, but that technology seems farther off. Requirements for such weapons are very different from those for tactical weapons. They would have to be ready to fire on a moment's notice at targets a few thousand kilometers away after sitting quiet for long periods. Light weight would be critical for orbiting battle stations and mirrors.

Strategic Point Defense. Conceptually, a cluster of high-energy lasers or particle-beam weapons could defend a "hardened" military target against nuclear attack. The defended target probably would be a cluster of missile silos or a military base. As warheads converged on the target, the beam weapon would zap one after the other, disabling them or triggering explosions of nuclear warheads far enough from the target that they wouldn't damage it.

A Revolution in Defense Strategy

It's only appropriate that the obstacles to developing beam weapons are high because the stakes involved are very high. The science-fictional scenario of orbiting antimissile battle stations would cause nothing short of a revolution in defense strategy. For some two decades we have been living with an uneasy balance of nuclear terror called "mutual assured destruction" or "MAD." That balance is based on the knowledge that there is no effective defense against nuclear attack. If one side attacked, the other could launch a devastating counterattack-guaranteeing a nuclear holocaust. Under these ground rules a nuclear war cannot be won.

Opponents of beam weaponry warn that their most insidious danger is that they might make a nuclear war appear "winnable." That is, the side with a beam weapon system able to defend against nuclear attack might decide it could launch its own attack with impunity. Critics also warn of other dangerous scenarios in which beam weaponry could dangerously destabilize the balance of power even if the actual weapon system was ineffective. For example, one side might attack a weapon system under construction in space to make sure that it never became operational, thereby triggering an ultimate escalation to World War III.

This cartoon appeared in the *Arizona Republic* in April, 1982, during the week the Conference on Lasers and Electro-Optics was held in Phoenix. It was apparently in response to a General Accounting Office criticism of management of the Pentagon's laser weapon program, and appeared before the talk at the laser conference during which a recently resigned Pentagon official showed a film of a 1973 laser weapon test. (Reprinted with permission.)

Advocates of beam weapons make other points. One of the most telling is that such weapons could put the balance of power in the hands of the *defense* for the first time in a generation. That would end the need to threaten each other with nuclear holocaust in order to maintain peace. A few days after proposing to develop ballistic missile defense technology, President Reagan said the United States might eventually offer the technology it developed to the Soviets "to prove to them that there was no longer any need for keeping offensive missiles."[13] Advocates also hold that assuming beam weapons are technologically possible, they're also inevitable, and that the United States had better get them first, lest we be blackmailed by the Soviet Union. They also warn that new technology is making MAD obsolete.

Beam Weapons and Arms Control

Tactical beam weapons would merely be another step, albeit perhaps a significant one, in the continuing escalation of battlefield arms. But space-based beam weapons for use against ballistic missiles and antisatellite weapons both have direct implications for existing treaties. The Anti-Ballistic Missile (ABM) Treaty signed in 1972 appears to prohibit deployment of antimissile beam weapons and the use (although not the deployment) of antisatellite weapons against spy satellites. Development is permitted, however.

Treaties are subject to modification and interpretation, and they can also be broken. The Reagan Administration has not appeared comfortable with the ABM Treaty in particular; there have been serious discussions of building an antiballistic missile system to defend the MX missile,[14] a move that would require at least revision of the ABM Treaty.

Beam weaponry is a complex issue, and one that I feel deserves the in-depth treatment that's only possible in a book. The following chapters delve into the subject more deeply than is possible in this brief overview. I start with the history of the concept of directed-energy weapons both because it's fascinating in itself and because it carries some lessons to remember. Next I stop to clear the air of some common myths that have no relation to reality. From there I go into the technology involved, the military role for beam weapons, and their impact on military strategy and arms control. These chapters lay out the facts and point out the diverging opinions in the field. Readers whose main interests are in the impact of beam weapons rather than their technology may want to skip parts of Chapters 4-9, which describe the hardware being developed for such weapons, but they should at least skim through those chapters to recognize the very real uncertainties that lie behind the debate on the feasibility of beam weapons. I deliberately steer clear of some of the more esoteric arguments-anyone interested can explore the references cited. Likewise I try to avoid getting overwhelmed with the continual political infighting in Washington, which is still evolving rapidly enough to be the stuff of daily newspapers and the *Congressional Record,* not of a book.

The final chapter is something different, my personal opinions on beam weaponry. To a certain extent my own biases must pervade the book, but I have tried as best as I can to concentrate them in this one chapter. These opinions are based not on any supersecret information leaked expressly to me, but on years of watching the field and talking with people involved. I include them out of a feeling that it's my obligation to help the reader sort out-as an engineer would say-the signal from the noise.

References

1. "President's speech on military spending and a new defense," (transcript) *New York Times,* March 24, 1983, p. A20.
2. Benjamin Taylor, "Reagan defends arms budget, seeks new antimissile system," *Boston Globe,* March 24, 1983, p. 1.
3. Dusko Doder, "Andropov calls ABM proposal 'insane'," *Boston Globe,* March 27, 1983, p. 1.
4. Senate Appropriations Committee, *Hearings on Department of Defense Appropriations Fiscal Year 1983 Part 4,* p. 569; the statement was made by Robert Cooper, director of the Defense Advanced Research Projects Agency, during hearings on June 23, 1982.
5. *ibid,* p. 591; also in testimony by Robert Cooper on June 23, 1982.
6. The human eye cannot see infrared light, and the beam itself is invisible. You can see a glowing spot that has been heated by the beam, and at high enough intensities you can see dust particles in the air being heated, ignited, and/or vaporized. A visible laser beam can be seen but only if there's enough dust and aerosol particles in the air to scatter some light back to the observer.
7. This sort of public announcement is very unusual in what has otherwise been a highly classified program. The press conference was reported by Jim Bradshaw, "Laser tests at Kirtland a 'milestone'," *Albuquerque Journal,* pp. A-1.
8. For example, "Laser weapon flunks a test," *Laser Focus,* July 1981, p. 4; "Laser fails to destroy missile," *Aviation Week & Space Technology,* June 8, 1981, p. 63; "Air Force laser weapon flunks a test," *Boston Globe,* June 3, 1981.
9. "Washington Roundup," *Aviation Week & Space Technology,* October 5, 1981, p. 17.
10. Washington Roundup," *Aviation Week & Space Technology,* June 7, 1982, p. 13.
11. DARPA's official charter supports what the Pentagon considers "high-risk, high-payoff" research, meaning work that could result in major advances but has a high risk of failure. Less serious members of the military development community call it "blue sky but golden pot" research.
12. Craig Covault, "Space command seeks Asat laser," *Aviation Week & Space Technology,* March 12, 1983, pp. 18.
13. Benjamin Taylor, "US could offer ABM to Soviets, Reagan says," *Boston Globe,* March 30, 1983, p. 1.
14. Some of the Reagan Administration's plans are described in Stockholm International Peace Research Institute, *The Arms Race and Arms Control* (Oelgeschlager, Gunn & Hain, Cambridge, Massachusetts, 1982) pp. 87, 88.

2.
Turning Fiction into Fact: An Outline of History

Directed-energy technology is new, but the idea that a concentrated beam of light or energy could inflict damage is not. Ancient Greeks incorporated the concept into their legends and may even have put it to use. Within the last century science fiction writers picked up the idea and gradually shaped it to their own needs, although ironically some of the most modem visions are among the least realistic.

Military officials may have shared the same dreams of potent beam weaponry, but it wasn't until the last two or three decades that the technology matured enough for them to do more than dream. Some of the earliest attempts failed to bear fruit, apparently because the technology wasn't ready at the time. Now, however, President Reagan and Pentagon officials feel the time is coming. The United States had invested $2.5 billion in directed-energy weapons through the end of fiscal 1983, and the Reagan Administration had proposed spending over $500 million in 1984, even before President Reagan's controversial speech.

Even by Pentagon standards, that's a lot to spend in trying to turn science fiction into fact-$10 for every man, woman, and child in the United States. Progress has been slow over much of the program's history, reflecting major technological problems that had to be overcome, and many more issues remain unresolved. The program continues to be a controversial one, as some critics charge that foot-dragging military bureaucrats are obstructing progress, while others say that the whole effort is a waste of money. The controversy goes back a long way.

Ancient Greek Legends

In a very real sense the idea of directed-energy weapons may have started with the ancient Greeks. They armed Zeus, the king of their gods, with thunderbolts-nature's form of beam weapon. In addition, they were among the earliest people to realize the power of sunlight. Their mythology tells the story of Icarus, who was able to fly with wings made of feathers and wax, but who fell to his death in the sea when he flew too close to the sun and the sun's energy melted the wax. The idea of focusing the sun's energy to inflict damage also occurred to the Greeks. The playwright Aristophanes, who lived around 400 B.C., has one of the characters in his play *The Clouds* envision destroying a warrant for his arrest by focusing sunlight onto it.[1]

The most intriguing legend of how the ancient Greeks used the sun's energy is the story of Archimedes' defense of the city of Syracuse against a Roman siege. Supposedly, Archimedes and the residents of Syracuse used large polished mirrors to focus sunlight onto the sails of Roman ships in the harbor during the 212 B.C. siege, heating the fabric (or perhaps other inflammable material on the ships) to a point where it caught fire.

One way that the defenders of Syracuse could have aimed their mirrors at Roman ships in the harbor was by looking through a hole in the center of the mirror. Experiments have shown that enough soldiers with mirrors could have heated flammable materials to their ignition temperatures and thus set fire to the ships, but it is not certain that the incident actually happened. (Courtesy of Albert C. Claus, Loyola University, Chicago; reprinted from *Applied Optics,* October 1973.)

Ancient sources say that Archimedes devised an devised an hexagonal mirror, and at an appropriate distance from it set small quadrangular mirrors of the same type, which could be adjusted by metal plates and small hinges. This contrivance he set to catch the full rays of the sun at noon, both summer and winter, and eventually by reflection of the sun's rays in this, a fearsome fiery heat was kindled in the barges, and from the distance of an arrow's flight he reduced them to ashes.[2] Unfortunately, that was not enough to stop the siege, and the city finally fell to the Romans. Archimedes himself was killed when the Romans took the city, supposedly by a Roman soldier angered by the mathematician's refusal to take leave of a problem he was working on. That story is in keeping with others that portray Archimedes as a brilliant but colorful character-the type who today might have problems getting a security clearance.

Most modem observers have been skeptical of the account of Archimedes and the Roman ships. Typical of these is L. Sprague De Camp who in his fascinating book *The Ancient Engineers* mentions two problems cited by modem critics of laser weaponry: the inability to make big enough mirrors and difficulty in holding the beam steady on the target.[3]

The feasibility of Archimedes' legendary achievement was debated in 1973 and 1974 in a series of letters that appeared in *Applied Optics,* an otherwise sober scholarly journal published by the Optical Society of America.[4]

Albert C. Claus of the physics department at Loyola University in Chicago started it off by pointing out that Archimedes could have equipped a few hundred people with flat pieces of polished metal and had each person aim the sunlight at the target. A rebuttal quickly appeared, saying that the whole idea was impractical. That was followed in a few months by another letter pointing out that the burning mirror scheme had actually been demonstrated in 1747 by the French naturalist George Louis Leclerc, Comte de Buffon.[5] Buffon used an array of 168 mirrors, each about 8 x 10 inches, in the early spring, when the sun was not at its highest point in the sky. Using this array, often with fewer than the maximum number of mirrors, he was able to ignite planks as far as 46 m (150 feet) away.

Only a few weeks after the debate began in the pages of *Applied Optics,* there was another demonstration by a Greek scientist, lonnis Sakkis.[6] He gave one oblong 3x5-foot mirror to each of about 60 Greek soldiers, who used the mirrors to ignite a wooden ship 50 m (160 ft) away. Skeptics responded by questioning whether the ancient Greeks could have made enough mirrors for the job and by noting that some early writers made no mention of the incident. We'll probably never know exactly what happened in Syracuse 22 centuries ago, but the story itself-as well as its reception-are significant. It tells us how long ago man recognized the destructive power of directed energy. It reminds us how long the dream of immensely potent weaponry has been around. And most of all, it serves as a reminder that the only way to be *sure* an idea will or will not work is to try it out.

H. G. Wells and the "Heat Ray"

The idea of directed-energy weaponry went into eclipse for the next two millennia, as armies found what seemed to be more efficient means of mayhem. It wasn't until the 1890s that the concept of beam weaponry was resurrected by pioneering British science fiction writer H. G. Wells. He used it to arm the Martian forces invading the earth with "heat rays" in *The War of the Worlds.* The book, published in 1898, has a creaky sort of Victorian elegance in its prose. The concepts, however, are remarkably modern. Wells's description of the heat ray is strikingly prophetic:

> It is still a matter of wonder how the Martians are able to slay men so swiftly and so silently. Many think that in some way they are able to generate an intense heat in a chamber of practically absolute non-conductivity. This intense heat they project in a parallel beam against any object they choose by means of a polished parabolic mirror of unknown composition, much as the parabolic mirror of a light- house projects a beam of light. But no one has absolutely proved these details. However it is done, it is certain that a beam of heat is the essence of the matter. Heat, and invisible, instead of visible light. Whatever is combustible flashes into flame at its touch, lead runs like water, it softens iron, cracks and melts glass, and when it falls upon water, incontinently [sic] that explodes into steam.[7]

It reads almost as if Wells had slipped nearly a century into his future to see a high-energy infrared laser weapon at work. His vision of an intense beam of invisible infrared radiation-called heat radiation in his day–is chillingly similar to the reality of an infrared laser weapon, and the damage inflicted by a real laser weapon would be similar to what Wells envisioned. It's a remarkable job of prediction.

When the laser finally arrived on the scene six decades after Wells had written *The War of the Worlds,* the similarity to the heat ray was obvious. In his preface to a 1962 edition of *The War of the Worlds,* science fiction writer Arthur C. Clarke said, "the heat ray has remained on paper, and until recently seemed likely to stay there. We cannot count on this much longer . . . if you have not heard of infrared lasers, I am afraid you will soon do so."[8]

Wells was a remarkably gifted writer, one of the few able to combine action with stimulating ideas in prose that even today remains readable. His successful predictions were not confined to laser weaponry. *The War of the Worlds* also contains what was probably the first detailed description of highly mechanized warfare and its effect upon cities.

The "Ray Gun" Era

Science fiction eventually became a distinct genre of literature, at least in the United States, in the 1920s and 1930s. Today that period is called the "pulp" era because the magazines were printed on cheap pulp paper that deteriorates rapidly, and by now are likely to crumble in your hands. Wells was not a part of that era; although he lived until 1946, he abandoned science fiction before the advent of the pulps. Yet today Wells's books are more readable than most of the science fiction of the 1920s and 1930s, and his technological projections often seem much more realistic.

That's certainly the case for beam weapons. Wells was vague about the exact nature of the heat ray, but it was clearly closer to a piece of artillery than a handgun. Pulp science fiction writers, in contrast, equipped their heroes (and villains) with hand-held "ray guns" or "death rays." The effects of these fictional weapons generally had little or no relationship to the laws of physics. The mysterious rays they emitted could kill, paralyze, or disintegrate on contact. Such weaponry wasn't so much part of an effort to forecast the future as part of an effort to set the stage for the story. A tale set in the future needs futuristic-sounding weapons. It wouldn't do for Buck Rogers to use an old- fashioned six-shooter. Even in their efforts to forecast the future, pulp writers often missed the mark by extrapolating some trends too far and failing to predict breakthroughs. If their visions of the 1980s had been accurate, we'd all be flying personal helicopters equipped with vacuum-tube electronics. We don't because aviation technology reached limits, while semiconductor technology emerged to make the vacuum tube obsolete.

The ray gun has become a standard prop of science fiction. Despite the advances in technology in the last 50 years and tremendous improvements in the quality of science fiction as literature, modem writers have made only cosmetic changes to the ray gun. The standard term now seems to be "blaster," but a fair number of writers use "laser." That has helped to create a beam weapon mythology that is addressed in the next chapter.

Science fiction ray guns in action, as envisioned by Frank R. Paul, one of the leading artists of the pulp era. This illustration is from the February 1929 issue of *Amazing Stories,* showing a scene from the story "The Death of the Moon" by Alexander Philips in which scientists visiting the prehistoric earth from their home on the moon have to fight off an attack from Tyrannosaurus Rex. (Reproduced by permission of Forrest J. Ackerman, 2495 Glendover Avenue, Hollywood, California 90027. Agent for the Estate of Frank R. Paul.)

Origins of Directed-Energy Technology

The technological groundwork for beam weapons was being laid even while science fiction writers engaged in their flights of fancy. The work that would lead to high-energy lasers, particle beams, and intense microwave beams started in widely different areas. These fields, traditionally considered separate, are today lumped together because they share the common feature of producing an energy-containing beam that can be aimed at a target.

Ironically, the least advanced beam weapon technology is also the oldest. At about the same time H. G. Wells was writing *The War of the Worlds,* Guglielmo Marconi was investigating the properties of radio waves. Increasingly powerful radio wave generators were built, first for wireless telegraphy and later for commercial radio broadcasting. Physicists and engineers were beginning to study other applications of radio technology in addition to communications. Serious interest in the possibility of using radio waves as weapons first surfaced in the early 1930s. British government officials sent a memo to Sir Robert Watson-Watt, a pioneer in radio technology, asking if he could produce a radio wave weapon for use against aircraft. The answer was no; at that time there wasn't any radio frequency source that could generate enough power to raise a pilot's blood temperature to levels that might be caused by an average fever.[9]

However, Watson-Watt had another. Since the early years of the century, he'd been studying the use of radio waves to detect the locations of thunderstorms.[10] Instead of using radio waves as weapons, he proposed using them to spot enemy aircraft. The concept was quickly developed into radar (an acronym for *radio* detection *and* ranging), which a decade later would play a vital role in World War II. Radar pushed researchers in new directions. Its precision improves as wavelength decreases (or, alternatively, as frequency increases). The range increases with the power transmitted. Efforts to improve radar system performance thus led to higher-power radio wave generators and to increasing the frequency of radio waves. From the 1-m wavelength that corresponds to a 300-million-Hz frequency, frequencies moved into the billions of hertz (or gigahertz) range, where wavelengths are measured in centimeters. Such short radio waves are called "microwaves," although their wavelength is not microscopic. Development of higher-power and shorter-wavelength sources continued over the years, until impressively large powers could be attained. These high powers reactivated military interest in microwave beam weapons, which now supports the comparatively small program described in Chapter 9. That effort is a comparatively small one largely because microwave energy is much harder to direct and easier to defend against than lasers or particle beams. A huge antenna is required to focus high microwave powers onto a distant target. Because of such problems, the Pentagon is investing about 100 times more money in a much newer technology–the laser.

The Great Laser Race

The origins of laser technology date back to a prediction made in 1916 by Albert Einstein. It had already been recognized that atoms and molecules emit and absorb light spontaneously, without outside intervention. Einstein suggested another possibility: that an atom or molecule could be stimulated to emit light of a particular wavelength when light of that wavelength reached it, a phenomenon called stimulated emission. In 1928 R. Ladenburg showed that Einstein's prediction was right.[11] At the time stimulated emission seemed to be a very rare occurrence that was inevitably overwhelmed by spontaneous emission.

It would be many years before physicists learned how to create the right conditions to make practical use of stimulated emission in lasers, the physics of which are described in Chapter 4.

The first person to put stimulated emission to work was Charles H. Townes, a Columbia University physicist, who in 1951 was looking for a source of microwaves better than any available at the time. Sitting on a park bench in Washington early one spring morning before going to a physics meeting, he was struck by the idea of using stimulated emission. He wasn't sure it would work but was optimistic enough to put one of his graduate students, James P. Gordon, to work on the problem. Two years later Townes, Gordon, and Herbert Zeiger operated the first "maser," an acronym for microwave amplification by the stimulated emission of radiation.[12]

Within a few years interest turned to extending the maser concept to the much shorter wavelengths and higher frequencies of light waves. In the late 1950s Townes and Arthur Schawlow published a theoretical description of the laser, which at the time they called an "optical maser."[13] Parallel work was being performed at the Lebedev Physics Institute in Moscow by Nikolai G. Basov and Aleksander M. Prokhorov. All four men would later become Nobel laureates; Townes, Basov, and Prokhorov shared the 1964 Nobel Prize in Physics for fundamental work leading to the development of the laser and maser, and Schawlow in 1981 for helping to develop laser spectroscopy as a tool for studying the internal structure of atoms. A race began to build the first laser.

Most researchers thought of the laser as a comparatively low-power device, much like its older microwave-emitting cousin, the maser. One exception was Gordon Gould, a graduate student in physics at Columbia, but not working directly under Townes. Gould worked out his own conception of the optical maser, which he says was independent of Townes's work–an assertion with which Townes disagrees.[14] Gould wrote up extensive notes on the concept in which he coined the word laser from light amplification by the stimulated emission of radiation. Gould had always wanted to be an inventor, and instead of going the usual route of publishing his results in scholarly journals, he applied for a patent. Hurt by poor legal advice, Gould lost a series of patent battles but saw the term he coined displace the more cumbersome "optical maser." Eventually, after many years of litigation, Gould was finally granted two patents, one in 1977 and one in 1979. At this writing both are still under dispute in the courts.

Gould left Columbia and took his ideas on the possibility of high-power lasers with him to TRG Inc., a small company on Long Island that used them as the basis for a research proposal it submitted to the Department of Defense. TRG asked for $300,000, but the Defense Advanced Research Projects Agency was so excited by the prospects that in 1959 it gave TRG $1 million.

Complications rapidly ensued. Gould couldn't get a security clearance to work on his own project-a problem he says stemmed from a brief interest in Marxism he had during the early 1940s, stimulated by his first wife. He became disillusioned with both Marxism and his first wife at the time of the communist takeover of Czechoslovakia in 1948 and, at that point, separated from both of them. That explanation wasn't good enough in the Cold War era, however. Gould ended up isolated in a building separate from where the classified laser research was being conducted. TRG researchers with clearances could talk with Gould, but they couldn't tell him results of classified work, a restriction which needless to say hampered progress.[14]

Gould, Schawlow, Townes, and most other participants in the great laser race were trying to build gas lasers. That is, their goal was to deposit energy into a gas and get a laser beam out, a process described in more detail in Chapter 4. In theory, a gas looked like the best choice, and indeed today there is a large variety of gas lasers in regular use.

However, certain solid materials can also be made to produce laser emission, a fact that was recognized by the winner of the great laser race, Theodore H. Maiman, who was then working at the Hughes Research Laboratory in Malibu, California. He made the first laser from a small rod of synthetic ruby, a material that other physicists had decided wouldn't work as a laser material. Maiman, who had been studying the optical properties of ruby, wasn't convinced. He put mirrors at the ends of a rod, slipped it into the middle of a corkscrew-shaped flashlamp, and found that flashes of light from the lamp stimulated the emission of pulses of red light from the ruby.[15] While DARPA, the Pentagon's lead agency for risky but promising research, was putting $1 million into TRG, Maiman was working alone and without a government contract. In fact, Gould says that Hughes management told Maiman *not* to work on the laser.[16] Work on it he did, however, and on July 7, 1960, he announced demonstration of the world's first laser.

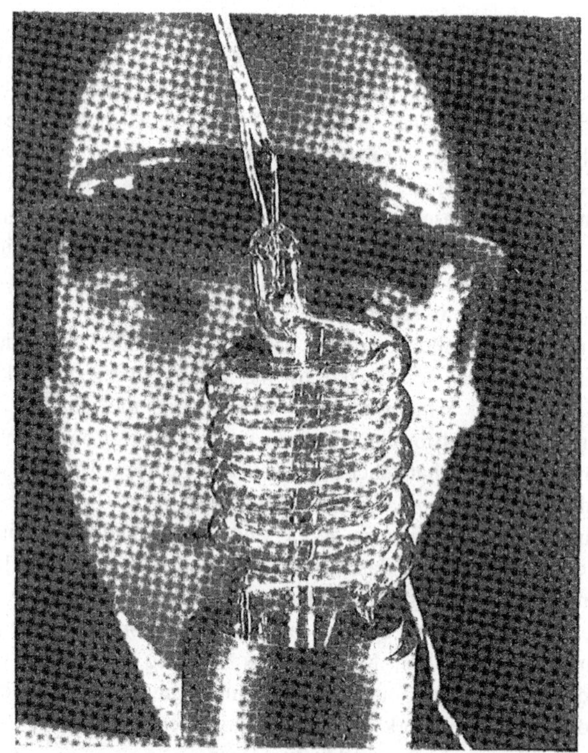

Theodore H. Maiman studies the first ruby laser, which he made at Hughes Research Laboratories in Malibu, California. The spiral tube is a flashlamp that produces a bright pulse of light, which excites the rod of synthetic ruby it surrounds to emit a pulse of laser light. (Courtesy of Hughes Aircraft Co.)

The months that followed saw a few other reports of different types of lasers, and within a few years there was a veritable torrent of reports of new lasers. Reports in the general press heralded the laser as a dramatic breakthrough, with a wide range of potential applications. Some of the reports got a bit carried away. One that appeared in late 1962, titled "The Incredible Laser," told about prospects for building laser cannons. This was too much for Arthur Schawlow, and soon a copy of the article was taped to his door at Stanford University-along with a note saying "for *credible* lasers see inside."[17]

Researchers working in the field were skeptical because the promise of high-power laser output remained elusive. Continuous power levels were generally measured in watts at most. To increase power, developers often resorted to the brute-force approach of enlarging the laser. The most powerful of the lasers developed in the early 1960s was the carbon dioxide type, which produced an invisible beam at a wavelength of 10 μm (micrometers) in the infrared; it was developed by C. Kumar N. Patel at Bell Laboratories.[18] But just increasing the length of the carbon dioxide laser wasn't good enough; the laser reached monstrous proportions before the output power reached useful levels. The logical conclusion of this line of development was a carbon dioxide laser able to produce continuous emission of 8800 W, which had a 750-ft beam path that passed through a series of cavities containing laser gas.[19] This was the sort of laser that prompted someone whose name is lost to history to crack that "a laser big enough to inflict militarily significant damage wouldn't even have to work–just *drop* it on the enemy."

Laser pioneer Arthur L. Schawlow put this sign on the door of his laser laboratory at Stanford University after the now-defunct Sunday newspaper supplement *This Week* published an article titled "The Incredible Laser" in late 1962 (inset at top). The article was accompanied by a picture of cannon-like lasers shooting down missiles with beams of light that so impressed Schawlow that he had another artist make a rendering of it for his own door. Schawlow writes: "Over the years, I gave away quite a few copies, and they turn up in all sorts of places. Even the author of the article eventually saw one of the posters, and wrote to say that he had not realized how appropriate the title had been." The *This Week* article was not the only one to talk about the possibilities of laser weapons in the 1960s; a 1964 issue of the German magazine

Kristall (lower inset) had as its cover story an article titled "Light Cannons Against Rockets." Schawlow's skepticism is typical of that of many laser pioneers, who understand very well the problems in getting a laser to work. (Courtesy of Arthur L. Schawlow.)

This early generation of lasers couldn't do the job of weapons, but it could help conventional weapons do their jobs. The Pentagon soon found that the ruby laser, and a similar laser in which the active medium was a rod of another synthetic crystal containing a small quantity of the rare-earth element neodymium, could help soldiers pinpoint their targets. One type of laser device, called a rangefinder, measures the time it takes a laser pulse to make a round trip from the soldier to the target and back, thus indicating how far away the target is, and where a gun or missile should be aimed. Another laser device, called a target designator, marks a potential target with a series of coded laser pulses; a homing device on a conventional bomb or missile can then home in on the laser-designated spot.

Such weapons found their first use in "smart bombs" during the Vietnam war. By the end of the 1970s, George Gamota, then assistant for research to the deputy undersecretary of defense for research and engineering, could call laser rangefinders and designators "the single most successful [Department of Defense] investment in the last decade."[20]

The Breakthrough to High Powers

The years around 1970 saw a series of breakthroughs that rekindled military interest in high-energy laser weaponry. These developments were centered in two areas, carbon dioxide and chemical lasers, the technology of which are described in more detail in Chapter 4.

The carbon dioxide laser's potential for high-power output was recognized soon after it was first demonstrated by Patel. However, it was not possible to realize that potential because of serious problems that were not overcome until a version called the "gasdynamic" carbon dioxide laser was invented in 1967. In the Pentagon's official words the gasdynamic laser "was the first gas-phase laser that appeared to be scalable to very high energies, and as such paved the way for serious consideration of a laser damage weapon system."[21] By the time the first public reports emerged from under the Pentagon's veil of classification,[22] continuous power levels had reached 60 kilowatts (kW) (60,000 W). Similar work was reported at about the same time by a Russian group,[23] which may have stimulated declassification of the American research.

Around 1970 important developments were also made in the area of chemical lasers, lasers in which the energy that emerges in the beam is produced by a chemical reaction. The basic idea, like that of the carbon-dioxide laser, dates back to the early 1960s, but it wasn't until 1969 that a series of developments pointed to the possibility of reaching high powers.[24] Further work led to even higher output powers from chemical lasers. Meanwhile, progress was being made on new high-energy variations of the carbon dioxide laser.

The Pentagon was quick to test the possibilities. The Air Force built what was officially described as a "moderate-power" gasdynamic carbon dioxide laser, which unofficial sources say produced a few hundred kilowatts. That laser was tested amid great secrecy in the early 1970s at Kirtland Air Force Base in New Mexico, the site of the Air Force Weapons Laboratory. Visitors tell of charred countryside as well as charred targets, silent but vivid testimony to the problems in getting the beam to its target. In 1973 the ground-based laser shot down winged drones at Kirtland's Sandia Optical Test Range by detonating fuel and cutting control wires. Details of the tests were kept secret for years, but in 1982 I was part of an unclassified audience that saw a film showing some of the tests at the annual Conference on Lasers and Electro-Optics.

The Army was the next armed service to demonstrate a laser "kill." Army scientists shoehorned a 30- to 40-kW carbon dioxide laser, which derived its energy from an electrical discharge, into a vehicle the size of a tank. Observers report that the fit was so tight that the only way to service the laser was to dissemble part of the vehicle. However, the system, called the Mobile Test Unit, did its job and in 1975 destroyed winged and helicopter drones in tests at Redstone Arsenal in Alabama.[25] Interestingly, it was not those tests but later results that convinced Army brass that lasers might be useful on the battlefield. The laser has since been removed from the tank, apparently becoming the first "Army surplus" laser weapon. It was given to the National Aeronautics and Space Administration for research on the use of lasers for rocket propulsion at its Marshall Space Flight Center in Huntsville, Alabama.[26]

The Navy had its turn in March 1978, when it used a "moderate-power" chemical laser (reportedly able to deliver 400 kW) to shoot down a TOW antitank missile in flight. The tests were part of the Unified Navy Field Test Program conducted at San Juan Capistrano, California.[27]

Military programs during the 1970s were not limited to these tests. There was extensive research into areas such as fundamental laser physics, optical components needed for high-energy lasers, beam-target interactions, pointing and tracking, and propagation of a laser beam through the atmosphere. By the end of fiscal 1978, total Pentagon spending on laser weaponry had passed the $1-billion mark. Spending accelerated since then, and the total reached roughly $2 billion at the end of fiscal 1982.

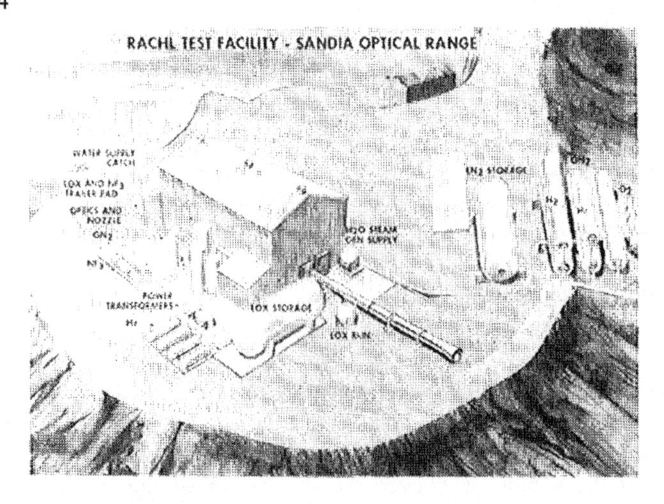

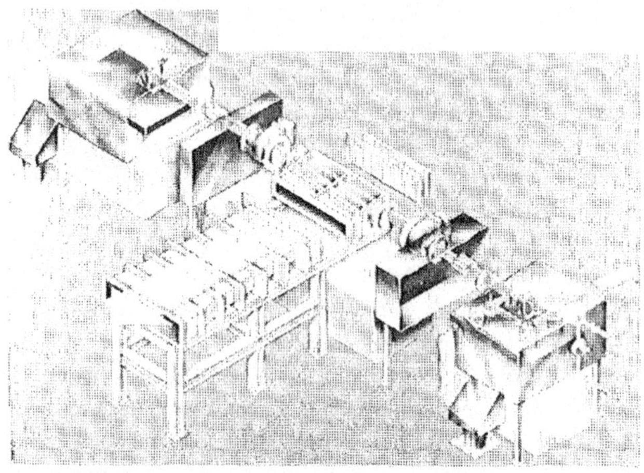

Artist's conceptions of a real high-energy laser, the Rocketdyne Advanced Chemical Laser (RACHL), first demonstrated in 1977. Built by the Rocketdyne Division of the Rockwell International Corp., this chemical laser can deliver power of about 100,000 W for up to 200 sec at a time. The top drawing shows the building housing the laser and the tanks of gases including hydrogen, helium, gaseous and liquid nitrogen, deuterium, nitrogen trifluoride, and liquid oxygen-required to operate the laser. The lower drawing shows a simplified view of the laser itself, which is inside the building near the center of the top drawing. The Department of Defense has used the infrared beam from this laser in a variety of weapons experiments. (Courtesy of Rocketdyne Division of Rockwell International Corp.)

Even by Pentagon standards, that price tag is a hefty one for a research program. Military scientists have learned some important lessons over the years, not all of them encouraging. During the 1970s, they found big problems with the gasdynamic laser. It seems adequate for tests of laser effects, and in fact the Air Force uses a 400-kW gasdynamic laser in its Airborne Laser Laboratory. However, the beam does not travel well through air. From a practical standpoint, a gasdynamic laser is "a ten-ton watch," in the words of one developer, a bulky and inefficient device that nonetheless requires extremely precise components.

The program has had only one widely publicized failure. In January 1981 overconfident Air Force officials called a press conference at which they announced plans to use the Airborne Laser Lab to shoot down air-to-air AIM-9L Sidewinder missiles while the modified KC-135 aircraft (a military version of a Boeing 707) housing the laser was in the air.[28] Two tests were conducted at the Naval Weapons Center in China Lake, California, but the missiles were not destroyed, resulting in the predictable headline mentioned in Chapter 1: "Laser weapon flunks a test."[29] Embarrassed air force officials insisted that they had learned a lot, and that the tests were not failures. However, there were no more public statements until mid-1983, when the Airborne Laser Lab did shoot down some Sidewinders.[30]

The Soviet laser weapon program is under tighter security wraps than that of the United States. Pentagon officials have said that the Soviet effort is three to five times larger than the U.S. program,[31] an estimate that seems to hold constant even when the American budget is being sharply increased. Weapons-related research is published in the open Russian scientific literature and seems to account for a larger share of papers there than in the open literature in the West. That does not necessarily indicate the relative size of the Soviet program, however, because of different classification and publication policies.

The past few years have seen a sharp increase in interest in nuclear defense laser weapons, a controversial idea whose initial public champion was Senator Malcolm Wallop (R-Wyo),[32] and which is described in more detail in later chapters. The field has also attracted interest from others, and at times the laser weapon budget-particularly the part devoted to space-based systems-shows signs of becoming one of the proverbial "political footballs." Two fundamentally new laser concepts have surfaced recently: the free-electron laser and the X-ray laser, which are described later. (The X-ray laser story is so strange it requires a chapter of its own.)

Particle-Beam Weaponry

The roots of particle-beam technology date back to the early days of subatomic physics, when physicists built the first "atom smashers." The idea was to accelerate charged subatomic particles-electrons, protons, and ions- to high velocities by passing them through suitably designed electromagnetic fields. Once accelerated to a high velocity, the particles would have enough energy to probe inside atoms.

The first particle accelerators were built about half a century ago, and by World War II the Department of Defense had begun looking at their potential use as weapons. At the time military officials decided that the technology wasn't ready. The years that followed brought major advances in the technology. In 1958 the Defense Advanced Research Projects Agency decided that the time was finally ripe and started a program code-named "Seesaw" at the Lawrence Livermore Laboratory. The goal was to use beams of electrons as a ground-based defense of strategic areas against ballistic missile attack. Altogether $27 million was spent on Seesaw before the program was halted in 1972, a decision made, according to the Pentagon, "because of the projected high costs associated with implementation as well as the formidable technical problems associated with propagating a beam through very long ranges in the atmosphere."[33] Large accelerators, in short, cost a lot of money, an unpleasant fact of life that has lately helped slow the progress of research in subatomic physics. What's worse, from the standpoint of the weapon system designer, is that it's very hard to make a bolt of electrons travel a predictable path through the atmosphere to hit a particular target-witness, for example, lightning bolts.

Seesaw had taught military researchers enough about the problems of particle-beam weapons that their next effort was more modest. In 1974 the Navy started a program called "Chair Heritage." This, too, involved electron beams, which in this case would be used to defend ships against attack by aircraft and missiles.[34] As a "point defense" system rather than one intended to protect a large area, the Navy application appeared more reasonable to military planners. One reason was that the beam wouldn't have to go as far, simplifying the problem of beam direction. Another was closely related: because potential targets would be closer to the beam generator, the system could be smaller and hence more reasonable in cost. Thus, Chair Heritage was never a massive program. During its five years of existence, from 1974 to 1978, it accounted for $21.2 million from the Department of Defense, miniscule indeed compared with the hundreds of millions of dollars spent over that interval on laser weaponry. In 1977 Congress shifted Chair Heritage from the category of weapon development programs to that of "technology base" (i.e., basic research) efforts because it did not seem likely to result in the near-term development of a weapon system.

Both Seesaw and Chair Heritage were comparatively small programs because they were able to draw heavily on other efforts to develop large accelerators. The Atomic Energy Commission, and later its successors the Energy Research and Development Administration and the Department of Energy, were pouring hundreds of millions of dollars into particle-beam accelerators for research in subatomic physics and other areas. Large accelerators were being built at places like the Argonne National Laboratory and Brookhaven National Laboratory. The technology was not directly applicable to directed-energy weapons because of problems such as the need for high currents to damage potential targets, while research accelerators generated only low currents. Yet the development of research accelerators helped create a strong technology base for military programs to draw upon.

By the mid-1970s, however, the golden age of particle physics research was beginning to run out. Or, to be more precise, the money was starting to run out. Physicists who asked the government to support bigger and better accelerators were increasingly being turned away empty-handed. The price tags were simply too large, and the benefits too nebulous to convince the people who paid the bills. Perhaps because they were getting hungry, the people who originally developed accelerator technology for use in pure re- search soon thought of other potential applications for particle accelerators. In addition to weaponry, particle beams might be useful in fusion, in particular in "inertial-confinement fusion." The basic idea is to hit a tiny target containing fusion fuel with a very short pulse of energy that makes the target collapse or implode. The implosion generates the high temperatures and densities needed to produce a fusion reaction, although these last only for a very short time. The plasma in which the fusion reactions take place is confined by inertial forces, hence the name. Initial studies focused on lasers, but research has been expanded to include beams of protons[35] and other ions[36] as well.

Although the other principal approach to fusion, magnetic confinement, is a civilian program, inertial confinement is a military one. The microexplosions of fusion fuel pellets are miniature versions of hydrogen bombs! Inertial-confinement fusion thus simulates the effects of nuclear explosions. For that reason some results are classified, and the entire program is operated by the Department of Energy's Division of Military Programs, which is responsible for the development of nuclear weapons.

Accelerator technology is part of a broader field called "pulsed power," which includes the generation and switching of extremely high electrical voltages as well as the production of energetic beams of ions or subatomic particles. (The high-voltage part of pulsed-power technology is important in high-power, electrically driven lasers as well as in particle beams.)

The potential of pulsed power has not been lost on the Soviet Union. A 1978 report from the Rand Corporation says that Soviet research in the field has been "gathering momentum" since the mid-1960s. It goes on to conclude that "the Soviet Union is engaged in a major effort to develop pulsed-power systems. The principal objectives of this effort appear to be (1) the achievement of controlled thermonuclear reactions as a national energy source and (2) military applications."[37]

The Soviet interest in military applications was first publicized a year earlier by retired Air Force Major General George J. Keegan. Shortly after retiring as head of Air Force intelligence in January of 1977, Keegan gave a press briefing at the American Security Council in which he warned that the Soviet Union was well on the way to development of particle-beam weapons for ballistic missile defense. He evidently had been unable to convince higher- ups at the Pentagon of his interpretation of intelligence information and had decided to "go public" with his analysis. In an interview with *Science* magazine, he said his goal was "to provoke and make enough people angry" about the situation.[38]

The controversy reached major proportions shortly afterwards, with the publication of an article in *Aviation Week & Space Technology* based on interviews with General Keegan.[39] The article opened with the charge that the "Soviet Union is developing a charged particle beam device designed to destroy U.S. intercontinental and submarine-launched ballistic missile nuclear warheads." Military editor Clarence A. Robinson, Jr., described evidence for the Soviet program, adding that "because of a controversy within the U.S. intelligence community, the details of Soviet directed-energy weapons have not been made available to the President or to the National Security Council."

Keegan was not the first to go to *Aviation Week* with an idea he couldn't sell to his superiors. Edited at the time by Robert A. Hotz, a long-time advocate of an aggressive military posture, the magazine is sometimes called *Aviation Leak* in the aerospace community because of its penchant for leaking classified material. [40] With a circulation around 140,000 the magazine is widely read and influential in the military and aerospace community and on Capitol Hill. Although generally considered authoritative, the magazine is far from infallible, an inevitable problem given the nature of its sources. For example, in 1958 *Aviation Week* published a report that the Soviet Union was flight- testing a nuclear-powered bomber, at a time when a United States program to build one was running into serious Congressional opposition. The U.S. program was finally killed in 1961, and since then no fleets of nuclear-powered Soviet bombers have appeared-nor has any other evidence that the Soviets were trying to develop such craft. Hotz later said that the information was supplied by U.S. government officials,[41] and that claim seems reasonable. What is not clear-and probably never will be-is whether the leak was a deliberate lie or a misinterpretation of intelligence data. It is not very likely to have been an accurate report of an ill-fated Soviet program that has been kept quiet for 25 years. Generally, the magazine tries to forget its mistakes, such as a late-1975 report – evidently based on misinterpretation of intelligence data – that a U.S. spy satellite had been blinded by a Soviet laser beam,[42] an event military analysts now attribute to natural gas fires in the Soviet Union. The article reporting that incident has not been cited in recent *Aviation Week* roundups of beam weapon technology.

Robinson's article succeeded in stirring up the desired hornets' nest, and many things started happening at once. Scientists debated the evidence for and against the tests, and many expressed doubt about the feasibility of the whole idea.[43] Congress conducted an investigation. The Defense Advanced Research Projects Agency commissioned the Rand Corporation study. A Swedish scientist reported detecting short-lived radioactive isotopes in the atmosphere, which he interpreted as possible indications of Soviet tests of a particle-beam generator.[44] Others questioned his interpretation and he is reported to have later decided that his observations did not agree with what would have been expected from the system described in *Aviation Week*.[45]

Exactly what the Soviet Union is up to remains a mystery. The latest edition of the Pentagon's "fact sheet" on particle-beam research says: "The Soviet effort is judged to be larger than that of the United States (particularly in the area of accelerators for fusion applications), and to have been in progress longer. However, no direct correlations between Soviet particle beam work and weapons related work has been established. "[46] Other observers inside and outside of the Pentagon are not so hesitant to make the connection.

With considerable prodding from Congress, the Department of Defense finally began increasing its particle-beam effort. For fiscal 1979 Chair Heritage was transferred to the Defense Advanced Research Projects Agency. Funding was gradually increased, from a total of $18 million for all military particle- beam research in fiscal 1979 to a request for just over $50 million in fiscal 1983.[47] Particle-beam research was combined with laser weapon development in a "directed-energy" program in the Pentagon. The current head of that effort is Air Force Brigadier General Robert R. Rankine, Jr., who recently succeeded Air Force Major General Donald L. Lamberson.

Inside view of the Pentagon's biggest particle-beam accelerator for weapon tests, the Advanced Test Accelerator at the Lawrence Livermore National Laboratory in California. Note the two workmen in light circle at upper right. (Courtesy of Lawrence Livermore National Laboratory.)

Today, development of particle-beam weapons is generally considered well behind that of laser weaponry, a fact manifest in a fiscal 1983 budget request for laser weapons that is over eight times larger than for particle- beam research. The technology and the current status of the particle-beam program are described in more detail in Chapter 7.

Controversy continues to surround beam weapon development, particularly efforts to build systems for defense against ballistic missile attack such as proposed by President Reagan. Reagan's March 23, 1983, speech stirred up a new round of political controversy. His request for preparation of a new report on the prospects for ballistic missile defense stimulated a new round of thinking on designs for missile defense systems. One new concept that seems likely to play a prominent role in the report-not yet completed as this book went to press-is leaving the large lasers on the ground and reflecting their beams off orbiting battle mirrors to their targets. Another is a "pop-up" weapon system that would be launched only when needed.

Reagan's involvement in the debate has led to a few inevitable puns on the pronunciation of his name ("ray gun"), but for the most part discussions have been taken more seriously than they have in the past. When General Keegan's charges about Soviet particle-beam weapon development were first reported in *Aviation Week,* then Secretary of Defense Harold Brown replied:

I won't go so far as to say that the suggestion that the Soviets have a new Manhattan Project capability breakthrough in the charged particle beam weapon is a piece of advance flackery for the new science fiction movie, "Star Wars."[48]

Six years later Reagan's proposal was quickly dubbed his *Star Wars* speech by some reporters-but I didn't hear any suggestions that Reagan was pro- viding advance publicity for the third movie in the *Star Wars* series, due out two months after his speech.

References

1. Quoted in O. N. Stavroudis, "Comments on: On Archimedes' Burning Glass," *Applied Optics* 12, A16 (October 1973).
2. Siculus Diodorus, *The Library of History* (Davis, London, 1814), XXVI, xviii, quoted in L. Sprague de Camp, *The Ancient Engineers* (Ballantine Books, New York, 1974), p. 170.
3. L. Sprague de Camp, *The Ancient Engineers* (Ballantine, New York, 1974) p. 170.
4. These letters make fascinating reading and provide an enlightening example of how armchair theorizing can diverge from reality. In sequence of publication, they are: Albert C. Claus, "On Archimedes' Burning Glass," *Applied Optics* 12, A14 (October 1973); 0. N. Stavroudis, "Comments on: On Archimedes' Burning Glass," *Applied Optics* 12, A16 (October 1973); Klaus D. Mielenz, "Eureka!" *Applied Optics* 13, A14-A16 (February 1974); E. A. Phillips, "Arthur C. Clarke's Burning Glass," *Applied Optics* 13, A16 and 452 (February 1974); D. Deirmendjian, "Re: Archimedes's Burning Glass," *Applied Optics* 13, 452 (February 1974); L. Simms, "More on That Burning Glass of Archimedes," *Applied Optics* 13, A14-A16 (May 1974); and Richard A. Denton, "The Last Word," *Applied Optics* 13, A16 (May 1974).
5. G. L. Leclerc de Buffon, *Memoires de l'Academie Royale des Sciences pour 1747* (Paris, 1752) pp. 82-101, cited in Klaus D. Mielenz, "Eureka!" *Applied Optics* 13(2), A14 & A16 (February 1974); the incident is also mentioned in Trevor I. Williams, ed, *A Biographical Dictionary of Scientists,* 3rd ed (Halsted Press, New York, 1982) p. 88.
6. *Washington Star-News,* November 13, 1973, p. A3 and *Time,* November 26, 1973, p. 60, cited in Klaus D. Mielenz, "Eureka!" *Applied Optics* 13(2), A14 & A16 (February 1974).
7. H. G. Wells, *The War of the Worlds* [Washington Square Press, New York, 1965 (edition published in one volume with *The Invisible Man)],* p. 174 (start of Chapter 6).
8. Arthur C. Clarke, "Introduction" to H. G. Wells, *The Invisible Man* and *The War of the Worlds* (Washington Square Press, New York, 1965), p. xvii.
9. Edward L. Safford, Jr., *Modern Radar: Theory, Operation and Maintenance,* 2nd ed (Tab Books, Blue Ridge Summit, Pennsylvania, 1981), p. 23.
10. *Ibid,* p. 17.
11. For a fuller description of the roots of the laser, see Bela A. Lengyel, "Evolution of Masers and Lasers," *American Journal of Physics* 34, 903-913 (October 1966).
12. This race is covered in Lengyel's paper (Ref. 7). There are also two interesting first-hand accounts: Arthur L. Schawlow, "Masers and Lasers," *IEEE Transactions on Electron Devices ED-23* 7, 773-773 (July 1976); Charles H. Townes, "The Early Days of Laser Research," *Laser Focus* 14(8), 52-58 (August 1978).
13. A. L. Schawlow and C. H. Townes, "Infrared and Optical Masers," *Physical Review* 112, 1940 (1958).
14. The struggle over Gould's laser patents and credit for inventing the laser is a story as convoluted as that of any soap opera. For an account through mid-1981, see Jeff Hecht and Dick Teresi, *Laser Supertool of the 1980s* (Ticknor & Fields, New York and New Haven, 1982), pp. 49-61; for an update to mid-1982, see "Laser Patent Litigation Reaches Crescendo," *Lasers & Applications* 1(2), 22-24 (October 1982).
15. T. H. Maiman, "Stimulated Emission in Ruby," *Nature* 187, 493 (1960); this remarkable 300-word description of a milestone in physics was rejected by the first journal to which it was submitted, *Physical Review Letters,* because the editors had decided there wasn't much significant going on in maser research.

16. Gordon Gould, private communication, during an interview I conducted for *Omni* (in preparation).
17. Arthur L. Schawlow (Stanford University), private communication. The article appeared in the November 11, 1962 issue of *This Week,* a now-defunct supplement to Sunday newspapers.
18. C. K. N. Patel, *Physical Review Letters* 12, 588 (1964).
19. F. Horrigan, C. Klein, R. Rudko, and D. Wilson, *Microwaves* 8, 68 (1968), cited in W. W. Duley, CO_2 *Lasers Effects and Applications* (Academic Press, New York, 1976), p. 16.
20. Testimony before Congress, quoted in Jeff Hecht and Dick Teresi, *Laser Supertool of the 1980s,* p. 103.
21. Department of Defense, "Fact Sheet: DoD High Energy Laser Program," February 1982, 3.
22. Edward T. Gerry, *IEEE Spectrum* 7, 51 (1970), cited in W. W. Duley, CO_2 *Lasers Effects and Applications.* (Academic Press, New York, 1976).
23. V. K. Konyukhov, I. V. Matrasov, A. M. Prokhorov, D. T. Shalunov, and N. N. Shirokov, *JETP Letters* 12, 321 (1970), cited in W. W. Duley, CO_2 *Lasers Effects and Applications.* (Academic Press, New York, 1976).
24. Early chemical laser development is recounted in Arthur N. Chester, "Chemical Lasers," in E. R. Pike, ed, *High-Power Gas Lasers 1975* (Institute of Physics, Bristol and London, 1975) pp. 162-221.
25. Philip S. Klass, "Laser destroys missile in test," *Aviation Week* & *Space Technology,* August 7, 1978, pp. 14-16.
26. "Postdeadline reports," *Laser Focus* 17(8), 4 (August 1981).
27. Philip S. Klass, *op. cit.,* (Ref. 25).
28. Jim Bradshaw, "Laser tests at Kirtland a 'milestone'," *Albuquerque Journal,* June 15, 1981, A1.
29. *Aviation Week* & *Space Technology,* befitting its advocacy of beam weaponry, was more circumspect in its headline, "Laser fails to destroy missile," (June 8, 1981, p. 63), but the June 3, 1981 *Boston Globe* simply used "Air Force laser weapon flunks a test." I couldn't resist that line either and used it on the "Postdeadline Reports" page of the July 1981 issue of *Laser Focus.*
30. "Airborne laser lab finally kills missiles," *Lasers* & *Applications* 2(9), 24 (September 1983).
31. Department of Defense, *Soviet Military Power* (U.S. Government Printing Office, Washington, 1981), p. 76.
32. See, for example, Malcolm Wallop, "Opportunities and imperatives of ballistic missile defense," *Strategic Review,* Fall 1979, pp. 13-21.
33. Department of Defense, "Fact Sheet: Particle Beam Technology Program," February 1982, p. 3.
34. *Ibid.*
35. Particle-beam fusion concepts, including approaches based on electrons as well as ions and protons, are described in R. B. Miller, *An Introduction to the Physics of Intense Charged Particle Beams,* (Plenum, New York, 1982), pp. 293-333.
36. *Ibid.;* see also John Lawson and Derek Beynon, "Heavy ions beam in on fusion," *New Scientist* 95, 565-568 (August 26, 1982).
37. Simon Kassel, *Pulsed Power Research and Development in the USSR* (Rand Corporation, Santa Monica, California, May 1978) p. v; this report is based on a survey of the open (i.e., unclassified and publicly published) Russian literature from the early 1960s through 1976.
38. Nicholas Wade, "Particle beams as ABM weapons: General and physicists differ," *Science* **196,** 407-408 (April 22, 1977).
39. Clarence A. Robinson, Jr., "Soviets push for beam weapon," *Aviation Week* & *Space Technology,* May 2, 1977, pp. 16-22.

40. Nicholas Wade, "Charged debate erupts over Russian beam weapon," *Science* **196,** 957-959 (May 27, 1977).
41. John Tierney, "Take the A-Plane," *Science* 82(1), 46-55 (January/February 1982).
42. Philip J. Klass, "Anti-satellite laser use suspected," *Aviation Week & Space Technology* (December 8, 1975), p. 12; cited in "Postdeadline reports," *Laser Focus* 12(1), 4 (January 1976).
43. Nicholas Wade, "Charged debate erupts over Russian beam weapon," *Science* **196,** 957-959 (May 27, 1977).
44. Lars-Erik De Geer, "Airborne short-lived radionuclides of unknown origin in Sweden in 1976," *Science* **198,** 925-927 (December 2, 1977).
45. Farooq Hussain, "Is the beam weapon the physicists' Flying Dutchman?" *New Scientist,* January 5, 1978, p. 30.
46. Department of Defense, "Fact Sheet: Particle Beam Technology Program," February 1982, p. 7.
47. *Ibid.*
48. "Brown comments on beam weapons," *Aviation Week & Space Technology* May 30, 1977, p. 12.

3. Beam Weapon Mythology

The long history of beam weapons in legend and fiction has helped create a kind of mythology, a large collection of widely held misconceptions about their power and potential uses. It isn't often that a new technology comes on the scene with a ready-made-but entirely fictional-history. Such a false background breeds misconceptions, which tend to take on a life of their own, especially in the more sensational segments of the general press.

Laser researchers have had to live with the problem for well over two decades, with reactions ranging from bemusement to annoyance. When an 1962 article, in a Sunday newspaper supplement, titled "The incredible laser" predicted the development of laser cannons, a copy soon appeared on Arthur Schawlow's door at Stanford University together with a note reading: "For credible lasers, see inside." The businessmen who manufacture lasers and laser systems tend to be less amused by the misconceptions that convey the false impression that all lasers are dangerous.

The fact is there is a large and healthy civilian industry building nonlethal laser systems. About $340 million worth of lasers were sold for nonweapon uses in 1982, and sales of systems containing lasers that year were about $2.6 billion, according to *Lasers & Applications* magazine.[1] Some of that hardware was used by the military, but total sales to the civilian world still ran well over $2 billion, and those sales continue growing. For all the words that have been devoted to beneficial uses of lasers in medicine, communications, and other fields, a distressing portion of the public still thinks first of death rays. Anyone who introduces him- or herself as working with lasers is liable to be grilled on the laser weapons program and to get strange looks when responding that he or she puts lasers to some innocuous use such as reading those cryptic striped symbols on cans in the supermarket.

The public has confused science fiction with fact. What's worse is that, despite a promising beginning in H. G. Wells, science fiction writers have largely blown their chance. Very few science fiction stories depict lasers as anything other than deadly weapons. And most science-fictional weapons bear little relationship to the concepts being considered by the deadly serious planners in the Pentagon and the Kremlin.

The reasons most science fiction writers missed the mark are tied in with the mechanics of writing fiction. Science fiction stories require futuristic props, just as historical novels have their own conventions. Weapons have been the traditional props of action-oriented stories, such as most early science fiction. The writer might invent new types of weaponry to depict omnipotent armaments, to create symbols of power, or simply to avoid arming 25th century characters with six-shooters. Some writers have tried to make their weapon ideas sound real by giving them realistic-seeming names or by borrowing real names, such as the laser. Few have made any real effort to think through how a laser would function as a weapon. That's not their job-they're storytellers, not weapon designers or generals. Stories, however, are the things of which myths are made, and myths and misconceptions flourish in the absence of hard facts. With much of the beam-weapon program shrouded in secrecy, there's been plenty of opportunity for a full-fledged mythology of sorts to develop unhindered.

Directed-energy weapons do have the potential to do things that other weapon systems have been unable to do, but they are far from omnipotent. Even if they work as well as advertised by their advocates, something which at this point is not at all certain, beam weapons will have only limited (although important) roles in warfare. The foreseeable future will *not* bring:

- Hand-held "ray guns," "disintegrator rays," or "death rays"
- Space-based weapons able to zap at will any target on the ground
- Weapons of mass destruction capable of nuclear devastation
- Planet-busting *Death Stars*

The rest of this chapter will explain why.

The Ray Gun Myth

Over the years the ray gun has become a standard prop of science fiction. It even shows up in the play of children who point fingers at each other, shout "Zap, you're dead," and imagine their fingers command some vast power of destruction that can be invoked at will. At first, fictional ray guns emitted beams of light or heat (invisible infrared radiation), but later a science-fictional arms race, created out of pulp paper and printer's ink, developed sources of mysterious rays with the ability to stun, paralyze, disintegrate, kill, or whatever else was needed to keep the plot going, or ultimately do away with the bad guys.

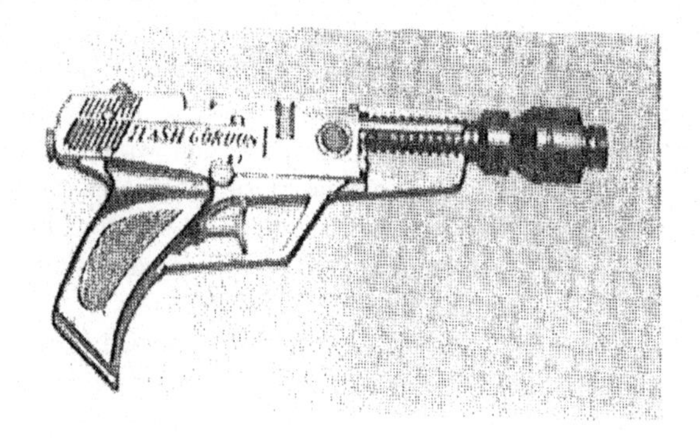

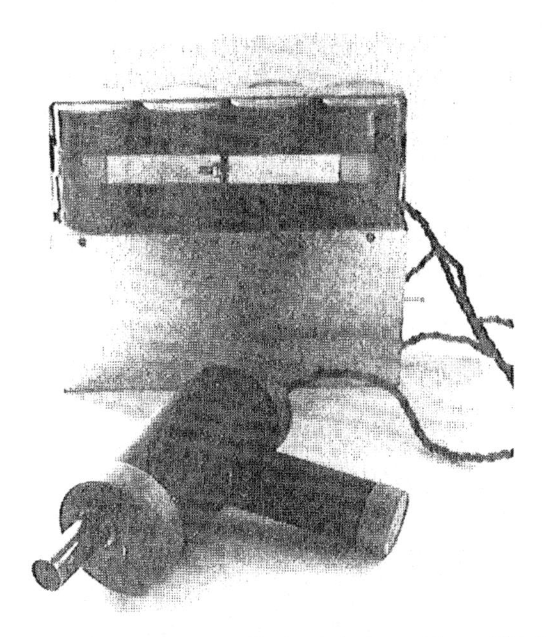

The toy ray gun at the top is just a squirt gun. The real portable laser is at the bottom. Powered by the 12-V battery in the background, the ruby laser is able to drill holes, according to Robert Ianinni, president of Information Unlimited Inc. in Amherst, New Hampshire, which makes the laser. But While the laser could punch holes in thin materials, or do serious damage to the human eye, it could not blast holes through people or tank armor. (Ray gun from the collection of Anthony & Suford Lewis; photo of laser courtesy of Robert Ianinni, Information Unlimited.)

A laser beam does carry light energy to its target, while a particle beam brings energy to its target in the form of fast-moving particles. A beam of intense microwaves can indeed heat its target, witness the microwave oven (which engineering legend has it was the ultimate outgrowth of the accidental cooking of a forgetful engineer's lunch by a microwave transmitter). However, the only way that a beam weapon could damage a target is by delivering energy to it. Paralysis, disintegration, and instant death are purely the stuff of fiction.

In fact, the concept of a beam weapon small enough to hold in your hand is far from reality. Powerful microwave generators are bulky, and more importantly, the beam spreads out much too rapidly to cook anything that isn't within a few meters (yards) of the source. Hand-held particle-beam generators seem out of the question. Today's compact ones come in boxes about the size and shape of a dumpster and require large amounts of electrical power. Large particle-beam generators include systems like the two-mile long Stanford Linear Accelerator.

Lasers can be made small, but the smaller they are, the less power they produce. Some hobbyists have built lethal-looking laser handguns, which could be hazardous to the eye in the same way that staring at the sun can be harmful because of the eye's extreme sensitivity to light. Their use in public places without special precautions is forbidden by general rules on laser safety issued by the federal Bureau of Radiological Health. The most serious laser eye hazard to the general public is not toy lasers-none of the models on the market have a laser in them. The real danger comes from rock musicians who like to scan multicolored beams through the audience, a practice strictly illegal in the United States.

The power level from a hand-held laser is so low that it couldn't kill anything larger than an insect. Rumor has it that a few engineers have used laboratory lasers as fly-swatters. It is possible to build a laser big enough to do serious damage to something larger, but it would be too large to carry. That is true for two reasons: (i) a beam of light isn't an efficient way to cause bodily harm and (ii) the laser itself is inherently inefficient in translating input power into a beam of directed energy.

An example of the magical rays of pulp science fiction. Artist Frank R. Paul is depicting a scene from "The Machine Man of Ardathia" by Francis Flagg in which rays are holding a man in the air for several minutes. The creature in the jar is a human being from 35,000 years in the future. The H. G. Wells story was a reprint, not an original for *Amazing Stories.* (Reproduced by permission of Forrest J. Ackerman, 2495 Glendower Avenue, Hollywood, California 90027. Agent for the Estate of Frank R. Paul.)

There's a big difference between the damage done by a projectile such as a bullet and the damage done by a laser beam. A bullet rips a hole through flesh because its momentum carries it through the tissue. It is the momentum of a bullet that causes a person hit by one to jerk in the direction of the bullet's flight and causes the recoil felt by a person firing a gun. A bullet kills by causing mechanical damage, tearing through tissue in vital organs. An ordinary laser beam doesn't pack that kind of punch; it can only cause damage by depositing energy in the form of heat. (Light does have momentum in theory, but that is not significant here.) To kill a person, a laser beam would have to carry enough energy to bum a hole through his body and do lethal damage to some vital organ, such as the heart. That would take a lot more energy than the 500 J (joules) carried by a typical 45-caliber bullet.

To get an idea of the energy required, I've calculated what it would take to drill a hole about 1 cm (0.4 in.) in diameter through a human body, which is roughly 20 cm (8 in.) thick. Although it might sound large, such a hole is roughly the size produced by a pistol-like laser on a target 10-100 m (30-300 ft) away. In addition, a fairly large hole would be needed so debris produced when the laser vaporized the tissue could escape without blocking the laser beam. It might also help in preventing the laser beam from automatically sealing off the sides of the wound-an effect that makes lasers useful in surgery but counterproductive in a weapon.

Drilling such a hole through the body would take at least as much energy as would be needed to vaporize 20 cm^3 of water because the human body is roughly 90% water. That would be about 50,000 J, 100 times more energy than is carried by a 45-caliber bullet. Actually, trying to vaporize a hole through a person would take much more energy because the process would not be 100% efficient. Some of the energy would be absorbed by the vaporized tissue, some would heat the tissue beyond the vaporization temperature, some would be reflected, and some would be lost in other ways.

A laser large enough to produce 50,000-J pulses would be much too large to hold. To get rid of excess heat produced by the laser, the material producing the beam would have to be a gas. The amount of energy that can be extracted from a gas depends in a complex way on its pressure, volume, and temperature; but if a laser weapon must be small enough to be held in a hand, the critical thing is volume. Any sort of practical laser "gun" couldn't hold more than about one liter (close to one quart) of gas, and with present technology you probably couldn't come close to producing 1000 J/l.

There's a further complication: the laser's inherent inefficiency. All the energy that goes to excite a laser does not end up in the beam, a problem discussed in more detail in Chapter 4. In practice, at best about 20% of the energy dumped into the gas by an electric discharge or chemical reaction emerges in the beam; the rest winds up as dissipated heat. Thus at least four times as much energy as was present in the beam would have to be dissipated, and a hand-held laser weapon would probably get very hot. Buck Rogers would do well to wear asbestos gloves!

Even tank-sized lasers might have power supply problems. This Pentagon artist's conception of a Soviet laser weapon shows the laser itself at left and the power supply in a separate vehicle at right; the two are connected by a thick power cable that would be the logical target in battle. (Courtesy of Department of Defense.)

Energy storage would also add to bulk because a laser itself merely transforms energy from one form into another. Some lasers are driven by the energy from chemical reactions and probably could get their energy from cartridges filled with suitable fuels. However, others require electrical power supplies, which are inherently bulky. This problem was highlighted around a dozen years ago in a cartoon published in a trade magazine showing a scientist holding a small laser pistol, which was attached by a thick power cord to an immense supply in the background. Even the thick power cord would not overcome the limit on how much laser energy can be produced from a small volume of gas.

Will future breakthroughs make compact, self-contained laser weapons possible? Don't bet on it in the foreseeable future. The problems of heat dissipation and packing the required amount of energy into a small enough volume are fundamental ones, and attempting to get around them may lead to even more serious difficulties. And lasers won't be cheap.

I've made some rough calculations to indicate how much energy could be extracted from one liter of gas under reasonable conditions. Assume that the laser gets its energy from a chemical reaction that produces a gas at a pressure ten times that of the atmosphere (or 10 atmospheres) and a temperature of 600 °K (roughly 330 °C or 630 °F). In practice, chemical lasers generally operate at much lower pressures, but because the number of atoms or molecules increases with pressure, assuming a high pressure gives a deliberately optimistic estimate. Also assume that the laser emits light at an invisible ultraviolet wavelength of about 300 nm (nanometers) (0.3 µm), about as short a wavelength as the atmosphere can transmit well. If each and every atom or molecule in the laser gas emitted one quantum of light energy at 300 nm in a pulse–an undoubtedly overoptimistic assumption–that pulse would contain 50,000 J of energy, the figure calculated earlier as the minimum energy needed to bum a hole in a person. That might make it seem as if a lethal laser handgun is almost within reach. However, my simple calculations haven't made any effort to account for many effects that would increase the lethal energy requirement and decrease the possible output energy, leaving a wide gap between the two. I would not flatly say that laser handguns are impossible, but the prospects definitely don't look good.

It is at least theoretically possible to obtain very high energies from very small X-ray lasers, as described in more detail in Chapter 6. Though potentially lethal, X rays don't penetrate the atmosphere well. However, the most serious objection is that the activation threshold for an X-ray laser is so high that a small nuclear explosion may be needed to get it to work. It's certainly not a weapon you would want to carry in your hip pocket!

Because the eye is very sensitive to light,[2] it should be possible to build a hand-held laser capable of causing permanent eye damage. Indeed, eye safety is a major concern in training soldiers to use low-power laser range-finders and target designators with conventional weapons; laser displays, light shows, and toy laser guns built by hobbyists are other potential problems. As mentioned in Chapter 13, there are some ways in which lasers could be used to temporarily or permanently blind soldiers, but it is unclear how effective this would be during an actual battle.

What about killing soldiers with larger lasers, much like the heat rays used by the Martians in *The War of the Worlds?* It's theoretically possible, but probably not likely for reasons discussed in Chapter 13. The basic problem, to borrow some military terminology, is that there are more cost-effective antipersonnel weapons. High-performance military rifles are far less costly than the large lasers that would be needed to kill people. One of the laser's major attractions for some missions-the ability of the beam to travel in a straight line-is a liability when it comes to seeking out soldiers hiding behind rocks or in trenches. If lasers are ever put on the battlefield, their job will be to hit targets such as fast-moving missiles or attack aircraft that are hard to destroy with conventional weapons.

Shooting at the Ground from Space

At first glance it might seem that anyone and anything on the ground is a sitting duck from the standpoint of an orbiting laser battle station. Fortunately, we can safely hide behind the same shield that protects us from dangerous radiation emitted by the sun-the atmosphere. We think of the air as perfectly transparent, but even at visible wavelengths it absorbs and bends light (which is why the stars twinkle instead of being steady pinpoints of light). These effects become much worse when you're trying to transmit a lot of light (in the form of a high-energy laser beam) through the air, as described in more detail in Chapter 5. The physics are complex, but the basic result is simple: the laser might not be able to deliver enough energy to the target to do it any harm.

Optical engineers and physicists have come up with a number of ingenious ways to compensate for atmospheric effects that work reasonably well when the air is close to the laser.[3] However, they can't overcome atmospheric scattering and absorption by clouds that might be in the way when shooting down at the ground from a satellite. The prospects for overcoming these limitations are poor. (It may be possible to send the beam from a ground-based laser up into space to be redirected by a battle mirror against potential targets, but the battle mirror would have to compensate for beam dispersion. Also, the laser could be put on a mountaintop with clear air, a site not likely to be picked for potential laser targets.)

There have been suggestions that space-based lasers could produce illumination intense enough to trigger explosions of flammable vapors around oil tankers.[4] Such tankers, as well as oil refineries and certain chemical plants, might be vulnerable to flash fires triggered by long-term laser heating to a critical temperature. However, it's hard to be sure *how* vulnerable they would be to laser attack. It seems easier to install a reflective aluminum umbrella to defend against laser attack than to protect against hostile terrorists equipped with bazookas. It's also worth noting that the price tag of a laser weapon system would be billions of dollars, far more expensive than equipping terrorists with bazookas and speedboats.

The atmosphere provides much better protection against many types of lasers because it does not let their beams through. Air strongly absorbs the wavelengths emitted by hydrogen fluoride chemical lasers, many excimer lasers, and X-ray lasers. In addition, air breaks up the type of particle beams most likely to be used in space, as described in Chapter 7.

Weapons of Mass Destruction

There is often an unfortunate confusion between *advanced* weapons and instruments of mass destruction. This may be a reaction to the threat of nuclear weapons, but though the concern is understandable, it leads to the wrong conclusions. In a sense laser and particle-beam weapons are the very antithesis of weapons of mass destruction because they focus their energy onto a small area. A true weapon of mass destruction, in contrast, strikes broadly and indiscriminately, like a nuclear bomb or nerve gas.

If a laser beam were spread out over a large area, the intensity would drop to low enough levels to be functionally harmless. If the beam from a 4-million-watt laser (a power level that has been considered for an orbiting battle station) were spread out over a spot 200 m (600 ft) in diameter, the resulting intensity would be roughly equal to that of the sun. If the beam were kept aimed at that area, it might make for a decidedly warm spell, but certainly not for mass destruction. An X-ray laser could make life much more unpleasant-but a beam from space would be rendered harmless by atmospheric absorption.

Microwaves can penetrate the atmosphere, but they face the same problem as lasers: in order to do damage rapidly, they must be focused onto a small area. Because of the long wavelength of microwaves, they would require a very large antenna. In theory, a 4-million-watt microwave source in a 400-km high orbit could focus its beam onto a spot 2 m in diameter on the ground and attain the power levels of 1000 W/cm^2 needed for heating.[5] Such a system would need a huge orbiting antenna 1 km (0.6 mile) across! If the satellite were in a higher orbit, above the tenuous remnants of atmosphere that would cause its orbit to decay, it would require an even larger antenna.

Fairly low levels of microwave exposure can be hazardous to humans, particularly over the long term. Health effects of microwaves have been the subject of intense controversy, and their exact nature remains unclear. The Soviet Union has stricter safety standards than the United States for microwave exposure,[6] indicating a belief in more severe hazards not shared by many U.S. specialists. Any effects appear to be both subtle and long-term in nature, and the evidence for them is so unclear that it's hard to imagine anyone spending the billions of dollars or rubles required to put a high-power microwave generator in orbit.

There is at least one possible scenario in which a particle-beam generator might serve as a weapon of mass destruction, although its feasibility and military usefulness are unclear. The idea, reportedly considered by the Soviet Union, is to point a powerful space-based particle generator down toward the earth and use the powerful beam to generate a lethal shower of high-energy radiation.[7] Energy would be transferred from the particles in the beam to atoms and molecules in the atmosphere generating radiation that would shower downward in a cone-shaped pattern. The Soviets raised this possibility in arms-control negotiation with the United States and proposed a treaty that would prohibit the use of charged or neutral particles to affect "biological targets," that is, people.[8] This could be taken as an indication of Soviet fears that the United States might develop such weapons or as a smoke screen to obscure Soviet efforts to develop them. Another suggestion is the use of a particle-beam generator on the battlefield to produce a similar effect; the beam would be scanned across the field to direct the secondary radiation at enemy soldiers.[9]

There is no doubt that secondary radiation exists; it's a well-known phenomenon produced when cosmic rays hit the top of the earth's atmosphere. It is unclear if it could be used as an effective weapon. The high energy of the secondary radiation would be much more deadly to soldiers than laser light. However, it is estimated that it would take a dose of about 18,000 J/m^2 to disable or kill a soldier. The beam would spread out rapidly once it hit the atmosphere because secondary radiation produced by the original particles would form a cone. It has been calculated that to be militarily useful such a weapon would have to produce over 10 billion watts for an hour, or a higher power for a shorter time.[10] That sort of performance is far beyond the state of the art on the ground, let alone in outer space. Even if the numbers aren't quite that bad, there certainly seem to be simpler and cheaper means of destruction.

No Planet Busters

At about the same time a major controversy erupted about particle-beam weaponry (see Chapter 2), George Lucas's film *Star Wars* opened, and the image of the planet-destroying *Death Star* has lingered in the minds of some people. Even people who should know better have invoked the *Death Star* vision in describing proposals for space-based beam weapon battle stations. The capabilities of the fictional *Death Star* are incredibly far beyond those ever envisioned for beam weapon battle stations. A very crude estimate indicates that it would take something on the order of 10^{30} J to vaporize the 6×10^{27} grams of material that comprise the earth. A 10-million-watt laser generates 10 million (10^{7}) J of energy per second. Advocates of laser propulsion have talked about lasers able to generate 10 billion watts in the far future, but that's still only 10^{10} J/sec. Even depositing energy at that fantastic rate, one laser would take three trillion years to vaporize the earth! In short, forget it. There are more than enough hydrogen bombs lying around to devastate the planet.

Beam Weapons and Science Fiction

Science fiction writers may have helped create many of the myths about beam weaponry, but they shouldn't be condemned for that. After all, creating fiction is their business. If blame is to be assigned anywhere, it should rest on the people who took science fiction too seriously.

Although science fiction writers have proved prophetic in some cases, such as H. G. Wells's vision of the high-energy laser, they've also missed the mark many times. The prevalent vision of 1930s pulp science fiction probably would have had the people of the early 1980s being waited upon by humanoid robots while they chatted by videotelephones. The technology didn't work that way. Predictions of mass television service were on target, but those of videotelephone service missed badly-something the American Telephone and Telegraph Corp. learned the hard way when it tried to market Picturephone to the general public. Real-world robots have turned out to be industrial devices with no resemblance to the creations of science fiction writers other than their metal exteriors. Science fiction has a real charm and fascination, but it never deals with the hard realities of beam-weapon technology described in the chapters that follow.

References

1. C. Breck Hitz, "The laser marketplace-1983," *Lasers & Applications* 2(1), 45-53 (January 1983).
2. For an authoritative description of laser hazards, see: David Sliney and Myron Wolbarsht, *Safety with Lasers and Other Optical Sources* (Plenum Press, New York, 1980).
3. Philip J. Klass, "Adaptive optics evaluated as laser aid," *Aviation Week & Space Technology* 115, August 24, 1981, pp. 61-65.
4. David Ritchie, *Spacewar* (Atheneum, New York, 1982), p. 115.
5. Ervin J. Nalos, "New developments in electromagnetic energy beaming," *Proceedings of the IEEE* 66(3), 276-289 (March 1978).
6. Eric J. Lerner, "RF Radiation: biological effects," *IEEE Spectrum* 17(12), 51-59 (December 1980).
7. "U.S. eyes Soviet offensive beam weapon plans," *Aviation Week & Space Technology*, November 13, 1978, p. 16.
8. Arms Control & Disarmament Agency, *Fiscal 1983 Arms Control Impact Statement* (U.S. Government Printing Office, Washington, D.C., 1982), p. 324.
9. M. Din, "The prospects for beam weapons," in Bhupendra Jasani, ed, *Outer Space: A New Dimension of the Arms Race* (Oelgeschlager, Gunn & Hain, Cambridge, Massachusetts, 1982), pp. 229-239.
10. Peter Laurie, "Exploding the beam weapon myth," *New Scientist*, April 26, 1979, pp. 245-250.

4.
High-Energy Laser Technology

The word "laser" was coined as an acronym for light amplification by the stimulated emission of radiation. Strictly speaking the term describes a physical process, but it has always been applied to devices that use the process to produce light. The earliest lasers bore a strong family resemblance to each other. However, over the past two decades the laser family tree has grown immensely, to include devices that seem at first glance to have little in common. A laser may be a tiny semiconductor chip similar in structure to the red LEDs (light-emitting diodes) used in some calculator displays, which can produce thousandths of a watt of invisible infrared light when an electric current is applied to it. Or the term can describe a building-sized behemoth full of complex piping that takes some of the energy from the burning of chemical fuels and transforms it into a beam of light with a billion times more power-well over a million watts.

The fundamental similarity of lasers remains, but the process by which atoms or molecules are made to emit light is on a level not easy to see. The same type of process takes place whether the light is produced by a solid, a liquid, or a gas and whether the power is measured in thousandths of a watt or millions of watts. Understanding that process is vital to learning how high-energy lasers work, and it requires a quick trip through the world of atomic physics.

The Quantum View of Atomic Physics

Laser physics, based on a quantum-mechanical view of atoms and molecules, is called "quantum electronics" by physicists working in that area. The field is an outgrowth of the revolution in our understanding of physics that took place during the first part of the twentieth century, and an explanation requires laying some groundwork-describing fundamental concepts of atomic physics and quantum mechanics.[1]

The best way to begin is with the simplest atom, hydrogen, in which a single electron circles a nucleus made up of a single proton. The laws of quantum mechanics limit the electron to only certain possible orbits around the nucleus, and each of these orbits has a specific energy. That is, when an electron occupies an orbit, it has exactly the amount of energy specified for that orbit. The orbit closest to the nucleus has the least energy, and the electron energy increases as it rises to orbits farther from the nucleus. (Physicists assign each of these orbits an integral quantum number, counting outward from the nucleus, to keep track of them.) The energy difference between successive orbits (often called "energy levels" or "quantum levels") decreases as the size of the orbit increases. It eventually reaches a nominal limit of zero at the point called the "ionization limit," where the electron has accumulated enough energy to escape from orbit around the nucleus and leave the atom altogether.

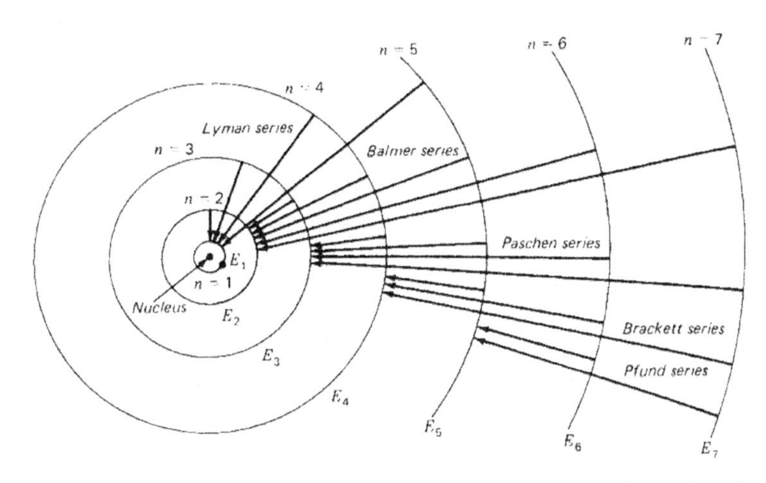

Energy levels or orbits in a hydrogen atom. As the electron moves farther from the atomic nucleus, its energy increases. If the electron absorbs a photon of light, it can jump to a higher energy level; if it emits a photon, it drops to a lower level. The transitions take place only when the photon energy is the same as the difference in energy levels of the electron orbits. The lines between orbits show transitions; the names are those used by physicists to identify series of wavelengths at which hydrogen emits or absorbs light because of those transitions. [Courtesy of Harvey E. White, from Frances A. Jenkins and Harvey E. White, Fundamentals of Optics (McGraw-Hill, New York, 1976, 4th ed.) p. 616.)

For an electron to make a "transition" from one orbital energy level to another, it changes its energy by an amount equal to the difference between the two energy levels. Moving to an outer orbit or a higher energy level requires an increase in energy, while a transition to a lower energy level or inner orbit requires a loss of energy. Typically, an electron gains or loses energy by absorbing or emitting electromagnetic radiation–visible, infrared, or ultraviolet light in most cases. The electromagnetic radiation is energy in a chunk called a photon. If the electron is dropping to a lower energy level, the photon it emits carries away the difference in energy between the two levels. To move to a higher level, an electron must absorb a photon carrying enough energy to boost the electron's total energy to the amount needed to reach the next rung on the energy level ladder.

It's important to stop here and stress that light and all other forms of electromagnetic radiation actually have a dual identity: they are both waves and particles. In certain cases light behaves like a wave, made up of electric and magnetic fields that are simultaneously oscillating at a definite frequency to make waves that have a measurable wavelength (the distance from the peak of one wave to the peak of the next). In other cases light behaves as if it is made up of particles, the photons (which are the "quanta" of electromagnetic radiation). Although it may seem contrary to logic and reason, whether you see light as a wave or as a particle depends on how you look at it.

Electromagnetic radiation comes in many forms: light, radio waves, infrared and ultraviolet radiation, microwaves, gamma rays, X rays, and a few other types that fall into the niches between the better known categories. There are three fundamental ways to identify each type: by the frequency with which the wave oscillates, the length of the wave (wavelength), or the energy of the photons. Actually, all three methods measure the same quantity, but use different yardsticks. Because electromagnetic radiation always travels at the speed of light, the frequency is simply the speed of light divided by the wavelength. The energy carried by an individual photon is the frequency of the electromagnetic wave multiplied by a constant. Thus, given one quantity, it's simple to calculate the other two. That means that the light emitted or absorbed by an electron making a particular transition has not only a certain amount of energy, but also a certain frequency and wavelength. For example, physicists might speak of a particular transition as having a wavelength of 3 μm (micrometers, or millionths of a meter), a frequency of 10^{14} Hz [100,000 GHz (gigahertz) or 100 trillion cycles per second], or an energy of 0.4 eV [electronvolt, the amount of energy an electron gains in passing through a potential of 1 volt; 0.4 eV is about 6×10^{-20} J (joules)].

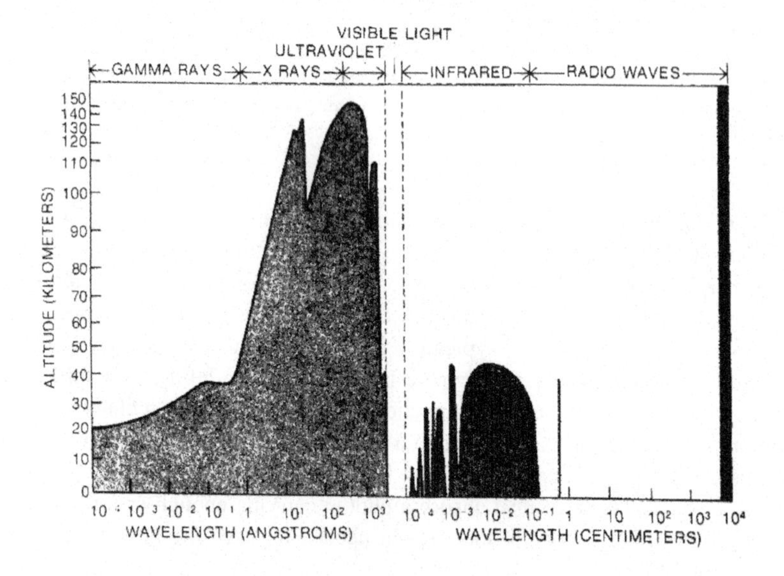

The types of electromagnetic radiation and how far they can travel through the atmosphere. The curve shows how far into the atmosphere radiation from space can penetrate before half of it is absorbed; wavelengths are indicated at the bottom (1 cm equals 100 million or 108 Å). Because this curve spans the whole electromagnetic spectrum, it does not show the atmosphere's many narrow absorption lines, which are particularly common in the infrared. [Courtesy of W.H. Freeman and Company, reproduced from Leo Goldberg, "Ultraviolet Astronomy," Scientific American, June 1969, p. 95.]

The simple structure of the hydrogen atom makes it fairly easy to calculate its energy-level structure and the nature of the light emitted as its single electron jumps between different orbits. However, hydrogen is by far the simplest atom. In atoms with more electrons, interactions can take place among the electrons themselves as well as between the electrons and the nucleus. Subtle interactions with the larger and more complex nuclei of these heavier atoms can split energy levels into sublevels. Other considerations of atomic structure enter the picture, leading to a complex array of electron orbits arranged in a series of "shells." With many electrons in an atom, it becomes essential to consider the Pauli exclusion principle, which states that only one electron can occupy each unique energy level.[2] In short, the energy level structures of most atoms are extremely complex.

There's also another layer of complexity to consider. Virtually all of the atoms in our world are bound together in molecules and/or in solid or liquid structures. Molecules can vibrate and rotate, and the vibration and rotation are quantized just as are the electron energy levels in an isolated atom. As a general rule, electronic transitions have higher energy than vibrational transitions, which in turn have higher energy than rotational transitions. The distribution of energy into vibrational and rotational modes creates additional energy levels, which sometimes can interact. For example, it's common for a molecule undergoing a vibrational transition to simultaneously change rotational state. It's easy for the molecule, but it's much harder on the physicist trying to keep track. To avoid unneeded complexity, most of the rest of this chapter will talk about energy-level transitions in general rather than specific types.

There are also what are called "nonlinear" effects, which in a sense put a joker into the deck. They can cause unusual things to happen in certain circumstances, such as a doubling of the frequency of light when high-intensity beams pass through certain materials.

What does all this have to do with the laser? Laser action is possible only when atoms and molecules have their energy distributed in certain ways. A complex set of rules determines what types of energy-level transitions are possible in atoms and molecules. 3 Only a small percentage of these possible transitions can serve as the basis for laser action. However, there are so many possible transitions that many types of lasers are possible, and even after more than 20 years of searching, physicists have come nowhere near exhausting the possibilities.

Stimulated Emission

One of the properties that makes laser action special is how it emits light. Virtually all of the light we see from "normal" sources such as light bulbs and the sun is spontaneous emission-light produced when an atom or molecule drops to a lower energy level without outside interference. About 65 years ago, Albert Einstein suggested that it might also be possible to stimulate emission of light from an atom or molecule in a high (or excited) energy state. The stimulation would require a photon from another source with an amount of energy corresponding to the difference between the excited energy level and some lower energy level. This external photon would, in essence, tickle the excited atom or molecule, stimulating the emission of a second, identical photon.

Stimulated emission sounds like a rather unusual process, and in the ordinary world it is. It works only when a photon with just the right energy is in the right place at the right time. Practically speaking, the most likely place to get such a photon is from another identical atom or molecule that has just undergone the same transition. It won't do much good simply to collect many identical atoms or molecules because of the workings of thermodynamics. Atoms or molecules normally tend to be in their lowest energy or "ground" states. At any temperature above absolute zero (the point at which all molecular motion would theoretically stop), some are in higher energy levels. But under normal conditions-what physicists call "thermodynamic equilibrium" because everything is in balance-the higher the energy level, the fewer atoms or molecules are in it.

Consider what would happen to a photon emitted when an atom makes a transition from a higher (or upper) energy level to one with lower energy. Because there are many more atoms in the lower level than the higher one (the population of the lower level is larger, in physicists' terms), the photon would be likely to meet an atom in the lower energy level that could absorb it before it could stimulate emission from an atom in the upper level. Stimulated emission could occur–and indeed was observed in the late 1920s–but it seemed doomed to be overwhelmed by spontaneous emission.

That line of reasoning was sufficiently obvious that for many years physicists considered stimulated emission an interesting phenomenon of no practical significance. The problem, in the words of Arthur Schawlow, was that "every scientist was trained to view the world as close to being in equilibrium. Really radical departures from equilibrium would be needed for stimulated emission to dominate, and that seemed unthinkable."[4]

What was needed was a nonequilibrium condition called a "population inversion," in which more atoms or molecules were in an upper energy level than in a lower one. Charles Townes produced one for his first ammonia maser (for microwave amplification by the stimulated emission of radiation) by physically separating energetic ammonia molecules from those in lower energy levels. His approach worked for microwave transitions, but different methods are needed to produce population inversions on the more energetic transitions needed to produce shorter wavelengths, such as light. Population inversions in lasers are produced by using an electric current, intense light, or a chemical reaction to generate atoms or molecules in an excited state. The idea is to selectively populate an upper energy level with many more atoms or molecules than would be in the lower energy level of the transition. Thus, a spontaneously emitted photon on that transition would be more likely to encounter an atom or molecule in the upper energy level (from which it could stimulate emission) than one in the lower level, which might absorb it. The result is light amplification by the stimulated emission of radiation, a net gain or increase in the number of photons, otherwise known as laser action.

Strangely enough, this sort of laser action is not simply an artificial, man-made process. After the laser was invented, radio astronomers discovered that clouds of hot interstellar gas generate stimulated emission in the microwave region. These gas clouds are called cosmic masers because they rely on physical processes that also occur in laboratory masers. What's even stranger is that the upper atmosphere of the planet Mars is acting as a laser. Carbon dioxide molecules far above the planet's surface are producing stimulated emission at 10 μm, the same infrared wavelength produced by man-made carbon dioxide lasers. Total power of the Martian laser is estimated at 5 billion watts (although for reasons described below, this 5 billion watts is not in a narrow beam).5 Both cosmic masers and the Martian laser are made possible by excitation of the gas by an external light source (a hot star or the sun) to produce a population inversion in the gas. It turns out that nature does not always believe in thermal equilibrium.

Generation of a Laser Beam

A population inversion is not the whole story behind what we know as a laser, however. Look at how stimulated emission works in a loose blob of gas such as a cosmic maser. An initial spontaneously emitted photon stimulates the emission of an identical photon from another atom ready to make the same transition (i.e., an atom in the upper energy level of the laser transition). The resulting two photons continue on their way in the same direction and can stimulate emission from other excited atoms as they pass through the gas. Eventually, their path takes them out of the gas, and nothing further happens.

As long as there is nothing in or surrounding the gas to deflect the light (or, in the case of cosmic masers, microwaves), it travels a straight line from where it was emitted. Along its path through the gas, it may be amplified by stimulated emission. However, the directions of the straight lines are essentially random, with no preferred orientation. The result is like a light bulb, but one that produces stimulated emission rather than ordinary light; its output would be concentrated at the wavelengths where stimulated emission occurs, but the light would go in every direction. The intensity of emission at particular wavelengths indicates that the source is stimulated emission Depending on the shape of the gas cloud and the locations of the surrounding stars, a cosmic maser may emit more intensely in some directions than in others, but it will not produce anything resembling the narrow, highly directed beam from a man-made laser. The Martian atmosphere operates in the same way, something for which we should be thankful. If the 5 billion watts thought to be emitted by that planetary laser was concentrated in a narrow beam, it would make a truly awesome natural laser weapon, a veritable "loose cannon" in the solar system.

Amplification by the stimulated emission of radiation occurs in both cosmic masers and the Martian laser. However, their diffuse output means that they can't concentrate high power onto a small area, a necessity for applications such as weaponry. Explaining how high-intensity beams are produced requires the introduction of two more concepts: laser gain and the laser resonator.

Laser gain is the degree of amplification experienced by light passing through a laser medium. It's usually measured per unit distance traveled. A typical value of gain is a few percent per centimeter, meaning that for each centimeter the beam travels through the laser medium, its power increases a few percent. Thus a 1-W beam would be amplified to 1.05 W after passing through 1 cm of gas with gain of 5% (0.05) per centimeter. The gain is cumulative, much like compound interest, so the further the beam goes, the higher its power.

It might seem simple to produce a highly directed beam from a long, cylindrical laser medium, but in general that approach provides neither high power nor a narrow beam. A few laser materials with very high gain can operate in that "superradiant" way, but their power is still limited, and the beam spreads out rapidly (at least for a laser). What is needed to obtain an intense beam of good quality from a laser of reasonable length is a pair of mirrors, one on each end of the laser. The mirrors reflect the light back and forth through the laser over and over again. Each pass amplifies the beam, which grows stronger on each trip through the laser. The two mirrors function as a resonator (sometimes called a laser cavity), preferentially amplifying light originally emitted along the line between them so that it will make many trips between the two. Light emitted in other directions leaks out of the laser and is lost before much energy is invested in amplifying it.

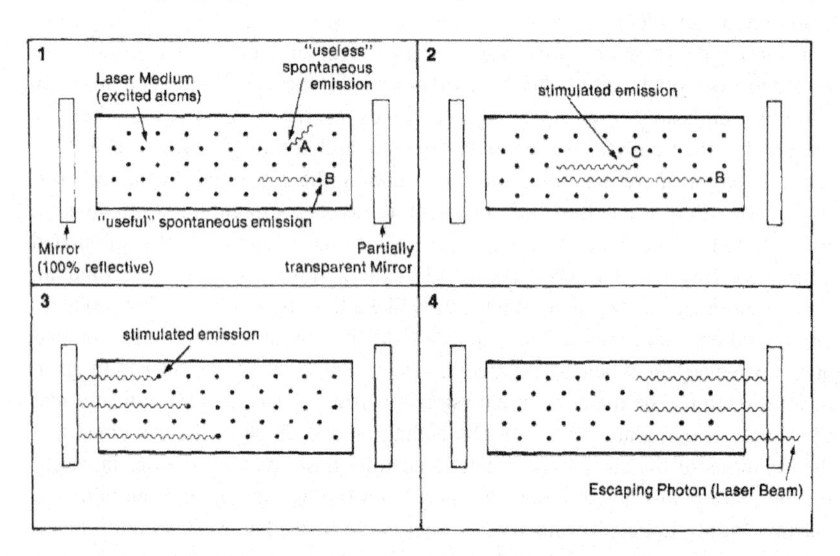

A simplified view of what happens inside a laser. Light spontaneously emitted along the axis of the laser resonator stimulates the emission of identical photons, generating light that bounces back and forth between a pair of mirrors. Some of the light leaks out one of the mirrors to form the laser beam seen by outside observers, while the rest of the light is reflected back into the laser medium to continue stimulating more emission. [From Laser: Supertool of the 1980s, by Jeff Hecht and Dick Teresi, Copyright 1982 by Jeff Hecht and Dick Teresi. Reprinted by permission of Ticknor & Fields, a Houghton Mifflin Company.]

The detailed design of resonators plays an important role in determining how well a laser operates, but those details are far beyond the scope of this book.[6] A couple of factors should be mentioned. In practice, one (or sometimes both) of the laser mirrors lets a small fraction of the light inside the laser leak out, forming the laser beam that is seen. Typical laser output mirrors let a few percent of the light leak out, but the percentage varies widely among lasers.

The way in which light is reflected back and forth between the mirrors also affects one characteristic critical to laser weaponry: the distribution of energy within the beam. The standard type of resonator produces a beam with most of its energy in the middle, but other designs produce different energy distributions. One type produces a ring or "donut"-shaped beam that carries little of its energy in the center, a configuration that may make it easier to get the laser beam through the air to its target.

The concentration of a laser's energy into a narrow beam is vital for weaponry because it makes it easier to damage a target with the limited power available from the laser. The damage produced by a laser beam usually depends on the intensity, that is, the amount of power or energy per unit area. In many cases it is only necessary to produce the damage in a small area, for example, puncturing the fuel tank of a missile and triggering an explosion. The degree of concentration in a laser beam is impressive. Small helium-neon lasers such as those used in high-school physics laboratories produce only a thousandth of a watt of red light, yet because that light is contained in a beam only 1 mm (1/25 of an inch) in diameter, its intensity is comparable to that of sunlight. An ordinary incandescent bulb, which requires about 100 W of electricity, produces about 1 W of light (the rest is lost as heat), which can illuminate a whole room, but at much lower intensity. Because the beam is highly concentrated, a power level that could be modest in other circumstances is powerful in a laser. A 1-kW (kilowatt) electric space heater should be only a few feet away to warm you with its diffuse output in a chilly room, but a 1-kW laser beam can be a powerful industrial cutting tool. A power level of 300 kW would suffice for laser weapon demonstrations, but it's equal to 400 horsepower, the capacity of some large automobile engines.

Laser beams are not perfectly directional; that's impossible even in theory. They spread out very slowly at an angle that depends on the design of the laser resonator or cavity and on the type of output optics (if any) that are used. For a typical small helium-neon laser the beam spreads out at an angle of 0.001 radian [1 milliradian (mrad) or about 0.06°].[7] This corresponds to the spreading of one part in a thousand, meaning that after the beam has traveled 1000 m, it will have expanded to 1 m in diameter. With sophisticated optics, it's possible to reduce the divergence to around one-millionth of a radian (1 μrad), meaning that it would expand to a diameter of 1 m only after traveling 1 million meters, or 1000 km. For many lasers the beam divergence is close to the theoretical minimum set by the diffraction of light, which is roughly equal to the wavelength divided by the diameter of the focusing optics.

The diffraction limit holds not only if the beam is small to start with, but also if it is large. In the case of a large beam it means not how rapidly the beam spreads out but rather how small a spot it can be focused to. Thus, a beam with a 1-μrad diffraction limit could be focused to a spot that formed a 1-μrad angle from the viewpoint of the laser. In theory this means it could be focused to a spot

1 mm in diameter at a distance of 1 km from the laser, even if the output beam was 1m in diameter. This turns out to be an important limit on the performance of long-distance laser weapons, and is discussed in more detail in Chapter 5.

Low- and High-Energy Lasers

The first lasers emitted comparatively weak pulses or low-power continuous beams, but physicists quickly started increasing the power that could be produced. An early landmark was the ability to punch a hole in a razor blade. Because the short, intense pulses able to perform that feat were hard to measure otherwise, laser developers began measuring output power in "gillettes," after the maker of razor blades, with one gillette being a pulse strong enough to drill through one blade, and so on. Before long, however, they started finding limits on how far they could go.

Limits were inherent in the nature of the first laser material-synthetic ruby crystals. Ruby is typical of several important types of lasers in which light from an external source (usually a flashlamp) energizes atoms embedded in a glass or crystalline rod. This produces a population inversion in the rod, and when mirrors are placed on each end, it serves as a laser. Unfortunately, only a small fraction, typically around 1%, of the light that goes into the rod emerges in the form of laser light. Most of it ends up in the form of heat, which must be removed lest its effects on the laser rod break up the beam or damage the rod itself.

Heat removal poses a serious problem in ruby lasers. Although the external light source efficiently deposits energy throughout the transparent rod, the excess heat is much slower in leaving the solid. This fact, combined with some complications due to energy levels in the material, limits operation of the ruby laser to no more than a few pulses per second except at very low power levels, and also sets upper limits on the practical output power. In general, the higher the pulse power, the longer you must wait between pulses. There are other crystalline and glass lasers that are more efficient or easier to cool than ruby, but the fundamental problem remains: crystalline and solid- state lasers can only get so big.

Crystalline or glass lasers can produce very high peak powers in very short pulses, but they are used in fusion research, not for weaponry. For example, the now-dissembled Shiva laser system at the Lawrence Livermore National Laboratory produced pulses that had a peak power of some 20 trillion watts, but lasted only about 0.2 billionths of a second. The problem from the standpoint of weaponry is that the laser could not be fired very often. The mammoth 20-beam laser, which occupied a four-story building, averaged about one shot per day during 1979.[8] That's fine for fusion research, but not if somebody is shooting missiles at you. It is possible to do somewhat better, and the Soviet Union has built some big crystalline lasers, but the fundamental limitations remain. Crystalline lasers may be helpful in weapon-related research, but at least in the United States nobody seems to be counting on them for use as weapons.

Pair of six arms of the 20-arm Shiva solid-state laser used for experiments in laser fusion at the Lawrence Livermore National Laboratory in California. The now-dissembled laser occupied the interior of a four-story building and produced extremely high powers in very short pulses. A new laser called Nova is being built in its place to produce higher powers. (Courtesy of Lawrence Livermore National Laboratory.)

Sustained operation at high power requires a laser material that can efficiently get rid of the waste heat inherent in laser operation. A gas is a logical candidate, if it can be made to flow rapidly and efficiently through the laser to remove the excess heat. Making gas flow fast and even turns out to be so important that big lasers tend to resemble jet engines. With the exception of the free-electron laser (described at the end of this chapter) and the X-ray laser (described in Chapter 6), all well-developed high-energy lasers rely on flowing gases.

The Carbon Dioxide Laser

The first high-power gas laser to come on the scene was the carbon dioxide laser.9 First demonstrated in 1964 by C. Kumar N. Patel at Bell Telephone Laboratories, the carbon dioxide laser produces invisible infrared radiation at wavelengths of around 10 μm, some 20 times longer than the wavelength of visible light. Its technology is fairly simple, and by laser standards it is an efficient device, able to convert up to around 20% of the input energy into laser output. It can produce either a continuous beam or short pulses of light, and its 10-μm output is transmitted reasonably well by dry air. These advantages have made carbon dioxide lasers workhorses in a variety of applications including metal cutting and welding. A wide variety of models are offered commercially, with power outputs running from about 1 to 20,000 W.

Patel produced a laser beam by passing an electric current through pure carbon dioxide. The fast-moving electrons in the discharge transferred energy to the gas molecules, causing them to vibrate energetically. As the carbon dioxide molecules drop to lower vibrational energy levels, they emit photons at a number of wavelengths centered around 10 μm. The electric discharge raises a large fraction of the molecules to high vibrational energy levels, creating a population inversion that makes laser action possible at this wave- length. (Actually, there are two vibrational transitions and a couple of dozen rotational transitions that can combine to make around 100 different laser transitions-and the same number of discrete wavelengths-possible. Some low-power carbon dioxide lasers are made to emit only one of these wave- lengths, but natural processes in high-power lasers concentrate the emission on just a few of the strongest transitions.)

The efficiency of energy transfer from electrons to carbon dioxide molecules can be increased by adding nitrogen to the gas. The nitrogen is an intermediary, collecting energy from the discharge and transferring it to the carbon dioxide, something that is possible because the two gases have some closely spaced energy levels.

Laser physicists have also learned to add helium to the gas mixture. The helium atoms play a different role, helping to knock carbon dioxide molecules out of the lower levels of the 10-μm transitions. This helps sustain the population inversion essential to laser operation by preventing an accumulation of molecules in energy states ripe to absorb 10-μm photons, which could stifle laser operation.

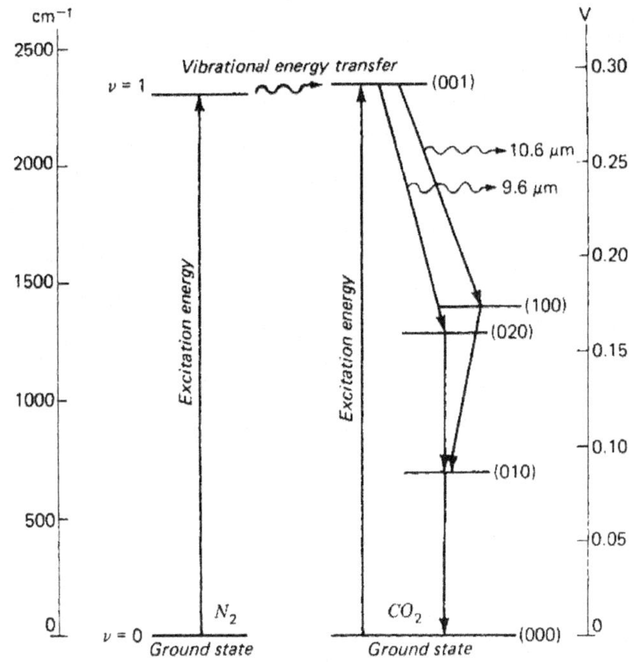

Energy levels in a carbon dioxide laser, showing how energy is collected by nitrogen molecules, then transferred to carbon dioxide, which is stimulated to emit light at two infrared wavelengths near 10 μm. [Courtesy of Harvey E. White, from Francis A. Jenkins and Harvey White, Fundamentals of Optics (McGraw-Hill, New York, 1976, 4th ed.) p. 646.]

Composition of the laser gas is thus an important factor in determining how well a laser will operate, and the ideal composition varies with the design of the laser. Gas pressure is also an important variable. The energy- and heat- transfer processes required for continuous laser operation work best at pressures between a fraction of one percent and several percent of normal atmospheric pressure. Pulsed carbon dioxide lasers can be made to operate at pressures to about ten times that of the atmosphere.

Patel and those who followed him in developing the first generation of carbon dioxide lasers put the gas in long tubes and passed both the gas flow and the electrical discharge along the length of the tube. They soon ran into limits. The long, narrow path limited the rate of gas flow, thereby slowing the removal of waste heat and contaminants produced by the discharge. At best, power was around 100 W/m of tube length, and as length increased, the law of diminishing returns also began to set in. The record power for such a laser apparently was 8800 W, set in the late 1960s by a group that used a collection of tubes totaling 750 ft (230 m) in length-not exactly a battlefield-ready laser.

The Gasdynamic Laser

The breakthrough to high continuous powers, which the Pentagon at one point officially defined as anything above 20,000 W,[10] came with the development of a variation of the carbon dioxide laser known as the gasdynamic laser. The name comes from the same "gasdynamic" process that produces the population inversion. Instead of passing an electric current through a cool gas, the gasdynamic laser gets its energy from a hot mixture of gases, generally produced by burning a fuel in oxygen or nitrous oxide. The hot gas, which is at thermal equilibrium and at a pressure typically well above that of the atmosphere, is then expanded through special nozzles into a near-vacuum. This expansion cools the gas, and if it occurs faster than the excited molecules can spontaneously emit their excess energy, it also produces a population inversion.

The gas flow is rapid, much faster than the speed of sound at atmospheric pressure (i.e., supersonic). Still the population inversion doesn't last long, appearing only after the gas has passed through the nozzles, and continuing only while the gas speeds across several more centimeters (a few inches) in the laser cavity. Energy can be extracted in the form of a laser beam by putting resonator mirrors on each side of the place where the population inversion exists in the flowing gas.

In practice, some fancy engineering is needed to make gasdynamic lasers work. The size and shape of nozzles are critical, as is the whole problem of gas flow. Gas composition is also important, with nitrogen used to aid energy transfer to the carbon dioxide molecules, and some "quencher" gas (such as water produced by combustion) needed to help remove molecules from lower excited levels that might otherwise absorb 10-μm laser radiation. Nonetheless, the powers that can be reached are impressive, particularly by the standards of 1970, when development of the gasdynamic laser was first publicly dis- closed. The first unclassified papers reported powers of 60,000 W.[11] Theoretical projections indicated that it should be possible to extract a few thousand joules per pound of gas flowing through the laser.

Researchers soon pushed power levels into the hundreds of thousands of watts. Gasdynamic lasers were used in some important laser weapon demonstrations. What may be the most powerful of these systems is a 400,000-W gasdynamic laser that today is at the heart of the Air Force's Airborne Laser Laboratory.

As gasdynamic lasers grew, their problems became painfully apparent. Their efficiency is lower than that of other carbon dioxide lasers, meaning that more waste heat must be dissipated. The atmosphere does not transmit the beam well. The gasdynamic laser is both bulky and complex, and there is little tolerance for error in design or manufacture of some critical components. One developer, speaking at an informal seminar I attended at the Massachusetts Institute of Technology, summed up the problem by calling the gasdynamic laser a "ten-ton watch." The problems are serious enough that the Pentagon has written the gasdynamic laser and other continuous carbon dioxide lasers out of its high-energy laser development plans, except for use as a demonstration tool.

Electrically Excited Carbon Dioxide

One type of carbon dioxide laser does remain in the running for some weaponry applications-an improved version of the electrically excited pulsed carbon dioxide laser. In this "transverse-flow" design the gas flow and the electrical discharge are perpendicular both to the axis of the laser resonator (the line between the two mirrors) and to each other. As in the axial-flow version described earlier, energy is transferred to the gas from electrons injected into it, either in an electrical discharge or in a beam of energetic electrons. The transverse-flow design makes it easier to get rid of excess heat and contaminants and permits much higher powers, around 10,000 W/m of resonator length, in commercial versions.[12] This approach is used in commercial lasers producing powers of a few thousand to about 20,000 W for metal working and other industrial applications, as well as in weapon development.

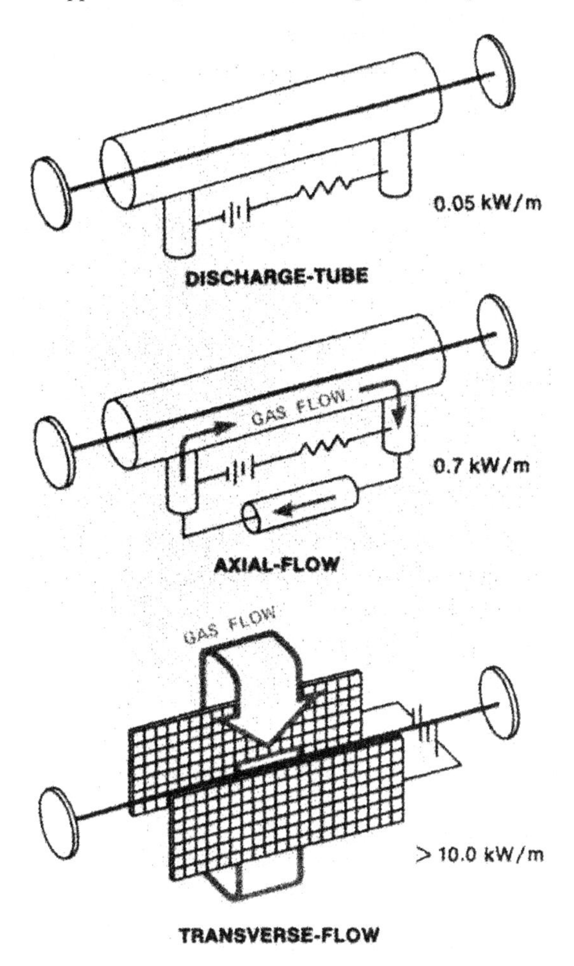

Types of carbon dioxide lasers, with output power per unit length indicated in kilowatts per meter. The discharge- tube devices shown at top produce comparatively feeble powers because they do not make the gas flow through the tube. Higher powers can be obtained by forcing the gas along the axis of the tube, but the highest powers require flowing the gas perpendicular to both the laser beam axis and the electric dis- charge (applied by the rectangular grids on either side), as shown in the bottom drawing. (Courtesy of Avco Everett Metalworking Lasers, Somerville, Mass.)

Gas flow is almost as important in these "transverse-flow" carbon dioxide lasers as it is in gasdynamic types. One manufacturer of high-power industrial carbon dioxide lasers, Avco Everett Metalworking Lasers of Somerville, Massachusetts, simply labels one component of its 15,000-W laser as a "wind tunnel."[13] However, electrical excitation avoids other important problems of gasdynamic lasers, particularly limited efficiency and some of the technical problems that make operation cumbersome. The Army has been considering electrically excited carbon dioxide lasers for some battlefield applications.

The biggest problem with the electrical approach is the need for a source of electrical power. Battlefields don't come equipped with power lines, so a laser weapon would have to carry its own power supply. Electrical power supplies are heavy and bulky, leading to one of the classic images of the laser weaponry field: a compact laser linked by a thick power cord to an immense power supply. Engineers have proposed, with tongue in cheek, squeezing a laser weapon into a small fighter aircraft and flying the power supply alongside the fighter in a jumbo jet. The possibility of a separate power supply on the battlefield does seem to be taken at least somewhat seriously; the first edition of the Pentagon's booklet Soviet Military Power included an artist's conception of a battlefield laser weapon which looks like two tanks-one housing the laser and the other housing the power supply-linked by a power cable.[14]

There is some military interest in developing compact power supplies that could provide electrical power in outer space for high-energy lasers and other power-hungry equipment. The Defense Advanced Research Projects Agency is sponsoring development of a 100,000-W nuclear generator that may eventually lead to generators capable of producing millions of watts of electricity for use in outer space.[15]

Carbon Monoxide Lasers

Another type of electrically excited laser is also worth mention, the carbon monoxide laser. This type is operationally similar to the carbon dioxide laser, also emitting on an array of combined vibrational and rotational transitions, but its wavelength is near 5 μm.

The big attraction of the carbon monoxide laser is its efficiency, in theory the highest of any type of electrically excited gas laser. Some impressive efficiency measurements have been made in the laboratory,[16] but the results were not calculated in the same way as for carbon dioxide lasers; the actual efficiency with which carbon monoxide lasers convert input power into a laser beam is much lower than the laboratory results would indicate. High powers have been produced, but the carbon monoxide laser suffers from two flaws that appear fatal for a laser weapon. Water vapor and other gases strongly absorb the laser's 5-μm wavelength, a problem not present with visible light or the 10-μm output of the carbon dioxide laser. In addition, to operate efficiently the laser must be kept at temperatures of 77 to 100 °K (roughly -230 to -200 °C or -380 to -330 °F), the so-called "cryogenic" temperatures at which gases such as nitrogen and oxygen turn to liquids.[17] The need for such cooling presents serious problems, and military researchers appear to have abandoned carbon monoxide lasers.

Chemical Lasers

Most demonstrations of laser weapon concepts in the works involve the chemical laser.[18] As its name implies, it derives its energy from a chemical reaction, the combination of hydrogen and fluorine to produce molecules of hydrogen fluoride in a vibrationally excited state. Strictly speaking, the hydrogen fluoride laser is only one member of a much larger family of chemical lasers, which derive their energy from a variety of chemical reactions. However, in practice, the term "chemical laser" usually refers to hydrogen fluoride lasers, including types in which hydrogen's heavier natural isotope, hydrogen-2, or deuterium, is substituted for the much commoner hydrogen-1 to get a wavelength more readily transmitted by air.

The reaction in a chemical laser can be triggered by an electric discharge, but it is simplest to view the process as being purely chemical. The starting point is a fuel containing hydrogen and an oxidizer containing fluorine. Sometimes pure hydrogen and fluorine are used, but those two gases are difficult to handle, particularly the toxic and highly reactive fluorine. The hazards led the builders of one large chemical laser to half-seriously talk about the "fire of the week," because every time there was a gas leak there would be a fire. Fluorine also has a reputation for such nasty habits as making stainless steel tanks explode unexpectedly. Because of these problems, developers would prefer easier-to-handle compounds that would release hydrogen and fluorine when needed. Thus, the fuel might be a hydrocarbon compound (such as alcohol) and the oxidizer nitrogen triflouride.

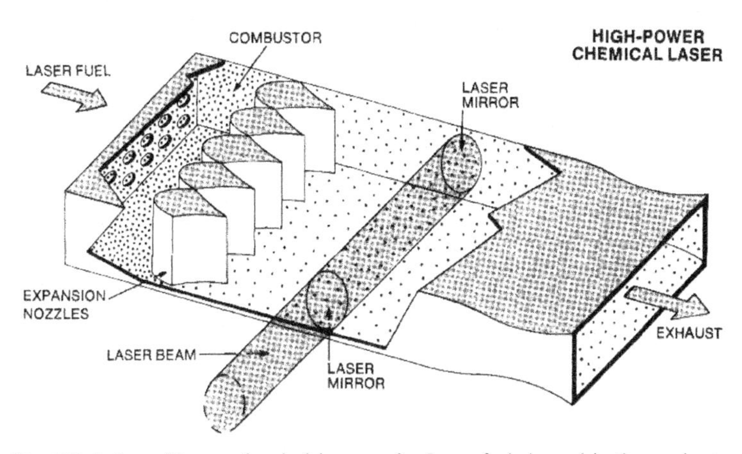

Simplified view of how a chemical laser works. Laser fuels burned in the combustor are passed through expansion nozzles, then the laser energy is extracted from the hot gas as it passes rapidly between a pair of laser mirrors. (From Laser: Supertool of the 1980s by Jeff Hecht and Dick Teresi, Copyright 1982 by Jeff Hecht and Dick Teresi. Reprinted by permission of Ticknor & Fields, a Houghton Mifflin Company.)

The first step in chemical laser operation is getting the fuel and oxidizer to release free atoms of hydrogen and fluorine, respectively. This is just a preliminary reaction that involves only a small fraction of the fuel and oxidizer; its role is to provide enough of the highly reactive atoms to trigger a continuing chemical reaction. The fuel and oxidizer gases are then passed through separate sets of nozzles in an arrangement that looks similar to those used in gasdynamic lasers. After passing through the nozzles, the fast-moving gases mix and react to form vibrationally excited hydrogen fluoride molecules. If the reactants are pure hydrogen and pure fluorine, the result is a self-sustaining chain reaction:

$H + F_2$ --> HF* + F
$F + H_2$ --> HF* + H
$H + F_2$ -=> HF* + F
ad infinitum

The chemical reaction produces about 500,000 J of energy per pound of chemicals, much-but by no means all-of which can be extracted in a laser beam.[19] The reaction produces a population inversion in the rapidly flowing gas, which passes between a pair of laser mirrors, where the laser beam is produced. The rapidly cooling gas passes quickly by the mirrors and ultimately into an exhaust system. As long as the gas keeps flowing, the laser can produce a continuous beam.

There are a number of operational complications. The cavity into which the laser gas flows must be kept at a very low pressure, typically 1% or less of atmospheric pressure. There must also be a way to get rid of the gas after it is passed by the laser mirrors. In the vacuum of space it could simply be vented to the outside. That doesn't work on the ground because the pressure differential works the other way: instead of letting the hydrogen fluoride out, a vent would let the air in. Instead, a vacuum pump is needed to suck up the gas and compress it into another container. That's probably just as well, because hydrogen fluoride is toxic in concentrations of three parts per million.[20] For battlefield use the Army is working on ways to pack the waste hydrogen fluoride into canisters, lest the laser weapon inadvertently wage chemical warfare against the soldiers running it.

Many of the critical components in a high-energy chemical laser influence gas flow. In this view of components of the 100,000-W RACHL chemical laser built by the Rocketdyne Division of Rockwell International Corp., the combustors, nozzles, and diffuser/ejectors all control gas flow. Some of the optics are shown at the lower right. A quick glance at the photo of the laser at center shows just how prominent a role gas plumbing plays. (Courtesy of Rocketdyne Division of Rockwell International Corp.)

Like the carbon dioxide laser, the hydrogen fluoride chemical laser can emit light at many wavelengths corresponding to a family of vibrational and rotational transitions. High-power chemical lasers emit a broader range of wavelengths than carbon dioxide types. Because the vibrational energy levels depend on the mass of the atoms that make up the molecule, the wavelengths emitted by the laser depends on the atomic mass. Thus, different wavelengths can be produced by using different isotopes of hydrogen (which have different masses). If the commonest isotope, hydrogen-1, is used, the hydrogen fluoride laser emits at infrared wavelengths between about 2.7 and 3.0 μm. If the rare stable isotope deuterium (hydrogen-2), which weighs twice as much as hydrogen-1 , is used, the resulting deuterium fluoride laser emits at somewhat longer infrared wavelengths between about 3.6 and 4 μm.

That difference in wavelengths turns out to be very important for use in the atmosphere. Deuterium fluoride wavelengths fall in an atmospheric "window," where the air is highly transparent, as to visible light. However, the shorter wavelengths produced by hydrogen fluoride lasers are strongly absorbed by the air. Deuterium is rare–making up only 0.015% of natural hydrogen, with the rest hydrogen-1–and hence much more expensive, but the difference in transmission is critical for the feasibility of laser weapons in the atmosphere. Hydrogen fluoride is fine for uses in space, but if the laser beam has to go through the air, deuterium fluoride is the clear choice.

Fortunately for system developers, the lasers that use hydrogen fluoride are very similar to those that use deuterium fluoride. In some low-power lasers, the two gases are regarded as interchangeable. It's probably not quite so easy for high-power laser weapons, although operational details of such big lasers are classified.

From the military standpoint chemical lasers have a couple of major advantages over carbon dioxide types. One is that energy can be stored more compactly in chemical rather than in electrical form. That's important both on the battlefield, where portability is at a premium and supply lines can be cut, and in outer space, where light weight and small size are absolute musts. Chemical fuels can be stored in canisters much easier to handle than bulky electrical power supplies.

The other important advantage is a much shorter wavelength, which as long as atmospheric absorption can be avoided is generally a good thing. The smallest spot that can be produced on a target depends on the laser wavelength divided by the diameter of the output optics. If a 3-μm chemical laser was used with the same output mirror as a 10-μm carbon dioxide laser, the shorter- wavelength laser would produce a focal spot with only 30% of the diameter and 9% of the area of the carbon dioxide laser. This would increase the highest power density on the target by a factor of 11, something which is important because, usually, the higher the power density (even on a smaller area), the more likely the laser is to do lethal damage to its target. (This is not always the case, though, and for some targets it is better to illuminate a larger area with a lower intensity.) There being no such thing as a free lunch in optics or military systems, there is a tradeoff–shorter wavelengths require more precise optics and better pointing control–but that price may be worthwhile. Most chemical laser research has concentrated on devices that emit a continuous beam. However, a series of short, intense pulses may be better able to damage targets and to overcome some of the atmospheric transmission problems discussed in Chapter 5. The technology is similar to that for continuously operating chemical lasers, but with some important differences.

Typically, pulses would be ignited with a burst of light from a flashlamp or a bolt of electrons from an electron beam generator. The laser gas would start at room temperature and a pressure of 1 atmosphere, but the reaction would heat it to about 1000 °F and raise the pressure to 6 atmospheres after the laser pulse, which would last a few millionths of a second. Up to 50 shots per minute have been produced in demonstrations.[21]

Pulsed chemical laser technology is not as mature as that involved in either carbon dioxide or continuously emitting chemical lasers. One major problem is the generation of acoustic waves, which can disrupt the laser beam, the gas flow, and the alignment of the optics–all critical factors in the performance of a laser weapon.

Shorter-Wavelength Lasers

Although Pentagon planners are willing to work with chemical lasers, they would rather have a laser with a much shorter wavelength. This is particularly true for use in outer space, where the ability to form a small spot on a target thousands of kilometers away is critical to some important potential missions. Their ideal laser would preferably emit light in the visible region (wavelengths of 0.4 to 0.7 μm) or the ultraviolet (0.4 to around 0.001 μm).

The main attraction of chemical lasers is that they are well developed and reasonably well understood. Power levels of a few million watts have been demonstrated for intervals measured in seconds, long enough to be of military interest. The technology has a long way to go before it is ready for the battlefield or for use in a satellite, but it is available.

Short-wavelength laser technology is newer, and its capabilities are not yet clear. The Pentagon has spent many years and millions of dollars on the effort, but the general consensus seems to be that short-wavelength technology is 5 to 10years behind chemical lasers. This work has produced four candidates for short-wavelength laser weaponry: the chemical oxygen iodine laser, the excimer laser, the free-electron laser, and the X-ray laser. Each type has its own advantages and disadvantages, but none has yet matched the performance of chemical lasers. The rest of this chapter will cover the oxygen iodine, excimer, and free-electron lasers; the X-ray laser is fundamentally distinct in nature and will be covered in Chapter 6.

Chemical Oxygen Iodine Laser

The chemical oxygen iodine laser (sometimes called "COIL") is a promising newcomer among chemical lasers. Its 1.3-μm near-infrared wave- length is transmitted moderately well by air. Although not as short as some developers might like, the iodine laser's wavelength is short enough to permit use of optics one-third to one-half the diameter needed for hydrogen fluoride or deuterium fluoride lasers. Developers expect the chemical iodine laser's efficiency and power level to be comparable to that of today's high-power chemical lasers, but outputs have yet to reach hundreds of kilowatts.

This type is actually a variation on an iodine laser system that was developed earlier. In that system intense ultraviolet light breaks up molecules containing iodine, producing excited iodine atoms that can emit light at 1.3 μm. Such an iodine laser, which derived its energy from intense flashlamps, was used in the mid-l970s to generate pulses lasting 1 billionth of a second with peak power of about a trillion watts for laser fusion experiments at the Max Planck Institute in Garching, West Germany.[22]

Flashlamps can produce only limited amounts of power; they are also too bulky, fragile, and inefficient for use in laser weapons. However, over the past few years researchers have found another way to produce excited iodine atoms: by transferring energy from excited oxygen molecules produced by a chemical reaction. Such chemically energized iodine lasers can produce a continuous beam or generate pulses as short as a billionth of a second.[23]

The chemical iodine laser is not exactly what the Pentagon was looking for. However, the short wavelength and the prospect of being able to work with liquid fuels much easier to handle than hydrogen and fluorine make the system very attractive to military planners. It is now getting considerable support, particularly from the Air Force, with goals including finding the best types of fuels and resolving uncertainties about beam propagation in air.

The technology seems to be coming along well. In 1980 the Air Force

Weapons Laboratory at Kirtland Air Force Base in New Mexico commissioned two major developers of high-energy lasers–TRW Inc. and Bell Aerospace–to build 50,000-W iodine lasers by mid-1982.[24] Results of those programs evidently were encouraging, because in late 1982 the weapons lab disclosed plans to spend nearly $20 million to develop the technology needed for an oxygen iodine laser with supersonic gas flow.[25] Spending on that scale is probably a prelude to the construction of a large laser.

The Soviet Union is also believed to be working on iodine lasers. Leaks of intelligence reports describing Soviet development of particle-beam generators have often mentioned construction of iodine lasers as another possible interpretation of the available intelligence information. [26]

Excimer Lasers

The excimer laser is somewhat related to chemical lasers in that it relies on a reaction between two atoms, but strictly speaking it isn't a chemical laser at all. Its energy comes not from a chemical reaction but rather from an electric discharge or beam of electrons directed into the gas. (Other types of energy transfer, such as excitation by microwaves, are also possible.)

The most unusual thing about the excimer laser is its namesake and active medium–an unusual molecule called an "excimer."[27] An excimer is a pair of atoms that are bound together only when the molecule is in an excited energy level. When the molecule drops down to what should be its lowest energy level, it falls apart. This means that there can be no excimer molecules in the lower level of the laser transition, and hence that a population inversion exists as long as there are excimer molecules around. The conditions turn out to be right for producing high-power pulses at ultraviolet wavelengths.

The best-known and most important family of excimers are molecules that contain one halogen atom such as chlorine or fluorine and a rare-gas atom such as xenon or krypton. (You may have been taught to think of the rare gases as "inert" because their outer electron shell is normally filled, but they can be excited so that one of the outer electrons is in a shell by itself, making the atom highly reactive.) Excimer molecules are formed when a mixture containing the laser gases is excited by electrons from a discharge or beam. The most important excimers for weaponry are two ultraviolet types, the krypton fluoride laser emitting at 0.25 μm and xenon fluoride emitting at 0.35 μm. Efficiencies, defined as percent of absorbed energy emerging in the laser beam, are up to 10% and 5%, respectively.

Excimer laser technology has made tremendous strides since the first member of the family was demonstrated in the mid-1970s, but some important problems remain. Because the lifetime of the excimer molecule is short, the laser tends to produce only short pulses, which may not be useful for weapons. Pulsed operation creates acoustic waves, which can disrupt the laser beam; excimer lasers are particularly subject to such problems because their short wavelength makes small aberrations more significant. Ultraviolet optics are a problem because high intensities at such wavelengths "could tear things apart" in the words of one observer. The whole problem of increasing output power to chemical laser levels is a tricky one, with highest average powers from excimer lasers still less than one-thousandth the best obtained from chemical lasers.

The Pentagon's initial interest in excimer laser weapons was for use in outer space, but lately another possibility has emerged: basing the laser on the ground and bouncing its beam off an orbiting focusing mirror to the target. The idea itself is not dramatically new, but had been bypassed because of concerns about problems in getting the beam through the atmosphere to the mirror. It has been getting renewed attention not just because of laser weapons, but also because of another laser program: using blue-green laser beams to convey messages to submerged submarines. Faced with serious problems in building a blue-green laser that could function reliably in space, developers in that program turned to the possibility of leaving the laser on the ground and bouncing its beam off an orbiting mirror.[28]

Basically the same idea could be applied to weapons, only using a higher-power laser. The laser itself would be built on a high mountain, where there were few clouds and a stable atmosphere, perhaps in the mountains of Arizona or New Mexico. Leaving the laser on the ground would simplify life in many ways. It would avoid the need to put a massive power supply in orbit, a clear drawback for the electrically excited excimer laser. It would also leave the laser within easy reach of repairmen and fresh supplies of chemicals. The main problem remains in coping with atmospheric absorption and scattering, which would be far more serious for a laser weapon than for a communication laser. These atmospheric effects become severe at short ultraviolet wavelengths; while the xenon fluoride laser might work well from the ground, strong absorption would probably rule out the shorter-wavelength krypton fluoride laser.

The Free-Electron Laser

The mid-1970s saw the development of the free-electron laser[29] as well as the excimer laser. The free-electron laser is so different from the types of lasers described so far that many observers don't even call it a laser. However, it may offer the most hope for producing high powers at short wavelengths.

The free-electron laser draws heavily on particle accelerator technology, which is described in more detail in Chapter 7. The basic idea is to accelerate a beam of electrons to a high velocity, then pass them through a specially tailored magnetic field. This magnetic field is formed by an array of magnets arranged in a line but alternating in polarity, so the electrons passing through it experience regular variations in the magnetic-field strength and direction. As the electrons pass from the influence of one magnet element to the next, the magnetic field bends their paths, causing them to emit and absorb light. If the magnetic field is properly designed, the beam of electrons will emit more light than it absorbs. Proper placement of laser mirrors then creates a free-electron laser.

The idea was conceived in 1971 by John M. J. Madey, a Stanford University physicist who went on to demonstrate it experimentally. In 1976 he amplified light from a carbon dioxide laser that was passed along a beam of electrons in a suitable magnetic field. The following year he demonstrated a free-electron laser that generated its own beam at a wavelength of 3.4 µm. In both experiments he used the beam from a 20-m accelerator at the Stanford Linear Accelerator Center.

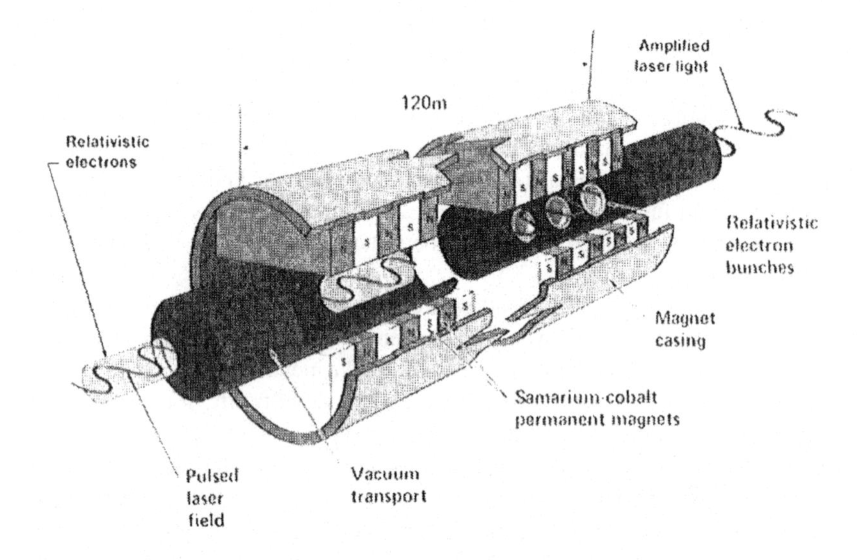

Structure of a free-electron laser. As the electron beam produced by an accelerator or a storage ring (not shown) follows a curved path through the array of permanent magnets, it transfers energy to the laser beam. The magnet array is called a "wiggler" because it makes the electrons wiggle back and forth as they pass through it. Laser wavelength depends on spacing of the magnets and energy of the electrons. In this version the spacing of magnets changes slightly along the beam path to enhance energy transfer. (Courtesy of Lawrence Livermore National Laboratory.)

It took other physicists a couple of years to assimilate Madey's results. There was a great outpouring of theoretical studies, most incomprehensible to anyone outside of a handful of other theorists, leading me to recall the old fable of the blind men trying to describe an elephant. The promise of the technology for generating high powers was evident to the Pentagon. Disenchanted by years of failure to produce a working X-ray laser, the Defense Advanced Research Projects Agency (DARPA) dropped that idea in favor of free-electron lasers. After the theoretical dust settled, DARPA sponsored experiments to assess the prospects for building high-energy free- electron lasers. Three major parallel experiments were conducted at the Los Alamos National Laboratory, Mathematical Sciences NorthWest Inc., and TRW.[30] The results were impressive, with energy extracted from the electron beam with 20 times greater efficiency than Madey reported in his first experiments. Charles A. Brau, leader of the free-electron laser group at Los Alamos and one of the fathers of the excimer laser, stated: "I've never seen an experiment that was so clearly in agreement with theory as this one."[31] At this writing a second round of experiments is being prepared.

The free-electron laser has several characteristics that are attractive for high-energy laser weaponry. The electron beam travels through a vacuum, so the problems of turbulence in laser gas, which can impair the quality of an ordinary laser beam, are not present. Because the electrons are free–not bound to a particular atom or molecule–they don't have to make fixed transitions and hence can be tuned to produce light at a wide range of wavelengths, from microwaves all the way into the ultraviolet. No single device would be able to operate across that entire range, but individual lasers could have their wavelength changed by adjusting the magnet spacing and the energy of the entering electrons. Theoreticians also predict that a free-electron laser should be able to operate at around 20% overall efficiency–even counting losses in the power supply generally not counted in other estimates of laser efficiency–by adding devices to recover the energy left in the electron beam after it passes through the magnetic field. The combination of high efficiency and short wavelength is particularly attractive for space-based laser weaponry. Developers of free-electron lasers can benefit from existing accelerator technology. Now they are borrowing other people's accelerators, but once they start building their own they will be able to draw upon technology developed over the past few decades for research in particle physics. Developers believe this will help make the free-electron laser inexpensive (at least compared with other high-energy lasers) and will make it easier to reach high powers.

There remains a large gap between the low powers that have been demonstrated experimentally and the millions of watts that designers of antimissile laser systems would like to have at their command. Madey estimates that existing accelerator technology could be used to generate free-electron laser powers to about 100,000 W.[32] Developers in general seem very optimistic about long-term prospects for generating high powers, but those power levels are clearly several years away.

Despite the prevalent optimism, there are problems as well as promises with the free-electron laser. The hoped-for high powers and high efficiencies are still undemonstrated. So is the production of intense ultraviolet light, and of ultraviolet optics that will withstand it. Accelerator technology is also inherently bulky–remember, Madey conducted his first experiments with a linear accelerator 20 m (66 ft) long. Such behemoths will be hard to put in orbit, to say the least. After hearing a developer of the gasdynamic laser call that system a "ten-ton watch," another laser physicist attending an informal seminar I attended at the Massachusetts Institute of Technology stood up and asked if in those terms a free-electron laser in space would be a "thousand ton television tube." It took a minute or two for the laughter to subside.

The problem of accelerator bulk seems to be one of the main factors generating Pentagon interest in using orbiting mirrors to reflect beams from ground-based lasers. Except for the problems in getting them off the ground, free-electron lasers seem in many ways an ideal long-term solution to the problems of developing laser weapons. Designers of free-electron lasers would be able to select operating wavelength, rather than being stuck with a particular wavelength produced by the laser medium. This would let them avoid any wavelengths where atmospheric absorption problems would be particularly severe. If the laser was on the ground, they would be able to rely on ground-based power, rather than having to wait for powerful nuclear generators to be developed for use in space. Plus, of course, the delicate thousand-ton television tube would never have to get off the ground, or be far from a repairman.

Other High-Energy Laser Approaches

Several other concepts for generating high-energy laser beams have got- ten serious attention over the years. Two that are at least worthy of passing mention are the nuclear-pumped laser and the solar-pumped laser, although neither seems to be getting much attention from today's weapon developers.

The basic idea of the nuclear laser is to transfer the energy from a nuclear reaction directly into a population inversion that could directly produce a laser beam. The ultimate goal would be to build a small nuclear reactor in which the power-generating core was a gas, not the solid rods used in conventional nuclear power plants. The gas core of the reactor would also serve as the laser medium.

After several years of effort, in late 1974 researchers at Sandia National Laboratories and at Los Alamos National Laboratory demonstrated nuclear-pumped lasers in separate but nearly simultaneous experiments. The enthusiastic head of Sandia's effort said that his group's results suggested it should be possible to convert more than half of the energy from nuclear fission reactions into laser output, although conceding that the efficiency he observed was far below that level.[33]

Alas, that optimism seems to have gone unrealized. The National Aeronautics and Space Administration, which had been the prime sponsor of nuclear-pumped laser research, stopped its program in 1981 after reaching a power level of 1000 W, citing the need for a special reactor beyond NASA's budget to demonstrate higher powers.[34] Laser weapon developers haven't expressed much interest in the nuclear-pumped laser concept, presumably because advantages of nuclear power would be offset by the need for shielding the reactor to prevent radiation hazards to friendly troops or control equipment. Many observers were also skeptical about the promise of nuclear pumping because performance fell short of desired goals.

Another possibility that remains under active consideration by NASA is using sunlight to directly or indirectly energize a laser. The long-term goals of both that and the nuclear laser effort are not in weaponry but in the use of high-power lasers to propel spacecraft[35] or to beam power from point to point, applications that would require much higher powers and total energies than laser weaponry.

Although such concepts so far have fallen short of theoretical expectations, laser developers are certainly not standing still, and the Pentagon is keeping its eyes open for new concepts for powerful lasers. We are still far from demonstrating any laser that is ideal for weaponry, that is, easy to operate, efficient, light in weight, and emitting a powerful beam at a useful wavelength.

Even the promise of many existing types of lasers has yet to be realized. Yet in many ways lasers are approaching the point of being "good enough" for use as weapons. The difficulty is that the laser is not all that is needed to make a weapon system. Even after you build a big laser gun, you have to find a way to use it to hit the target–and it looks like the latter problem is a much more serious one than building a big enough laser.

References

1. Laser physics is a discipline in its own right and can be treated in much more detail. For a fuller popular-level description see: Jeff Hecht and Dick Teresi, *Laser Supertool of the 1980s* (Ticknor & Fields, New Haven, 1982). For a deeper explanation see: Orazio Svelto, Principles of Lasers, 2nd ed (Plenum Press, New York, 1982).
2. Many books leave the impression that two or more electrons can occupy the same orbit or energy level. This is true only when electron spin is ignored. In reality each electron has a unique set of quantum numbers that uniquely determines its energy level. In practice, the energy level differences sometimes are too small to measure except with very sophisticated equipment.
3. See, for example, James D. Macomber, The Dynamics of Spectroscopic Transitions (Wiley- Interscience, New York, 1976).
4. Arthur L. Schawlow, "Masers and lasers," IEEE Transactions on Electron Devices ED-23 (7), 773-779 (July 1976.)
5. Michael J. Mumma, David Buhl, Gordon Chin, Drake Deming, Fred Espenak, Theodor Kostiuk, and David Zipoy, "Discovery of natural gain amplification in the 10-micrometer carbon-dioxide laser bands on Mars: a natural laser," Science 212, 45-49 (April 3, 1981).
6. The details of laser resonator physics are far beyond the scope of this book. For an introduction see Chapter 6 in Orazio Svelto, Principles of Lasers, 2nd ed (Plenum Press, New York, 1982).
7. A radian is a measurement of angle typically used by scientists and engineers. It is formally defined as the angle subtended at the center of a circle by an arc equal in length to the radius of the circle. From a practical standpoint, that means that an object l m long occupies a one radian angle when you look at it from l m away. If you look at the same object from 10 m away, it is one-tenth of a radian across. The radian is a particularly useful unit because it simplifies conversion between linear and angular measurements. One radian equals 57.3°.
8. Lamar W. Coleman, ed, Laser Program Annual Report 1979, Vol. 1, Lawrence Livermore National Laboratory, Livermore, California, pp. 2-12-2-19.
9. For more details see: W. W. Duley, CO2 Lasers Effects and Applications (Academic Press, New York, 1976) or Orazio Svelto, Principles of Lasers, 2nd ed (Plenum Press, New York, 1982), pp. 219-233.
10. IO. The definition appears in the July 1980 edition of the Department of Defense "Fact Sheet: DoD High Energy Laser Program," p. l; higher-power lasers were subject to security classifications.
11. Cited in W. W. Duley, CO2 Lasers Effects and Applications (Academic Press, New York, 1976), p. 31.
12. HPL Lasers (data sheet), Avco Everett Metalworking Lasers, Somerville, Massachusetts.
13. Ibid.
14. Department of Defense, Soviet Military Power (U.S. Government Printing Office, Washington, D.C. 1981), p. 75.
15. William J. Broad, "Nuclear power for the militarization of space," Science 218, 1199-1201 (December 17, 1982).
16. Mani L. Bhaumik, "High-efficiency electric discharge CO lasers," pp. 243-267 in E. R. Pike, ed, High-power Gas Lasers I975 (Institute of Physics, Bristol and London, 1976).

17. Orazio Svelto, Principles of Lasers, 2nd ed (Plenum Press, New York, 1976), p. 232.
18. Orazio Svelto, Principles of Lasers, 2nd ed (Plenum Press, New York, 1976), pp. 247-250; see also Arthur N. Chester, "Chemical lasers," pp. 162-221 in E. R. Pike, ed., High-power Gas Lasers I 975 (Institute of Physics, Bristol and London, 1976).
19. Richard Airey, presentation at Laser Systems & Technology Conference, sponsored by the American Institute of Aeronautics and Astronautics, July 27-28, 1981, Boston.
20. Robert C. Weast, ed, CRC Handbook of Chemistry and Physics, 62nd ed (CRC Press, Boca Raton, Florida, 1981), p. D-I05.
21. Peter Clark, presentation at Electro-Optical System Market Forecast Seminar, November 17-18, 1980, Boston, sponsored by American Institute of Aeronautics and Astronautics.
22. K. Hohla, G. Brederlow, E. Fill, R. Volk, and K. J. Witte, "Prospects of the high-power iodine laser," pp. 97-113 in Helmut J. Schwarz and Heinrich Hora, eds, Laser Interaction and Related Plasma Phenomena, Vol. 4A (Plenum Press, New York, 1977).
23. Gerald M. Hays and George A. Fisk, "Chemically pumped iodine laser as a fusion driver,"
24. IEEE Journal of Quantum Electronics QE-17(9), 1823-1827 (September 1981).
25. "Weapons laboratory aids beam effort," Aviation Week & Space Technology, August 4, 1980, pp. 56-59.
26. "Laser funding," Aviation Week & Space Technology, September 27, 1982, p. 16.
27. See, for example, "Soviets build directed-energy weapon," Aviation Week & Space Technology, July 28, 1980, pp. 47-50.
28. Orazio Svelto, Principles of Lasers, 2nd ed (Plenum Press, New York, 1982), pp. 237-239.
29. "Ground-based laser has the edge for sub communications," Laser Focus 18(1), 38-40 (January 1982).
30. For a comprehensible overview of the free-electron laser, see Charles A. Brau, "The free-electron laser: an introduction," Laser Focus 17(5), 48-56 (May 1981). For more detailed treatment see the August 1981 issue of the IEEE Journal of Quantum Electronics, which was devoted entirely to free-electron lasers, or the two-volume set Free-Electron Generators of Coherent Radiation edited by Stephen F. Jacobs, Gerald T. Moore, Herschel S. Pilloff, Murray Sargent, Marlon 0. Scully, and Richard Spitzer, (Addison-Wesley, Reading, Massachusetts, 1982).
31. H. Boehmer, M. Z. Caponi, J. Edighoffer, S. Fornaca, J. Munch, G. R. Neil, B. Saur, and C. Shih, "Variable-wiggler free-electron laser experiment," Physical Review Letters 48 (3), 141-144 (January 18, 1982); Charles A. Brau, private communication; Jack Slater, private communication.
32. Quoted in "Free electrons make powerful new laser" High Technology 3 (2), 69-70 (February 1983).
33. John M. J. Madey, private communication.
34. "Sandia and Los Alamos obtain lasing from nuclear-pumped CO and He-Xe," Laser Focus
35. 10(12), 13-16 (December 1974).
36. Lynwood P. Randolph, presentation at Laser Systems and Technology Conference, Washington, July 9-10, 1982, sponsored by American Institute of Aeronautics and Astronautics, quoted in "Postdeadline reports," Laser Focus, August 1981, p. 4.
37. Lee W. Jones and Dennis R. Keefer, "NASA's laser-propulsion project," Astronautics & Aeronautics 20, 66-73 (September 1982).

5.
Hitting the Target:
Beam and Fire Control

It takes much more than a powerful beam to make an effective weapon system. Buck Rogers can simply pull his ray gun from his holster and blast away, but in the real world things are far more complex. The beam must be aimed and focused through a generally uncooperative atmosphere or over long distances. It must concentrate high power on a small area long enough to do fatal damage, a requirement that typically means the beam must follow the target along its path. Most of the targets envisioned for beam weapons are fast enough that automatic tracking and identification are needed. In the case of systems for defense against nuclear attack, there should be a way to verify "kills" from far away. All in all the problem is much more complex than asking Luke Skywalker to "trust the Force" before blowing the *Death Star* to oblivion with "proton torpedoes."[1]

The Pentagon is well aware of the complexities involved in getting a laser or particle beam from weapon to target. Some of the lessons came through experience. Visitors to the New Mexico site where the Air Force tested high-power lasers in the early 1970s describe a swath of charred brush surrounding the target area, vivid evidence that laser beams could miss. Later tests showed that even if the beam did reach the target, it might not destroy it. Military researchers have explained why such things can happen, but explaining problems is not enough to solve them. Today, the critical issues in laser weaponry center on "lethality" and target vulnerability–how to get the laser beam to the target, and how to make sure it does lethal damage once it gets there. These issues are receiving much attention from military planners. For example, about two-thirds of the Pentagon's fiscal 1983 budget for the Space Laser Triad (the demonstration of chemical laser technology for use in space weapon systems described in Chapter 11) was allocated for developing optics for the laser and for work on pointing and tracking systems. Only one-third was set for building a large demonstration laser.

There are several ways to look at the problems in destroying targets with beam weapons, and the job is complicated by the very complexities involved. The problems are best defined for laser weapons, and for that reason this chapter will concentrate primarily on lasers (other than the X-ray types described in the next chapter, which involve quite different physics). There are several tasks involved:

- Identify the target (which generally will not be sitting by itself on a target range) and a vulnerable spot on it
- Track the target both until the weapon is ready to fire and while it

is firing.

- Point the weapon in the direction of a vulnerable spot on the target
- Focus the beam so it has the desired intensity (generally the highest possible) at the target
- Compensate for atmospheric effects that otherwise would tend to make the beam wander off target or disperse the beam's energy
- Maintain focus of the beam on target during the attack
- Make sure that as much as possible of the energy in the beam is deposited on the target, not deflected away from it
- Verify that the target has been disabled.

That is a hefty shopping list of goals, many of which are hard to achieve technically. To reach these goals, military developers are working in a number of areas, generally lumped under two headings: beam control and what in the military world is called "fire control." The precise borderlines are somewhat hazy, but there is a general division between the two. Beam control involves steering the beam and focusing it onto a target, in the process compensating for atmospheric distortions to the beam. Fire control is the aiming and firing of the weapon, a task which includes identifying and tracking targets (providing information which the beam-control system uses to point the beam), triggering shots from the weapon, spotting vulnerable points on the target, and making sure that the target is disabled.

Beam control is unique to directed-energy weapons, but fire control is a well-established military technology that is used in many kinds of weapon systems. The fire-control systems used with many modem missiles and "smart" bombs incorporate low-power lasers to mark potential targets with a laser spot that can be detected by a sensor in the bomb or missile. Low-power lasers can also measure the distance to a target to aid artillery fire-control systems in pointing weapons. But although fire-control equipment is used widely in conventional weapons, directed-energy weapons would demand more than current equipment can deliver. Indeed, many observers believe that beam and fire control are much more difficult problems than building a big laser. The difficulties arise because of the very demanding missions envisioned for beam weapons.

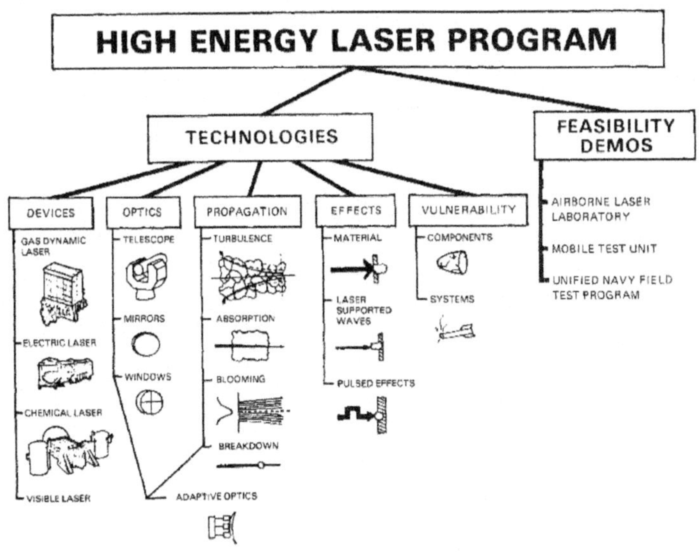

The Pentagon's high-energy laser weapon program goes far beyond the building of large lasers (labeled "devices") to cover laser and beam-focusing optics, beam propagation, laser effects, and target vulnerability. The feasibility demonstrations listed at right are those that had been completed or were in progress at the time this chart was prepared; others are in the works, as described in Chapters 11 and 13. (Courtesy of U.S. Department of Defense.)

Mission Requirements

The different missions proposed for beam weapons each impose distinct requirements on beam- and fire-control systems. The missions can be roughly divided into two families: "endoatmospheric" applications in the atmosphere and "exoatmospheric" uses within outer space, where there is no air to get in the way. Some proposed systems fall into a hazy intermediate category because they require sending a laser beam up from the ground into space.

Directed-energy weapons intended for use in the atmosphere typically have tactical missions on or above the battlefield. That means that they must be able to function over ranges measured in miles or kilometers. For all practical purposes a beam traveling at or near the speed of light can reach such a target instantaneously. However, the atmosphere can bend, distort, or break up the beam, and sophisticated compensation techniques are needed to concentrate the beam energy onto the right point of the target. Many missions, such as defense of a battleship against an onslaught of cruise missiles, would probably require extremely fast response to zap many targets before they could reach the ship. Speed might not be as critical in some other missions, such as destroying comparatively slow-moving helicopters, but in most cases the system would have to pinpoint enemy targets in a field that included friendly forces. Typical targets would require concentration of enough power at a critical point to physically disable them, but in some cases much lower powers would be sufficient to blind a sensor and thereby disable the target.

Ground- and air-based antisatellite weapons present rather different design constraints. Their range would be hundreds or perhaps thousands of kilometers or miles, depending on the orbit of the target satellite. Moderate powers reaching the satellite should be sufficient if the goal was to blind sensors or disable vulnerable sensing systems or electronics, probably the most effective way to kill current satellites. Seen from the ground, satellites would be slow-moving targets. Some compensation for atmospheric effects might be needed, but the task would be easier than if high powers had to be delivered to a small point on the target.

Systems to defend a country against nuclear attack would have very different requirements. In an all-out attack hundreds of targets would appear thousands of kilometers or miles away, and the weapon system would have to go after them as rapidly as possible at that distance. Except for X-ray lasers, any weapon system would have to destroy a series of many targets in succession. The degree of devastation caused by a nuclear bomb would make it important to kill as many targets as possible (ideally *all* of them) and to know which ones were not disabled so other weapons could shoot at them. If the lasers were in space, there would be no need to worry about atmospheric effects. If the lasers were on the ground, transmitting their beams to "battle mirrors" in space, compensation for atmospheric effects could be performed both by the optics on the laser and by the large mirrors in space, which together would form the beam-control system.

Additional differences come from the nature of the weapon itself. It takes different techniques to direct beams of visible light, X rays; charged particles, uncharged particles, and microwaves. Fire-control methods for such weapons would be more closely related to each other, although there would be some notable differences. This chapter will focus on technology for the most developed type of beam weapon–the high-energy laser. The different ways in which other types of beams would be controlled will be covered in the chapters that follow, which describe the principal directed-energy alternatives to lasers in more detail.

Wavelength Effects

The wavelength of a laser sets some fundamental constraints on the optics that can be used with the laser. By far the most important is the Fraunhofer diffraction limit, which determines how small a spot the beam can form. In this case the spot size is measured as an angle (as viewed from the laser) that is proportional to the ratio of wavelength to the diameter of the focusing optics. The theoretical formula for the ideal case is[2]

$$\text{Spot size} = 1.22 \times \text{Wavelength/optics diameter}$$

This formula gives spot size in radians, an angular measure equal to the diameter of the spot divided by the distance to it. (One radian equals 57.3°.) Laser physicists usually talk in terms of spot size (sometimes sloppily called beam divergence, although in this sense it isn't exactly that) in radians. However, the formula can be altered to give spot size in meters:

$$Spot\ diameter\ (meters) = \frac{1.22 \times wavelength \times target\ distance}{optics\ diameter}$$

Note that in making calculations with either formula, it is essential that everything be measured in the same units. Thus, if wavelength starts in micrometers, target distance in kilometers, and optics diameter in meters, they would all have to be converted to meters before making the calculation.

Spot diameter is important because it indicates onto how small an area on the target the laser's output can be concentrated. Dividing the area of the focal spot on the target into the laser power or energy gives the power or energy density at the target. Measurements of laser power or energy density on the target are useful in making rough approximations of the threshold for causing damage, but the actual mechanisms involved are quite complex and far beyond the scope of this book.

Another important factor is the beam "wander" or "jitter." That is, how precisely can the laser beam be kept on one spot on the target while it is depositing its lethal dose of energy. If the beam wanders all over the place, it won't stay at any one point long enough to do any damage. The usual assumption is that beam wander will have to be somewhat smaller than the spot size, but actual performance details are nowhere to be found in the public domain.

With this information in mind, you can calculate some very general requirements on a laser weapon system if you know the lethal power or energy density and the maximum range. For example, suppose that in the case of missile defense, it seems desirable to concentrate a laser power of about 5 million watts onto a spot about 1 m (about a yard) in diameter (a little over 6 million watts per square meter). If the target is 5000 km (3000 miles) away, in angular terms, the focal spot is 0.2 millionths of a radian (or 0.2 μrad in standard scientific terminology). That figure can be inserted into the equation that relates laser wavelength and optics diameter to spot size. Suppose, for example, the weapon system uses a space-based hydrogen fluoride chemical laser with a nominal wavelength of 2.8 μm (micrometers), or 0.0000028 m. Simple division shows that the output mirror must be 17 m (56 ft) in diameter. If that doesn't sound impressive enough, remember that the largest telescope in the United States, the 200-in. giant at Mount Palomar Observatory, has a main mirror only 5 m in diameter. The largest telescope mirror in the world is a 6-m one in the Soviet Union, which unofficial sources report hasn't been working very well. The largest mirror yet built for use in space is the 2.4-m (8 ft) mirror for NASA's space telescope.[3]

Use of a laser with higher power or shorter wavelength would allow use of a smaller mirror. If a 1.3-μm chemical oxygen iodine laser could be substituted for the hydrogen fluoride laser in the previous example, a mirror 8 m (26 ft) in diameter could produce a 1-m spot 5000 km (3000 miles) away. Increasing the laser's output power does not produce such dramatic reductions in required mirror size because damage depends on power density multiplied by illumination time, which increases faster with decreasing spot size than with increasing power. Thus, a 10-million-watt beam could be spread over twice as much *area* as a 5-million-watt beam, but as the area doubled, the spot diameter would increase only by the square root of two, a factor of 1.4. Thus, a 10-million-watt hydrogen fluoride laser beam could be focused onto a spot 1.4 m in diameter at a distance of 5000 km (3000 miles) with a mirror only 12 m (40 ft) in diameter, yielding the same power density as would be obtained when focusing a 5-million-watt laser over the same distance with a 17-m (56-ft) mirror onto a 1-m spot.

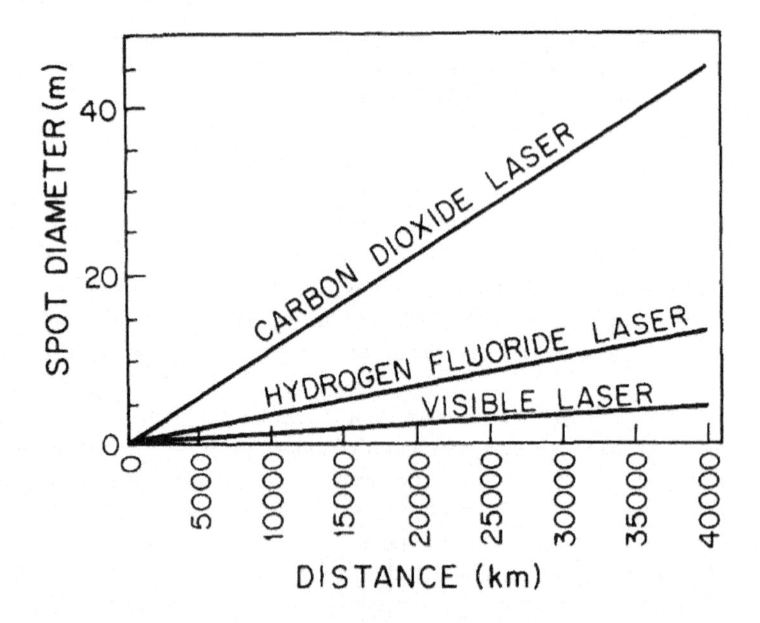

Laser spot size that would be produced at various ranges and wavelengths by a 30-m output mirror. The graphs show spot diameter for the 10.6-μm carbon dioxide laser, the 2.8-μm hydrogen fluoride laser, and a hypothetical 0.5-μm visible laser, assuming that the figure of the mirror is accurate to within 1/50th of the laser wavelength. (Drawing by Arthur Giordani based on calculations by Wayne S. Jones, Lockheed Missiles and Space Co.)

From a practical standpoint, shrinking the mirror diameter from 17 m (56 ft) to 12 m (40 ft) would be important. On the ground or in space, the weight of a massive mirror would present a problem. If weight was simply proportional to area (which in turn depends on the square of the diameter), reducing the diameter from 17 to 12 m would cut the weight in half. In practice, an even greater weight reduction would be possible, because the smaller-diameter mirror could be thinner and still have enough mechanical strength to maintain its shape.

These simple calculations demonstrate why the Pentagon is so interested in developing short-wavelength lasers. The prime allure of the ultraviolet is a wavelength about one-tenth that of the hydrogen fluoride laser, making it possible to use a mirror much smaller than needed for an iodine laser, although other considerations described below might weigh against picking the smallest possible mirror diameter.

It may be possible to put a laser weapon for use against missiles or satellites on the ground, but if the targets were in space, the focusing mirror would have to be there, too. Putting a large mirror into orbit is not going to be an easy job. Current proposals for space-based lasers envision either a 5-million-watt chemical laser with a 4-m (13-ft) mirror or a more potent weapon using a 10-million-watt laser and a 10-m (33-ft) mirror. Mirror sizes would have to be the same to deliver the same power to the same size spot, with the same laser wavelength, even if the laser was on the ground. Military contractors seem to think the task is achievable. The Coming Glass Works, Perkin-Elmer Corp., Itek Corp., and Eastman Kodak have proposed a plan for a 4-m (13-ft) glass mirror.[4] The United Technologies Research Center has offered to build a 10-m (33-ft) lightweight mirror, using a graphite fiber reinforced glass matrix for the body of the mirror and vaporized silicon for the reflective coating. Reports published in 1981 say that the company offered to build the mirror for a fixed price of $87.5 million, with completion to be five years after the Department of Defense gave the go-ahead.[5] However, neither of those mirrors would fit intact into the Space Shuttle. The 2.4-m (7.9-ft) mirror used in NASA's space telescope is close to the limit of the shuttle's capacity. Putting a bigger mirror into orbit would require a larger booster rocket to launch it, assembly of components in space, and/or a mirror that unfolded to large size after being squeezed into the shuttle's cargo bay. Although such problems are formidable, they do not appear to deter the Pentagon or the aerospace industry. Probably the most ambitious concept to have received serious study was a 30-m (100-ft) mirror that NASA was considering for space use. The giant mirror would be intended for use with large lasers for two long-term NASA programs: laser propulsion of spacecraft and the use of laser beams to transmit power for use far from the laser.[6]

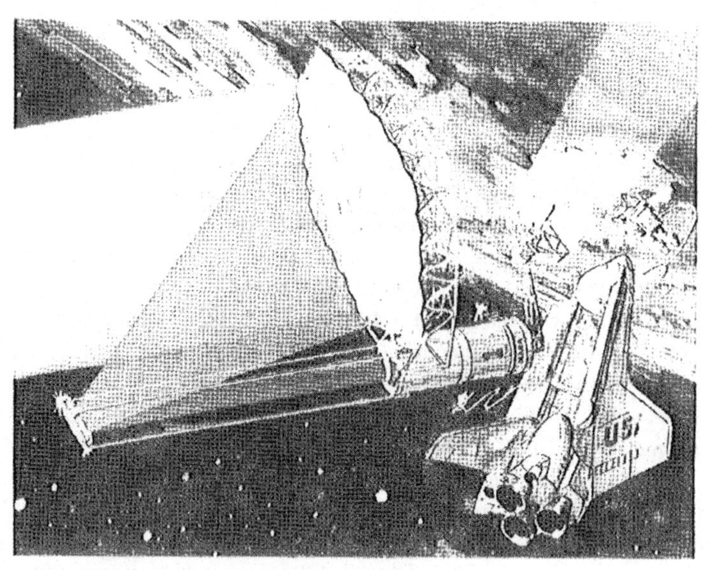

Artist's conception of a 30-m mirror being assembled in space to direct laser beams for powering and propelling spacecraft. The concept was developed by the Itek Corp.'s Optical Systems Division under a NASA contract. (Courtesy of Itek Optical Systems Division, from Ref. 7.)

In 1975 engineers at the Itek Corp. evaluated three approaches to building such mirrors: developing structures that would inflate in space, designing "variable geometry" mirrors that would be launched in compact form but would unfold to full size, and assembly of structural elements in space. All three designs were based on the "existing aerospace state of the art" to assess their feasibility. They concluded that the most practical approach was to assemble components in space, even though that would require astronauts to work in open space.[7] Their final proposal recommended that the mirror surface be made of glass or ceramic materials that changed little in size as the temperature changed. They urged that the supporting structure be made from a graphite-epoxy composite matched to the mirror surface material in thermal characteristics. Size of the output mirror would be much less of a problem for laser weapons used against targets on the ground, at sea, or in the air. That is because they would be much closer to their targets, with 20 km (12 miles) probably a reasonable upper limit to assume. The lasers themselves would probably be lower in power, but still it would be possible to use mirrors much smaller than needed in space. Suppose, for example, that a ground-based laser emitted 500,000 W, one-tenth the power of the space-based laser in the earlier example. Assume that it was necessary to achieve the same power density as in space at a distance of 20 km (12 miles) from the laser, but that the spot would be smaller, roughly one-tenth the area on the target and hence about 0.3 m (1 ft) in diameter rather than the 1-m spot required in space. Assuming that a 3.8-μm deuterium fluoride laser would be used rather than the 2.8-μm hydrogen fluoride laser that could be used in space (but whose beam would be absorbed by the air), the ground-based laser would need a mirror about 0.3 m (1 ft) in diameter.

In fact, for a variety of reasons, the short-range laser would probably require a larger mirror. The most important reason is to prevent laser damage to the mirror itself. If the mirror was indeed 0.3 m (1 ft) in diameter, it would be exposed to the same flux of laser energy per unit area as the target. (This would not be the case for the larger mirrors used in space, which would be several times the diameter of the focal spot on the target.) It is possible to polish the laser mirror to very high reflectivity and to cool it by passing a refrigerant through a network of holes fashioned in the mirror. However, the mirror would be exposed to the laser power for much longer than any target, and such high laser power densities are a way of asking for trouble because all mirrors absorb a small fraction of the incident light. Expanding the mirror to 1 m (3 ft) in diameter would reduce the power density on the mirror surface by a factor of 10, leaving a vital safety margin. Given the bulk of the laser itself, expanding the mirror diameter to provide such a safety margin shouldn't pose serious problems.

Bulk is not the only problem with a large mirror, however. The ability to focus beam energy onto a spot size depends critically on the quality of the optics, that is, on how accurately the surface of the mirror matches its ideal theoretical shape. Deviations should be no more than a small fraction, perhaps a tenth or a twentieth, of the wavelength of the laser light, or they could scatter part of the light out of the beam. That means that the surface of a mirror for a hydrogen fluoride laser should deviate by no more than a couple tenths of a micrometer from the ideal value. Because the tolerances are proportional to the wavelength, they are tighter for a shorter-wavelength laser. For an ultraviolet laser mirror, tolerances would be a couple of a hundredths of a micrometer.

Those tolerances are by no means trivial to achieve, but neither are they as difficult as you might think if you're not familiar with optics. Many years of experience in making precision optics have taught industry how to make very good components. For a couple of hundred dollars a major optics supplier can ship you a mirror one-tenth of a meter in diameter with surface accurate to one-tenth of the wavelength of visible light–a precision of 0.05 μm.[8] Such off-the-shelf mirrors are nevertheless not practical for use with laser weapons. One problem is that they couldn't stand the high power densities produced by a high-energy laser. Another problem is size. The difficulty in achieving tight control over the shape of a mirror surface increases rapidly with the size of the mirror. In part, this reflects the simple mechanical problem of maintaining high accuracy over a large area. It also is due to the problem in trying to maintain uniform conditions throughout a large object. Small temperature differentials, for example, can warp a large mirror enough to seriously reduce its surface accuracy.

The mirrors best able to stand up to high laser powers are made of solid metal, typically honeycombed with holes through which coolant flows. The Department of Defense has spent millions of dollars developing ways to produce mirrors that absorb less than 1% of the incident laser light, and which can efficiently conduct away what heat they do absorb. Much of the effort has gone to development of machines that use diamond-edged tools to cut mirror surfaces directly into metal blocks, vastly simplifying the traditional time-consuming process of making optics. Diamond turning, as the technique is called, also makes it possible to produce mirrors with surface shapes impossible to obtain by conventional grinding and polishing methods.

High-energy laser mirrors are high on the Pentagon's list of strategically important technologies. Thus, government officials became very upset when a metal mirror able to withstand high laser powers found its way from Spawr Optical Research Inc., a small company in Corona, California, to the Soviet Union, through the company's West German sales representative. The government threw the book at the company and its president, Walter Spawr. The incident came at the beginning of the government's crackdown on the export of sensitive technologies, and Spawr appears to have been singled out as an example to other companies inside and outside of the laser field. Both Spawr personally and the company were convicted of violating export regulations, but at this writing the case was under appeal.[9]

The Spawr case was not the only one involving Soviet interest in U.S. metal mirror technology. Concern about the transfer of technology for using powdered metal to fabricate mirror substrates apparently was one reason the United States, in 1980, cancelled a $134 million contract Dresser Industries Inc. of Dallas signed to build a plant in the Soviet Union. The plant was intended to produce tungsten carbide components for oil well drilling, but there were fears that the powdered-metal technology could be used to make armor-piercing weapons as well. The possibility of using the technology to make laser mirrors only surfaced recently. It appears that the Dresser deal did result in the Soviets learning at least some of the manufacturing technology involved.[10]

The Atmospheric Propagation Problem

However, even if they were optically perfect, ordinary mirrors wouldn't always suffice to get a laser beam through the atmosphere. Air looks much more transparent than it really is. Alexander Graham Bell learned that lesson the hard way a century ago when he tried to transmit conversations through the air using beams of light and a device he called the Photophone.[11] Beams of light might have seemed aesthetically more pleasing than telephone wires, but the latter proved more practical at the time. Eighty years later another generation of researchers learned the same lesson in the same way when they tried to develop communication systems that relied on laser beams going through the air.

The problems were magnified many times when efforts were made to get a high-power laser beam through the atmosphere. Around 1970 the Air Force set up a high-energy laser target range in the New Mexico desert. Visitors recall considerable areas of charred brush in the general direction of the target. The air kept bending the beam away from the target, often in the process reducing the power level, for some complex reasons.

One important factor was turbulence. Air is continually undergoing small random fluctuations in density, caused by winds, temperature gradients, passing objects, air currents generated by buildings or local topology, precipitation, and other factors. Such effects make stars twinkle on seemingly clear nights. Similar effects are visible in a swimming pool with a patterned bottom. When a high-power laser beam goes through air, these random fluctuations combine to cause the beam to wander.

The high power of the laser beam itself causes other problems, most notably a phenomenon called "thermal blooming." Although air looks transparent, it actually absorbs a tiny fraction of the light energy passing through it (much like a seemingly perfect mirror actually absorbs a tiny fraction of light). This absorbed energy heats the gas, causing it to expand and become less dense. This warmer, less dense air has a lower refractive index than the cooler gas that surrounds it, making it look like a negative or dispersing lens to the laser beam. This lens effect makes the laser beam spread out or "bloom," dispersing its power and making it harder for it to damage the target.

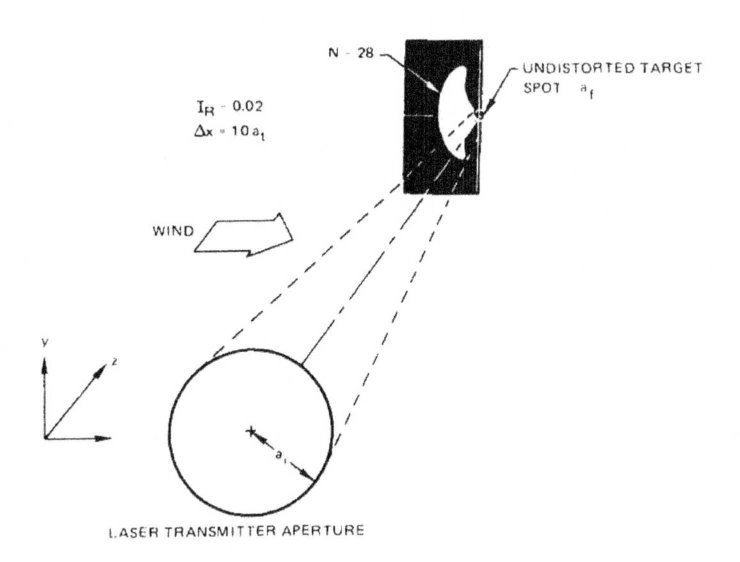

The effects of thermal blooming on a continuous laser beam when the air is moving, If there were no air in the way, the beam could be focused onto the small circle labeled "undistorted target spot." Thermal blooming and motion of the air combine to spread out the beam and bend its path so it covers the much larger area shown in white. The numbers shown refer to the laboratory simulation test that yielded the data used in the figure. (Courtesy of David C. Smith, from Ref. 17.)

Once motion of the laser beam through the air is considered, thermal blooming becomes a very complex question. That's actually the real situation, since the laser typically has to move to track its target, and the air itself is never perfectly still. Motion does help the blooming problem, but it also complicates it. As the air moves through the beam, it is heated and expands, thus decreasing in density. Refraction bends the light away from this rarified air, much as it bends light passing through stationary air away from the beam axis. The result for the simple case of a steady flow of air relative to the beam is that the beam bends into the direction of the wind. The beam also spreads out, producing a focal spot that is longer in the direction perpendicular to wind flow than in the direction parallel to the wind.[12]

The theory of how laser beams propagate through the atmosphere is complex, and physicists working on military grants have erected elaborate mathematical artifices to try to explain what happens. Books have been written on the topic in both the United States and the Soviet Union.[13] Papers on atmospheric propagation, often full of abstruse mathematics, appear regularly in both American and Russian scholarly journals, where they may serve as a quiet reminder of how much research in the field is ultimately derived from military directed-energy programs.

Virtually all of the published research on atmospheric transmission relies on simplified assumptions such as the air being clear. That isn't always the case. Various weather conditions can get in the way of the beam. As military systems have come to rely increasingly on optical and infrared systems, engineers have grown more concerned about the weather. Snow, rain, fog, or clouds can strongly attenuate both visible and infrared light because water droplets or snowflakes are larger enough than the wavelengths of light that they can block transmission. Longer wavelengths, such as microwaves, can penetrate through dust and precipitation because the particles in the air are much smaller than the wavelength. In some cases longer infrared wavelengths are transmitted much better than visible light because of the difference in their wavelengths.[14]

The natural heating effects of high-energy laser beams could help to overcome some weather-related transmission problems. A high-energy carbon dioxide laser, for example, can bore its way through fog by heating the tiny water droplets that obscure vision enough to make them evaporate. Military researchers have even studied the possibility of clearing away fog banks with lasers. Much the same could be done with clouds. Heavy rain would be harder, particularly in the realistic case where the beam is being scanned quickly through rapidly falling rain. If the beam scans across the drops too quickly, it may only partly evaporate them before moving on to illuminate other drops.

There are also additional complications likely to be encountered on the battlefield. There's likely to be lots of smoke, either incidental to the firing of guns and artillery, or laid down deliberately to block the enemy's vision. Partly burned diesel fuel, known as "fog oil," is a particularly effective smoke screen that blocks both visible and infrared light. Clouds of dust are also bound to be kicked up by moving vehicles and explosive blasts, and these, too, could block the transmission of laser beams.

Foul weather, dust, and smoke can do more than just block the beam from the high-energy laser. They can make it impossible to find targets visually or with infrared optical systems, making it necessary to rely on microwave radars that can penetrate the obscuration. However, even if the high-power beam could bum its way through, the limited resolution of microwave radars may not be able to pinpoint the target accurately enough for the laser beam to hit a vulnerable spot.

It is worth noting that thermal blooming produces a result that seemingly contradicts logic: beyond a certain point, increasing the laser power can decrease the amount of laser energy that reaches the target. This happens because the thermal distortions caused by the laser beam grow faster than the laser power. The more laser energy poured into the beam, the larger the percentage of energy that is bent away from the target. It's nature's reminder to the weapon system designer that bigger isn't always better, and it sets an ultimate limit to the size of battlefield lasers. Note that this is not the case if the beam doesn't have to travel through much air, so advocates of space- based laser weapons are free to propose lasers as big as they like–if they can build them and get them into orbit.

Trying to Beat the Atmosphere

It is possible, at least in theory, to adjust the output characteristics of a laser in ways that reduce some propagation problems. Transmitting pulses rather than a continuous beam can reduce thermal blooming because it takes time for the gas heated by the laser beam to thin out and act like a beam- spreading lens. The density changes are acoustic waves that travel at the speed of sound, typically taking a time measured in millionths of a second.[15] If the laser pulse stops passing through the air before the acoustic wave can create the lens effect, beam dispersion will be reduced.

The solution is not as simple as packing tremendous amounts of energy into ultrashort pulses. Packing too much energy into a short pulse can cause other problems, such as stripping electrons away from air molecules to create an ionized gas that can block the beam by absorbing light. It has even been suggested that intense pulses might ionize the air enough to create a ready path for lightning.[16] The vision of a bolt of lightning–nature's version of a particle-beam weapon–descending from the heavens to zap the user of a laser weapon system is certainly intriguing, although many observers tend to be skeptical that the effect would occur.

The best compromise between the problems encountered with a continuous beam and those caused by ultrashort pulses is to use a series of short, repetitive pulses that together would carry about as much energy as a continuous beam. Each individual pulse would be short enough to avoid thermal blooming effects, and the interval between pulses would be long enough for the air heating effects produced by one pulse to dissipate before the next pulse came along. This makes the optimum pulse repetition rate dependent on the size of the beam, the relative motion of beam and air, and on how much energy the air absorbs from each individual pulse.

It's also possible to influence thermal blooming by "tailoring" the distribution of energy within the laser beam. Energy is concentrated at the center of most laser beams, a type of "beam profile" that would tend to accentuate thermal blooming. Other energy distributions, such as one in which most of the energy is concentrated in a ring near the edge of the beam, can somewhat lessen thermal blooming.

Because the temperature differentials that cause thermal blooming depend on power density rather than on the absolute power level, spreading the beam out over a larger area can reduce distortion somewhat while helping to avoid too high a power density on the output mirror. The usefulness of this approach is limited by the problems encountered with large mirrors. Too large a mirror would be costly and hard to make, bulky to handle, and vulnerable on the battlefield. There would be some advantages to using this approach to transmit the beam from a ground laser to a beam-focusing mirror in space, but then the problem would become putting a large enough mirror in orbit.

Adaptive Optics

There turns out to be a better way to deal with distortion caused by atmospheric turbulence and thermal effects–corrective optics. Because the effects are dynamic and continually changing, the corrective optics would have to be adjustable or, in the jargon of the field, "adaptive." The idea is to use the optics to adjust the wavefront leaving the laser in a way that would compensate for the distortions the beam will encounter as it goes through the atmosphere. [18]

Ordinary transmissive optics are out of the question. The lens of the eye is deformable or "adaptive," but essentially all other transparent optical materials are rigid. What's more, transparent solids can be damaged by high-power laser beams because they, like the air, absorb a small fraction of the optical energy they transmit. There is some research into the possibility of developing gaseous optics to make the required corrections, but it is far from clear if that concept can be made practical for laser weaponry. The most promising solution is to use a mirror in which the reflective surface is deformable, sometimes called a "rubber" mirror by engineers, although it really doesn't contain rubber. There are three basic types being developed. One is the segmented mirror in which there are many discrete segments, each of which is moved back and forth mechanically by a separate piston-like device called an "actuator." Another is a mirror in which a reflective coating has been laid down on a base material that changes its shape when signals are applied to it. (In practice, the base material is generally a piezoelectric material, which changes size when an electrical voltage is applied to it. An array of electrodes applies different voltages to different parts of the mirror base.) A third concept is similar in that the mirror has a continuous flexible surface, but in this case the precise contour of the surface is shaped by an array of individually controlled mechanical actuators lying beneath it.[19] All these concepts have been demonstrated, although not in the sizes necessary for practical weapon systems.

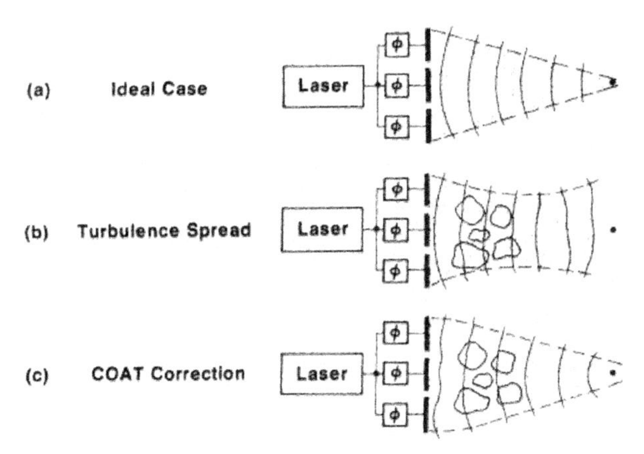

The atmospheric transmission problem, and how it can be overcome–at least in theory. In a vacuum a laser beam could be focused tightly onto a small spot (top). In the atmosphere turbulence, thermal blooming, and other effects spread out the beam over a much larger area (middle). By using adaptive optics to adjust the wavefront of the light emitted by the high-energy laser, the spreading can be reduced (bottom). In these drawings, <f>, the Greek letter phi, indicates control systems that adjust the shape of the output mirror to control the laser wavefront; they are compensating for atmospheric effects only in the bottom drawing. (Courtesy of James E. Pearson, from Ref. 18.)

Performance requirements are stringent for high-energy lasers. The mirror must be able to withstand the laser's high power, a need often met by forcing a liquid coolant through holes in the body of the mirror. Getting fine enough control over the wavefront of the beam requires many separate and ultraprecise control elements in the mirror. A mirror 16 cm (6.5 in.) in diameter requires at least 60 separate actuators, and proportionately higher numbers are needed for larger mirrors. The 16-cm mirror, together with its mount, should hit the scales at about 1000 kg (2200 lb.) and the 60 actuators should weigh no more than 800 kg (nearly 1800 lb.).[20]

The shape of the optical surface must be precisely controlled. The mirror surface should be able to move back and forth over a total range of at least four times the laser wavelength. When the surface control is operating, the surface should be within one-twentieth of the laser wavelength of the ideal shape. As if that isn't enough, adjustments in mirror shape have to be made about 1000 times a second to keep up with fluctuations in the atmosphere. Because the optical tolerances depend directly on laser wavelength, they get tighter at shorter wavelengths. This helps to offset the advantage of being able to use smaller optics at such wavelengths.

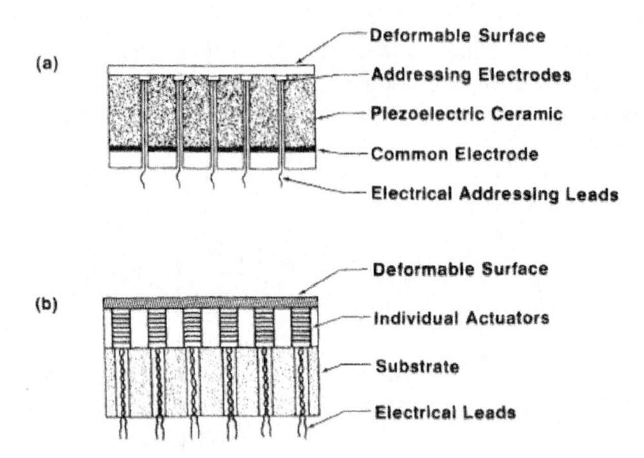

Two types of active mirrors shown in cross section to indicate how the surface shape is controlled. In the mirror at top the flexible surface layer rests on a block of piezoelectric ceramic, which changes its height when an electrical voltage is applied across it. Applying different voltages across different parts of the mirror alters its shape because the height of the piezoelectric material changes unevenly across the surface. In the mirror at bottom the flexible surface layer covers an array of piston-like actuators, which move back and forth in response to electrical signals, thus changing the shape of the mirror. (Courtesy of James E. Pearson, from Ref. 18.)

Adaptive optics can help compensate for effects other than atmospheric distortions that might defocus a laser beam. Some turbulence is inevitable in the laser itself, as gases flow rapidly through the laser cavity and react to release energy. Corrections applied through deformable mirrors can help in precisely tracking targets and in finely focusing the laser beam onto a distant target, although gross mechanical motion of the mirror would be needed to provide full compensation for anything beyond small movements.

Phase Conjugation and Optical Black Magic

Adaptive optics is more than simply deformable mirrors. A control system is needed to determine how much to adjust the mirror's shape. Extensive theoretical work has been done on the propagation of high-energy laser beams through the atmosphere, but theory is not enough. Some of the most important distortions are caused by random atmospheric turbulence that theory cannot predict. The effects influencing light along the path the beam is going to travel must be measured, and that information must be converted into a control signal. This means that the control system must receive light returning along the beam path and analyze what has happened to it. This is by no means an easy process, and the details are well beyond the scope of this book.

After the control system has measured the effects that the beam will be subjected to, the type of compensation required must be determined. This process is called phase conjugation. It is a complex operation in which the measured effects of turbulence are used to create a laser wavefront that will undo what the turbulence did, making it possible to produce a tight focal spot on the target. The precise method by which phase conjugation works is too complex to describe here; suffice it to say that in theory the technique can be used to compensate for aberrations inside the laser and in the atmosphere.[21] The critical corrections are made by adjusting the relative phase of different parts of the laser beam–that is, by making parts of the laser beam slightly out of step with each other, instead of staying in the normal lockstep of the light waves in a laser beam. Interestingly, changes in the intensity pattern of the laser beam are less important in compensating for atmospheric effects than the more subtle phase shifts.

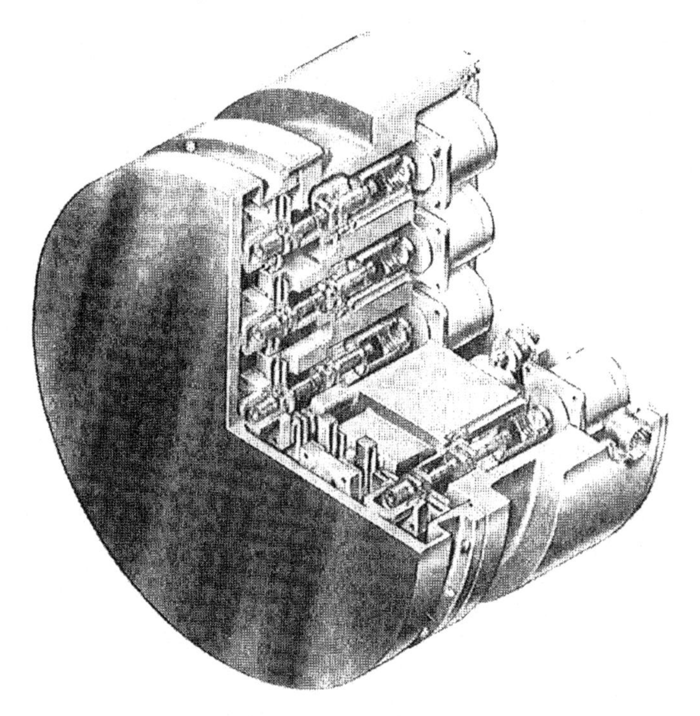

Cutaway of a 19-actuator deformable mirror built by Rockwell International's Rocketdyne Division shows the complexity of adaptive optics. This mirror is 16 in. (40 cm) in diameter and weighs 100 lb. (45 kg). (Courtesy of Rocketdyne Division, Rockwell International Corp.)

Current adaptive optical technology clearly has some limitations in practice. J. Richard Vyce, director of advanced program development at the Itek Corp.'s Optical Systems Division, says that adaptive optics can compensate for atmospheric turbulence in some, but not all, uses of high-energy laser weapons.[22] He believes it may sometimes be cheaper to increase the laser power or the size of the optics than to use adaptive optics, particularly with long-wavelength (i.e., chemical and carbon dioxide) lasers. That is especially true if the beam need not be tightly focused, such as when attempting to blind the sensors on a spy satellite.

He believes that thermal blooming close to the laser may be "tractable" using adaptive optics for compensation. But he is much less optimistic about correcting for turbulence and thermal blooming near slow targets far from the laser. The problem is twofold: the slow motion of the beam would accentuate thermal blooming, while information on what is happening to the beam would take a long time to reach the optical control system, probably too long for it to do anything useful. Those problems have a serious practical impact, making it very hard to hit targets on the ground with a laser beam from space, the archetype of slow targets far from the laser.

Because of the important role adaptive optics can play in laser weaponry, the Pentagon seems increasingly inclined to keep it under security wraps. Many of the approximately 120 papers withdrawn from an August 1982 optics meeting at the behest of Department of Defense security officials dealt with adaptive optics.[23] Researchers say they have made significant progress in the technology recently, but because of security restrictions they can't talk about it publicly.

The best systems that have been described in public are still far from what would be needed for an actual weapon system. These include:

- A 37-actuator, 16-cm (6.3-in.) mirror developed at Itek, which can make corrections accurate to 0.1 μm up to 2000 times/sec. It is not actively cooled, however, as it would have to be to withstand the beam from a high-energy laser.
- A 61-actuator mirror 20 cm (8 in.) in diameter developed by Hughes Aircraft that is actively cooled and can change its shape up to 10,000 times/sec.
- A 69-actuator mirror 16 cm (6.3 in.) in diameter built by United Technologies that can respond 3000 times/sec.[24]

There is a long road to travel before these laboratory demonstrations can be translated into practical military hardware. Adaptive optical systems are expensive, complex, bulky, and unable to cure all the ills of laser weaponry. However, there seems little alternative but to try to live with these problems, especially if the goal is development of laser weapons for use within the atmosphere. Beam distortions are not so serious in space, where there is no air to get in the way, but the tremendous distances involved present other problems.

An impressive amount of scientific manpower and Pentagon money is going into development of adaptive optics. The topic is a common one in scholarly journals such as *Optics Letters, Applied Optics,* and the *Journal of the Optical Society of America.* It turns up regularly on the programs of optical and laser conferences. It has even attracted some interest from some civilians facing somewhat similar problems with a much skimpier budget: astronomers who want to reduce the distortion the atmosphere causes in the light they receive from stars and distant galaxies.

Target Effects

Once the beam reaches the target, it deposits part of its energy there. This involves a complex interaction between beam and target that depends strongly on the nature of the beam and of the target, and which ultimately determines how much of the energy in the beam is transferred to the target. Only after the energy is transferred to the target can it do any damage.

No one is seriously thinking of using a laser beam to completely vaporize any military targets. Instead, a continuous laser beam would cause physical damage by heating a target until the beam melted through the skin and lethally damaged some internal components. The actual type of damage would depend on the target and where it was illuminated. Drilling a hole in a fuel tank could cause an explosion. Disabling the device called a "fuze," which triggers the explosion of a warhead, would prevent a bomb from exploding or alternatively might trigger a premature explosion of the warhead, in a place where it would not damage the intended target, but could cause considerable damage to other objects and people. Knocking out the control or guidance system could make a missile land far from its intended target.

A continuous laser beam can't do damage instantaneously. Heating the target to the required temperature would probably take a few seconds, depending on the laser power and the nature of the target; exact requirements are classified by the government. The illumination time is long enough for the beam to wander off the target spot and let the heated area cool off. Techniques called "countermeasures," described in Chapter 9, could be used to reduce the amount of energy that the beam could deposit on the target. Other types of interactions could also help protect the target.

There is considerable interest in substituting a rapid series of short laser pulses for a continuous beam. As mentioned earlier, this might simplify the task of getting the beam through the air to its target. The abrupt heating and cooling could cause thermal shock, sufficient to shatter materials such as glass. A short, intense pulse could also rapidly evaporate a burst of material from the surface, generating a shock wave that would travel through the target and could cause mechanical damage. (Evaporation caused by a continuous beam would be more gradual and would not cause a shock wave.) The combination of thermal and mechanical damage and heating effects caused by a series of short, closely timed laser pulses does a better job of breaking through sheet metal than either laser heating or laser-produced shock waves can do by themselves.

Physical damage is not the only way a laser beam can disable a target. The beam could also attack sensors that guide weapons to their targets, blinding or disabling them by means discussed in more detail in Chapters 9 and 12. If the target is a spy satellite, disabling the sensor is tantamount to destroying the target, because the satellite can't do its job without its sensitive electronic "eyes."

Particle beams, microwaves, and X rays have their own distinct ways of producing damage, described in more detail in the chapters that follow.

A quick scan of the scientific literature in the United States and the Soviet Union shows what at first seems to be a rich collection of unclassified information about target effects. A more careful examination shows sharp limits to what is being described in public. Research that falls in the generic area of fundamental physics seems to be published freely. It is even finding some practical civilian uses in showing engineers promising uses of lasers in metal-working.

Specific information on the vulnerability of real military targets is conspicuous by its absence. Aiming a laser beam at a sheet of metal in a carefully controlled laboratory environment is much different from shooting at a moving missile a few kilometers away. There's every reason to believe that such experiments are being conducted. Over the past few years the Pentagon has issued a number of contracts to assess the vulnerability of various military hardware to laser attack and, at the behest of Congress, is stepping up those efforts.[25] It is not clear how much is known at this point; realistic experiments are hard to conduct, and may not yield definitive information. Whatever results do exist are unlikely to see the light of day outside of classified publications. The Pentagon and the Kremlin devote much effort to keeping each other guessing about the vulnerability of their hardware.

Fire Control

Another technology that has received little public airing is fire control. Unlike beam control, fire control is a discipline with broad application to different types of directed-energy weapons. The fundamental problems of identifying a target, tracking it, aiming and firing the weapon, and verifying a kill are similar for ordinary and X-ray lasers and for particle beams. (The concepts might be similar for microwave weapons as well, but it's too early to be sure what form such weapons might take.) The nature of the task, however, helps keep fire control quiet. By definition, its applications are almost exclusively in the military realm, making it a logical candidate to keep under security wraps.

By far the most work has been done on fire control for laser weaponry because the technology is considered the closest to application. Fire-control requirements for beam weapons are going to be demanding. The Pentagon's official position is reflected in its public description of high-energy laser technology:

> Fire control for laser weapons will have to be especially capable. It must be able to recognize and classify a host of potential targets, and determine which to engage first. In addition, to realize the firepower potential of a laser weapon, the fire control must be quick to recognize that the target being engaged has been damaged sufficiently that it can no longer perform its mission, so that the laser beam can then be moved to the next target.[26]

In its own vague way, that statement is somewhat informative, but understanding the nature of the problem requires some background in the nature of fire-control techniques and how they interact with the missions of beam weapons.

The developers of laser and particle-beam weapons generally believe that the fire-control system will have to be able to spot its own targets. Automatic target identification, often called IFF for identification of friend or foe, is one of the major thrusts in Pentagon research. The concept is believed to have a role not just in directed-energy weapons but in all sorts of military systems, reflecting the growing importance of sophisticated technology on the battlefield.[27] This broad trend will be described in more detail in Chapter 10.

Current military systems rely on soldiers to spot targets, although once the target has been identified it is often possible to home in on it automatically. From the standpoint of military planners the soldier is not always a desirable part of the system. On the battlefield he could get scared and duck when someone starts shooting at him, losing track of a potential target. Outside of the comparative security of a tank, he's lightly armored at best and very vulnerable to many weapons. His reaction time and discrimination ability are also limited, problems that become increasingly pressing as weapon systems become able to move faster and operate over longer ranges in more difficult circumstances. Missile defense is such a demanding application that the universal assumption seems to be that control will have to be completely automated to cope with the possibility of a rapid, massive attack.

Beam weapons generally are intended to react rapidly against hard-to- hit targets. On the battlefield they may have missions such as disabling incoming cruise missiles before they can destroy a battleship. In space they may have a few minutes to destroy a fleet of nuclear-armed missiles thousands of kilometers away. Typical targets would be fast moving and abundant, not leaving a soldier the luxury of time to scratch his head and figure out what something on the radar screen could be. The problem is much more complex than ray guns at 40 paces, and it requires sophisticated equipment with fast response times.

There are two basic approaches to target recognition: active and passive sensing. Active sensors are radar-like systems that send out a signal and watch for a return signal, which is then used to identify the target. The signals could be microwaves, millimeter waves, or visible or infrared radiation, the last two probably from a laser. The returned signal would require some form of interpretation. Each type of plane, for example, has a characteristic radar signal (or "signature") that it reflects back to the source. The signature varies somewhat with the orientation of the plane and other conditions, and interpreting the returns now generally requires a skilled operator. That is one task the Pentagon would like to automate.

The classical passive sensor is a soldier with a pair of binoculars, scanning the sky for enemy aircraft. Today passive sensors include electronic systems that look for infrared radiation from other sources and convert it into a visible image. Passive sensing still requires interpretation, however. Military researchers are working on a variety of automated techniques for pattern recognition, some using optical techniques and some using computers, in hope of replacing human operators.

For both active and passive systems it is the ability to *discriminate* among objects that is critical. A system that can't tell the difference between an enemy tank and a friendly one is downright dangerous if it is used to control firing on a battlefield where both forces are present. A system that can't tell tanks from barns isn't as bad, except from the viewpoint of the owners of the incorrectly identified barns. However, at best it can waste much expensive antitank ammunition on targets that could be destroyed much more readily–and which probably have little military significance anyway.

Once the target has been spotted in a field of view that generally contains other objects, it has to be tracked. Standard diagrams used by beam weapon planners show two stages of tracking: coarse and fine. The coarse tracking would be used to spot the target, establish its trajectory, and follow it by moving a large primary mirror. Once the general path was known, the system would switch to more precise tracking to pinpoint a vulnerable spot on the target and make small, rapid movements with a smaller, lightweight mirror. Under the direction of the fire-control system, the beam-pointing optics would follow the path of the target. When the optics were properly aligned with the target, the fire-control system would turn the laser on and keep the beam directed at the target long enough to disable it. Ideally, the system would then verify that the target had been disabled, a task that would be much easier if the laser caused some visible event, such as an explosion of fuel or warhead than if it simply killed vital electronic components hidden inside the target.

Demands on the fire control differ widely for missions on the battlefield and those in space. Unsophisticated fire control can do some simple tasks on the battlefield. As far back as 1975, the Army used a hand-aimed pointing system to shoot down helicopter drones with its vehicle-mounted Mobile Test Unit laser. In 1978 the Navy shot down TOW missiles with a 400,000-W chemical laser using a pointer-tracker system developed by Hughes Aircraft.[28] An advanced pointer-tracker was to be used in tests of the 2-million watt MIRACL laser planned at the National High-Energy Laser Test Site at White Sands, New Mexico, to evaluate ship and air defense possibilities as part of the Navy's "Sea Lite" program. Although beam and fire control are generally considered to lag behind laser development, the fact is that at least rudimentary capabilities have been demonstrated.

The problem of target tracking is much more severe in space because the distances involved are up to 5000 km (3000 miles) instead of a mere 10 or 20 km (6 to 12 miles). The feasibility of the required technology will be tested by the Defense Advanced Research Projects Agency in a program called "Talon Gold," part of the Space Laser Triad described in Chapter 11. Tests involving a low-power pointing laser plus a scaled-down version of a target tracking and pointing system are planned on the Space Shuttle. These experiments will track high-altitude aircraft and spacecraft to learn how well such a system can spot realistic targets against the type of backgrounds likely to be encountered. According to DARPA, "This experiment will establish the feasibility of achieving the fire-control performance levels required for operational missions and will provide the necessary database for designing a first generation laser weapon system" for use in space.[29]

Talon Gold will offer what DARPA calls "a significant improvement over current capabilities for pointing and tracking." However, that will not be good enough for actual use in a weapon system intended to destroy ballistic missiles and strategic bombers. DARPA says such a system will require "acquisition, tracking and pointing performance levels beyond those that are currently projected for the Talon Gold demonstration. Substantial improvements in pointing precision and the development of rapid acquisition and retargeting capabilities will be necessary."[30] The agency is sponsoring other research aimed at developing the needed technology.

A key technology in advanced tracking systems will be laser radar in which the shorter wavelengths (and higher frequencies) of light substitute for the radio waves and microwaves normally used in radar. It is one of a number of military efforts to develop alternative forms of radar. Although conventional microwave-based radars are adequate for many battlefield uses, they also present some problems. One is that new types of missiles can home in on radar emissions to destroy microwave antennas. The Pentagon is working on ways to deal with this problem, but some observers feel it is so serious that in many cases radars may prove more harm than good.[31] In addition, all radars cannot resolve details smaller than the wavelength used, so moving to wavelengths shorter than the centimeter (0.4 in.) or more of microwaves makes it possible to see finer details in radar images.

The Pentagon has an extensive program aimed at developing radar systems using millimeter waves, which as the name implies have a wavelength of around a millimeter (0.04 in.). There is also interest in "submillimeter" radars, but at present there are few suitable sources of high powers at wavelengths between about 0.02 and 1 mm. At shorter wavelengths in the infrared, lasers are available, and these are receiving serious attention from radar system developers who would like to use laser light rather than microwaves or millimeter waves.

The leading contender for use as a laser radar source is the carbon dioxide laser. With a wavelength of 10 μm (0.01 mm), this laser can be made to operate efficiently at low powers, as well as the high powers mentioned in Chapter 4. A variety of carbon dioxide laser radars intended for use over ranges typical of those encountered on the battlefield have been built and demonstrated.[32]

Large, sophisticated laser radars are capable of excellent performance over longer ranges, although they don't meet the requirements for use with space laser weapon systems. The Massachusetts Institute of Technology's Lincoln Laboratory has assembled a system called "Firepond," which in 1981 was considered the most sophisticated laser radar in existence.[33] The heart of the system is a carbon dioxide laser that can produce peak pulse power of 15,000 W, with average power over a number of seconds of 1400 W. The system can detect and track satellites with an accuracy of one-millionth of a radian, approaching the accuracy that would be needed in a space-based laser weapon. Extensions of the technology used in the Firepond system will be used in Talon Gold, but even that experiment won't meet the needs of laser weapons for space use.

It's worth stressing that the pointing accuracy of the fire-control system, and hence the tracking capacity of the target-spotting system, will have to be comparable to that of the high-energy laser beam. That means that a laser radar in a space-based laser weapon system will also require a large mirror, both to transmit its beam and to receive the returned signal, although the mirror would not have to withstand high power. It is unclear if that would require a second mirror, or if the mirror used to direct the high-energy beam could do double duty.

Other Fire-Control Components

A fire-control system is basically a special-purpose control system. Like other types of control systems, it has to have some kind of information-processing capability–that is, a computer. It would have to have some way to transmit signals to and from various elements of the weapon system, that is, a communication network. Some tasks required are quite sophisticated; reliably recognizing a potential target no matter what profile it presents and what lies in the background, for example, is a job that computer technology (even some exotic, special-purpose variants) has yet to master.

Further complications come from the need for very fast reaction times on the part of a beam weapon system. Military planners have talked about firing rates of one shot every few seconds from laser weapons, and perhaps many shots per second from particle-beam weapons. Yet those times are short compared with what large computers now need to handle automatic target-identification problems. There's much work ahead in both speeding up and shrinking the size of computers, and these are the major thrusts of research efforts such as the Pentagon's Very High Speed Integrated Circuit (VHSIC) program. Beam-control systems will also require fast and sophisticated computers to convert information on atmospheric conditions into signals to adjust the entire surfaces of adjustable mirrors at rates of thousands of times per second.

The details of the electronic and other requirements of fire- and beam-control systems go beyond the scope of this book and well into the realm of classified information. Clearly, there are many technical problems that remain to be solved, as there are throughout the area of fire and beam control.

Indeed, the whole problem of beam and fire control is extremely difficult. To many observers the barriers in that area seem harder to overcome than those to building multi-million-watt lasers. Developers have made notable progress in building some components for beam and fire control, but in other areas wide gaps yawn between demonstrated capabilities and the requirements for weapon systems. Ultimately, the components have to be put together in a system able to get the beam to the target, and that is expected to be a formidable task in system integration. Difficulties abound, and unofficial reports circulating in the laser weapon community indicate that some of them have already surfaced in planned demonstrations. Work on the 2.2-millionwatt MIRACL laser for the Navy's Sea Lite test was said to be well ahead of that on the tracker and pointer before Congress cancelled Sea Lite. In a similar vein reports on progress of the ALPHA laser for the Space Laser Triad demonstration tend to be more encouraging than those on the companion Talon Gold and Large Optics Demonstration Experiment (LODE) tests.

References

1. The port through which Luke dropped the torpedoes was "ray shielded" so no "energy beams" could be used. George Lucas, *Star Wars* (Ballantine, New York, 1976), p. 181. Like most science fiction writers, Lucas doesn't bother to explain the principles behind the weaponry in his story.
2. This formula actually defines the first point at which the intensity falls to zero. There are a series of bright rings surrounding the central spot, falling off in intensity as the distance from the central spot increases. The formula is valid for a circular output mirror and a perfectly uniform laser beam, which does not exist in practice, but which does give a rough approximation for real lasers. It actually gives the sine of the angular spot size for small spots such as would be produced by a laser beam, which is virtually identical to the angle in radians. The formula is a fundamental one and comes from Donald H. Menzel, ed, *Fundamental Formulas of Physics,* Vol. 2 (Dover, New York, 1960), p. 416.
3. John N. Bahcall and Lyman Spitzer, Jr., "The space telescope," *Scientific American* 247 (I), 40-51 (July 1982).
4. "Senate directs Air Force to formulate laser plan," *Aviation Week & Space Technology,* May 25, 1981, pp. 52-53.
5. "Laser battle station mirror proposed," *Aviation Week & Space Technology,* May 25, 1981, p. 64.
6. For more information on such concepts see, for example, papers in Kenneth W. Billman, ed, *Radiation Energy Conversion in Space* (American Institute of Aeronautics and Astronautics, New York, 1978).
7. R. R. Berggren and G. E. Lenertz, *Feasibility of a 30-Meter Space-Based Laser Transmitter* (Itek Optical Systems, Lexington, Massachusetts, October 1975).
8. The mirrors are listed in *Optics Guide 2* (Melles Griot, Irvine, California, 1981), p. 186; prices are from the company's 1982 price list.
9. "Report from Rochester: industry, government confront technology export," *Optics News* 8 (4), 4-5 (July/August 1982).
10. "U.S. damaged by transfer of mirror technology to Soviets," *Lasers & Applications* 1(4), 24 (December 1982).
11. Forrest M. Mims III, "The first century of lightwave communications," *International Fiber Optics and Communications Handbook and Buyers Guide 1981-1982,* pp. 6-23.
12. This description is based largely on David C. Smith, "High-power laser propagation: thermal blooming," *Proceedings of the IEEE* 65 (12), 1679-1714 (December 1977).
13. See, for example, J. W. Strongbehn, ed, *Laser Beam Propagation in the Atmosphere* (Springer-Verlag, New York, 1978), and V. E. Zuev, *Laser Beams in the Atmosphere* (Plenum Press, New York, 1982).
14. Albert V. Jelalian, "Laser and microwave radar," *Laser Focus* 17 (4), 88-94 (April 1981).
15. Description based on an explanation in David C. Smith, "High-power laser propagation:

thermal blooming," *Proceedings of the IEEE* 65 (12), 1679-1714 (December 1977).

16. Leonard M. Ball, "The laser lightning-rod system: thunderstorm domestication," *Applied Optics* 13 (10), 2292-2296 (October 1974); Leonard M. Ball, private communication.
17. David C. Smith, "High-power laser propagation: thermal blooming," *Proceedings of the IEEE* 65 (12), 1679-1714 (December 1977).
18. For a detailed technical review of the concept and of some representative systems as of the late 1970s, see James E. Pearson, R. H. Freeman, and Harold C. Reynolds, Jr., "Adaptive optics techniques for wave-front correction," in Robert R. Shannon and James C. Wyant, eds, *Applied Optics and Optical Engineering,* Vol. VII (Academic Press, New York, 1979) pp. 246-340.
19. R. H. Freeman and James E. Pearson, "Deformable mirrors for all seasons and reasons," *Applied Optics* 21 (4), 580-588 (February 15, 1982).
20. *Ibid.,* p. 581.
21. For more details, see James E. Pearson, R. H. Freeman, and Harold C. Reynolds, Jr., "Adaptive optics techniques for wave-front correction," in Robert R. Shannon and James C. Wyant, eds, *Applied Optics and Optical Engineering Vol VII* (Academic Press, New York, 1979), pp. 246-340.
22. J. Richard Vyce, presentation at Laser Systems and Technology Conference, Washington, July 9-10, 1981, sponsored by American Institute of Aeronautics and Astronautics. An excellent summary of this presentation and an interview that followed is Philip J. Klass, "Adaptive optics evaluated as laser aid," *Aviation Week & Space Technology,* August 24, 1981, pp. 61-65.
23. " 'Remote censoring': DOD blocks symposium papers," *Science News* 122, 148-149 (September 4, 1982).
24. Philip J. Klass, "Adaptive optics evaluated as laser aid," *Aviation Week & Space Technology,* August 24, 1981, pp. 61-65.
25. "House-Senate Compromise," *Lasers & Applications* 1 (3), 20 (November 1982).
26. Department of Defense, "Fact Sheet: DOD High Energy Laser Program," February 1982, p. 2.
27. For a critical history of the early stages of this trend, with particular emphasis on the Vietnam war, see Paul Dickson, *The Electronic Battlefield* (Indiana University Press, Bloomington, 1976).
28. J. Richard Airey, presentation at Laser Systems and Technology Conference, Boston, July 27-28, 1981, sponsored by American Institute of Aeronautics and Astronautics.
29. Defense Advanced Research Projects Agency, *Fiscal Year 1983 Research and Development Program,* March 30, 1982, p. III-47.
30. *Ibid.,* p. III-44.
31. Thomas A. Amlie, "Radar: shield or target?" *IEEE Spectrum* **19** (4), 61-65 (April 1982).
32. Brian E. Edwards, "Design aspects of an infrared laser radar," *Lasers & Applications* **1** (2), 47-50 (October 1982); Aris Papayoanou, "CO_2 lasers for tactical military systems," *Lasers & Applications* **1** (4), 49-55 (December 1982).
33. Albert V. Jelalian, presentation made at Laser Systems and Technology Conference, Boston, July 27-28, 1981, sponsored by American Institute of Aeronautics and Astronautics.

6.
The Strange Saga of the X-Ray Laser

By far the most controversial proposal for beam weaponry is the X-ray laser battle station. With a small nuclear bomb at its core, it would orbit the earth or sit atop a missile on the ground or in a submarine, waiting to be launched into orbit. If a nuclear attack was made, any X-ray lasers on the ground would "pop-up" into space. Then in each X-ray laser the bomb would explode, transferring its energy to an array of perhaps 50 X-ray laser rods, which would channel the energy from the bomb into ultrapowerful pulses of X rays that would destroy the missiles long before they could reach their targets. Advocates such as controversial nuclear physicist Edward Teller see the concept as permitting the United States to take a bold leap forward in missile defense, to a point well ahead of the Soviet Union.

Critics say that development of X-ray laser battle stations would be a dangerously destabilizing strategy-and probably a futile effort as well because of the massive technological barriers. One major concern is that building X-ray lasers would involve breaking three major arms-control treaties going as far back as 1963.[1] The limited test ban treaty of 1963 prohibits nuclear tests in the atmosphere or outer space, which would almost certainly be required to test an X-ray laser weapon system. The outer space treaty of 1967 expressly forbids putting nuclear weapons in orbit, a necessity if the X-ray lasers are to stop a nuclear attack. And building new types of missile defense systems is prohibited by the 1972 SALT-I Treaty. Deployment of other types of missile defense systems using conventional lasers or particle beams would require changes only in SALT-I, but just trying to develop X-ray laser weapon systems could lead to a whole new round of atmospheric nuclear testing by breaking the two-decade old limited test ban treaty.

There are also grave questions about the feasibility of X-ray laser battle stations. Although only one bomb might be needed to power an array of 50 lasers, it might take 50 separate pointing and tracking systems to aim the beams at the targets; and, as indicated in the previous chapter, such systems are likely to be difficult and expensive to build. It may be possible to devise attack strategies that could make an X-ray laser battle station comparatively ineffective. The very technical feasibility of the X-ray laser itself is uncertain. Researchers have encountered serious trouble trying to tame X-ray physics to build an X-ray laser in the controlled environment of the laboratory; building a reliable weapon system is an even tougher job.

The story of X-ray laser research is almost as bizarre as the physics involved. Years of futile research were interrupted by one premature report of an X-ray laser that wasn't. Progress was agonizingly slow, and some researchers and their military sponsors finally gave up. Just when the field seemed almost dormant, developers were awakened by an apparent breakthrough in a British laboratory. Within a year word of another, more dramatic, breakthrough by a group from the Lawrence Livermore National Laboratory leaked through a heavy veil of government secrecy. It was reports of this Livermore experiment that stimulated the current round of proposals for X-ray battle stations.

These serious technical and strategic uncertainties create plenty of controversy when added to the political debate in Congress over the direction of the space laser weapon program. Enthusiasm for short-wavelength lasers in general–and X-ray lasers in particular–seems to be concentrated in the House Armed Services Committee, with a prime mover there being staff member Anthony Battista.[2] There is much less enthusiasm in the Senate and the Department of Defense, even among advocates of other types of laser systems for ballistic missile defense. One proponent of chemical lasers for missile defense labeled the X-ray laser as "Teller's toy."[3] President Reagan's March 1983 speech advocating development of missile defense systems in general (he did not publicly indicate any preference as to the type of weapon) intensified the controversy.

The time delays inevitable in book publishing make it impossible to keep up to date with the rapidly shifting political winds. The rest of this chapter will go into the technological and strategic issues that lie beneath the political debate. To explain the seriousness of the technical issues, it's necessary to start by describing some of the basics of X-ray physics and how they differ from the laws of physics that affect other lasers.

X-Ray Laser Physics

Although X rays and visible light are both forms of electromagnetic radiation, from a practical standpoint they are very different. Visible light has wavelengths of 0.4 to 0.7 μm (micrometers), or equivalently 4000-7000 Å (angstrom units). X-ray wavelengths are shorter than 100 Å, and some of the most interesting X-ray wavelengths are in the range of 1 to 10 Å. Because the energy carried by a photon increases as the wavelength decreases, X rays have 100 to 1000 times more energy than visible light. That energy difference means that X rays and visible light interact with matter in very different ways, making them appear very different to us.

From the standpoint of atomic physics, X-ray transitions are simply electronic transitions that are much more energetic than those that produce visible light. Generally, X-ray transitions occur between one energy level near an atomic nucleus containing several protons and a second energy level much further from the nucleus. The energy difference is large because the positively charged nucleus strongly attracts the negatively charged electron, and the force involved is much stronger when the electron is close to the nucleus.

Like longer-wavelength transitions, X-ray transitions can emit light spontaneously or by stimulated emission. However, there are some important qualitative differences. The likelihood that a given atom will give up its energy by stimulated rather than spontaneous emission is proportional to the cube (third power) of the wavelength. Because the probability of stimulated emission declines sharply as wavelength decreases, so does the likelihood of amplification by the stimulated emission of radiation–laser gain. This does not mean that X-ray laser action is impossible–just hard to produce. However, theoretical calculations indicate that laser gain is not possible at wavelengths shorter than about 0.01 Å in material media.[4]

Because of the high energy of X-ray transitions, it takes extremely high powers to raise the energy levels of enough electrons to produce a population inversion on one. Even once the electrons are raised to a high energy level, they don't stay there long, tending to drop back down to a lower level very quickly. Quantitatively, the lifetime of excited states is proportional to the square of the wavelength, with one rule of thumb being that the lifetime in seconds is roughly 10^{-15} times the square of the wavelength in angstroms.[5] Thus, an atom would stay in an excited state ready to emit a 10-Å photon for only about 10^{-13} sec (one tenth of a trillionth of a second), and the situation would get even worse at shorter wavelengths. Excited-state lifetime is important in lasers because it indicates the time available to stimulate emission before the excited atoms drop back down to the lower energy level by spontaneous emission. If stimulated emission is not produced during that fleeting interval, the X-ray laser does not work.

The combination of high excitation energy and extremely short excited-state lifetime make it necessary to use extremely high peak powers in a pulse that can be very brief to energize an X-ray laser. Both requirements get more demanding as the wavelength gets shorter, rapidly compounding problems. Physicists have calculated that pumping energy of about 2 W *per atom* would be needed for a 1-Å laser.[6] To put that into perspective, it means that a trillion-watt pulse would be needed to excite a trillionth of a gram of carbon at that wavelength. The pumping energy requirements become much less formidable at longer X-ray wavelengths, but they are still considerable.

The same effects that make it necessary to put energy rapidly into an X-ray laser make it possible to get the energy out very rapidly. In fact, the X-ray laser would emit a pulse roughly as short as the energizing pulse, and with a high peak power. The overall peak power produced by the laser would inevitably be less than the peak power needed to get it to work because of the inefficiencies inherent in laser physics. However, the laser beam would probably concentrate the energy so that much higher intensities could be obtained in the narrow beam than from the source of the laser energy.

A side effect of the intense pumping would be the vaporization of the X-ray laser material. This isn't as bad as it might sound at first. Even if the pumping energy could be transferred instantaneously to the atoms, it would take time for the freshly excited atoms to speed away. The atoms would emit laser light very fast, faster, in fact, than they could move. The laser light would then speed through the material, being amplified along the way by excited atoms. Because the X rays travel at the speed of light, they would be able to get out of the material before it was disrupted by vaporization. The fact that an X-ray laser inevitably self-destructs (or at least the material emitting the laser beam) does not prevent it from working (once), although it does impose limits on the way in which it can be used.

There's another complication that also deserves mention: X-ray lasers probably wouldn't have mirrors or resonators. Some materials can reflect X rays incident at a glancing angle, but nothing efficiently reflects X rays in the way an ordinary mirror reflects light. Although progress is being made, it probably wouldn't matter that much anyway. If X-ray mirrors existed, the intense power carried by an X-ray laser pulse would probably vaporize them, if the laser pumping energy hadn't done the job already. Nor would it do much good to keep bouncing X rays back and forth through the freshly vaporized laser medium, especially since most of the atoms would have dropped back to a lower energy level.

Instead of a resonator, an X-ray laser would rely on what is called amplified spontaneous emission. Spontaneously emitted photons would stimulate emission of other photons as they passed through the laser material. With no mirror at the end of the laser rod, the photons just keep on going. This process produces diffuse "cosmic maser" emission at much longer wavelengths from gas clouds in interstellar space. It could be made to produce an X-ray laser beam by shaping the laser material into a long, thin fiber or cylinder. Photons passing along the length of the fiber would be amplified much more strongly than those headed in other directions because they would pass through more laser material. The result would be a laser beam shaped not by a resonator cavity but by the dimensions of the laser material itself.

What would an X-ray laser look like? Theorists have talked about fibers around 1 cm (0.4 in.) long and 1 μm (one millionth of a meter) in diameter, 10,000 times as long as it is across. With a wavelength around 1 Å, they predict that such a laser would produce a beam that spread out at an angle of a fraction of a thousandth of a radian,[7] that is, somewhat less than an ordinary helium-neon laser. That would mean that at a distance of 1000 m (3300 ft), such an X-ray laser would produce a spot a fraction of a meter (about a foot) across. It would produce a pulse that lasted around one-trillionth of a second, or perhaps less. During that very brief interval, the peak power would be very intense, probably well into the trillions of watts.

Ways to Excite X-Ray Lasers

Producing the conditions needed for an X-ray laser to operate is no easy task. A population inversion must be produced with a density of 10^{18} excited atoms/cm^3 for a laser 1 cm long. (The required density of excited atoms is inversely proportional to laser length, and hence is higher for shorter lasers and lower for longer ones.)[8]

There are two basic approaches to powering an X-ray laser. X rays produced by any of several possible sources could excite electrons to high energy levels. Alternatively, energy could be transferred by collisions between atoms (or more likely ions) and electrons. Either mechanism would probably involve dumping so much energy into the laser medium that it would become ionized. Thus, X-ray laser action would probably take place in a plasma, a mixture of ions and electrons at high temperature, although only an instant before that plasma might have been a solid.

Many X-ray laser experiments have relied on an extremely short pulse from a longer-wavelength laser to deliver the required power. Such a short, powerful pumping pulse could excite an X-ray laser in different ways. The pump pulse could heat a small target to temperatures so high that it emits thermal X rays analogous to the infrared or "heat" radiation that other objects emit at room temperature, or the white light from the hot filament of an incandescent bulb. These X rays could, in theory, be used to excite an X-ray laser.

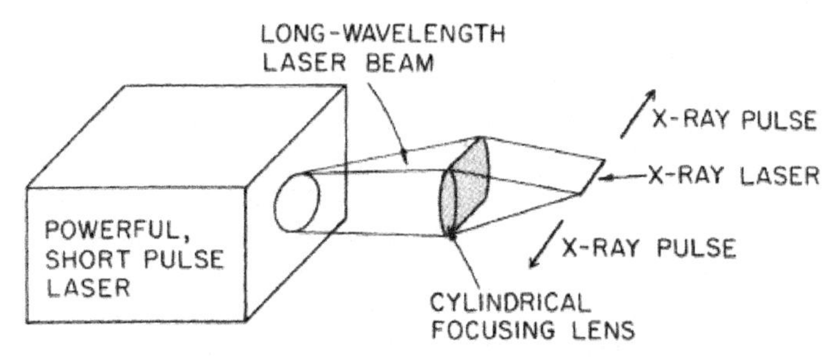

To get an X-ray laser to operate, a large amount of energy must be deposited very rapidly into the laser material. One common approach is to use a very short pulse from a powerful laser emitting at a longer wavelength, usually one designed for laser fusion experiments. Focusing of the beam onto the X-ray laser medium produces the required high-power excitation. In this diagram the beam from such a powerful laser is passed through a cylindrical lens, which focuses it onto a straight line rather than a point focus as with an ordinary "spherical" lens. The X-ray laser medium, a thin fiber, is vaporized by the pulse from the other laser, but before it evaporates produces a rapid burst of X rays. Because there are no mirrors, this sort of X-ray laser produces two beams, emerging from opposite ends of the fiber. This arrangement is similar to one used at the University of Hull. (Drawing by Arthur Giordani.)

Rapid deposition of energy on a small target also can ionize part of it, producing a very hot plasma, which cools rapidly as it expands. As electrons in the plasma recombine with ions, they can produce a population inversion. This approach produced the most successful results yet reported openly in experiments at the University of Hull in Britain.

In the Hull experiments carbon fibers only a few micrometers in diameter were zapped with laser pulses lasting less than one billionth of a second. This produced a hot plasma, which cooled rapidly as it expanded. Carbon ions that had lost all of their electrons then began capturing electrons, which tumbled into low energy levels and produced stimulated emission at a wavelength of

182 Å (often called X rays, but more properly considered part of the "extreme ultraviolet" part of the spectrum).[9]

There are other alternatives. Using intense laser pulse to heat a target is only one way to generate a pulse of X rays suitable for exciting an X-ray laser. A much more powerful source is a nuclear explosion, which generates prodigious quantities of X rays. The possibility of using small nuclear bombs to drive X-ray lasers is being investigated in a supersecret program at the Lawrence Livermore National Laboratory, described in more detail later in this chapter.

The experimental problems of X-ray laser research do not stop with generating the X-ray laser beam. Researchers have to make measurements to check what they have done, but the characteristics of X-ray laser pulses are hard to measure. The pulses are so intense they can overwhelm sensitive detectors and so short that they require sophisticated measurement techniques–in some cases beyond the capabilities of existing instruments. Another problem is the limited resolution of measurement methods needed to show that X-ray emission is in a narrow range of wavelengths and in a well-collimated beam, characteristics expected from a laser beam.

The quirky and unrepeatable nature of X-ray effects can also frustrate researchers. Tantalizing bits of evidence that might indicate the presence of an X-ray laser beam could also indicate something else. Experiments simply might not work reliably. One Soviet group reported generating pulses at a wavelength of about 600 Å but noted that "shot-to-shot reproducibility has not been achieved because of the difficulty to provide [sic] the needed plasma parameters."[10] Western observers privately remain skeptical about that experiment because the published reports do not describe diagnostic measurements adequate to rule out effects other than X-ray laser action.

X-Ray Interactions

Although their characteristics can be hard to measure, X rays can produce some very obvious effects when they interact with matter. Energetic X rays can ionize atoms deep inside solids, including semiconductor electronics. The resulting burst of electrons might knock out a semiconductor device's memory and render it temporarily useless or permanently disabled. Intense enough bursts of X rays could physically weaken some materials, perhaps evaporating surface layers and generating damaging shock waves. Low levels of X rays can cause cancer and mutations in humans, and high-level exposure can be fatal. Most of these effects are of military interest, particularly the ability to disable electronic guidance and control systems.

How well X rays travel through materials depends on their wavelength. In general, the shorter the wavelength, the deeper it will penetrate. For example, 2.7-Å X rays can penetrate about 6.9 μm (millionths of a meter) into iron, while 0.56-Å X rays can penetrate 64.1 μm into the metal, nearly 10 times further.[11] (The high contrast and penetration of X-ray photographs of the human body are possible because the light elements, hydrogen, carbon, and oxygen, that make up most of the body's soft tissue are fairly transparent to X rays. Calcium, a major constituent of bone, strongly absorbs X rays, and in fact at some wavelengths is nearly as opaque to X rays as lead.)

The same general trend holds for X rays in the atmosphere; the shorter the wavelength, the farther they travel. However, X rays don't travel as far in air as you might think. At sea level, 1-Å X rays can travel through a little over 2 m (about 7 ft) of air before half of them are absorbed. For 10-Å X rays the situation is much worse–half are absorbed before they make their way through 1.5 mm (0.06 in.) of air.[12]

The strong atmospheric absorption limits the use of X-ray laser weapons to space, where there is no air to absorb the beam. If the beam from an orbiting X-ray laser was pointed down at the earth, the air would simply soak it up. A 1-Å beam would be half-absorbed by the time it reached an altitude of 60 km above the earth's surface, while a 10-Å beam would be half-absorbed by the time it reached an altitude of 110 km.[13] The same atmospheric effects protect us from solar X-rays and short-wavelength ultraviolet radiation.

Even in space, propagation effects would tend to point toward the use of shorter X-ray wavelengths. One reason is that the shorter wavelengths would more effectively penetrate the metal skins of potential targets. Another is that residual traces of the atmosphere reach altitudes well over 100 km. Those traces might be enough to hamper transmission of a long-wavelength X-ray beam if it had to skim through them, as seems likely if battle stations in low earth orbit have to shoot at missiles being launched far away.

Weaponry is not the only potential application of X-ray lasers. Prospects for high-resolution holographic images of living cells have lately been receiving attention.[14] Other potential uses include recording very fine patterns for the manufacture of semiconductor integrated circuits, medical diagnosis and treatment, metallurgy, radiation chemistry, research in atomic physics and spectroscopy, studies of the structure of materials, and diagnostic tools for the observation of laser fusion reactions.[15] However, none of these appear important enough to justify support for a large research program.

Theorists have begun to look beyond the X-ray region to the even shorter wavelengths of gamma rays. The obstacles to building gamma-ray lasers (or "grasers" as some call them) appear to be formidable. In the words of a team of three theorists from the United States and Soviet Union who recently wrote a comprehensive review of the topic: "The gamma-ray laser presents as difficult a challenge as any that Man has ever undertaken."[16] It should be added that the authors are optimistic and think that a gamma-ray laser may ultimately be possible.

The Early Days of X-Ray Laser Research

It is clear now that any X-ray or gamma-ray laser will be the outcome of a long effort–one that's already been under way for over two decades. The leap from the microwave frequencies of the maser to the much higher frequencies of visible light in the laser stimulated the emission of a broad range of new ideas. Many of them focused on the possibility of taking another large leap to much higher frequencies, or, equivalently, much shorter wavelengths. There were even more proposals for new types of lasers emitting in the visible range.

Some of the ideas for new lasers worked. Enough worked, in fact, that for a time a kind of euphoria prevailed. One veteran of short-wavelength laser research recalled, with tongue in cheek, that there was a school of thought that held that "a telephone pole would lase if you zapped it hard enough." It wasn't quite that easy, however. Many of the bright new ideas didn't work and remain buried in musty archives of moldering scientific journals in the basements of university libraries.

Proposals for gamma-ray lasers in the United States date back to 1960, the year the first visible-wavelength laser was demonstrated. That proposal was not published in the open scientific literature until 1963, which some observers interpret as "evidence of concern that the applications might include military uses."[17] Soviet research also was not published until 1963, although a patent application was filed in 1961.

The pattern of the bold leap was not to be repeated, however. Instead, there was a gradual march to shorter wavelengths, first from the red of the ruby laser to shorter wavelengths, then into the ultraviolet. Slogging through the ultraviolet to progressively shorter wavelengths proved slow, hard work.[18] Despite intense interest and support from the Pentagon that contributed to a short-wavelength laser race, the X-ray laser remained elusive.

What at first seemed to be a dramatic breakthrough came in 1972 when John G. Kepros, then a junior-level researcher at the University of Utah, announced that he had demonstrated X-ray lasing.[19] He produced packets of X-ray film that showed small, well-aligned spots and cited other measurements consistent with X-ray lasing. The demonstration was dubbed "X-ray Jell-O" because the source of the emission was a mixture of copper sulfate in Knox Unflavored Gelatin. The report first made headlines, then it made trouble. Other researchers couldn't duplicate the experiments.

Exactly what happened has never been definitely agreed on by X-ray laser researchers.[20] *Laser Focus* magazine simply described it as "neither X rays nor a laser."[21] That much is probably true, but no attempt to provide a detailed explanation has ever satisfied the majority of observers. Ten years after the experiment, Ronald W. Waynant, a short-wavelength laser researcher at the Naval Research Laboratory in Washington, would shrug his shoulders while recounting the history of the field at a laser conference and say that the problem could simply have been the "hot, dry climate" of Utah.[22]

Kepros ended up taking the blame for what is generally considered to have been overenthusiastic interpretation of inconclusive results rather than deliberate fraud. After some early attempts to salvage his work by reinterpreting it, he eventually went on to other things, except for an abortive attempt to present his interpretation of the supersecret Livermore experiments at the 1981 Annual Meeting of the Optical Society of America.[23] However, the incident left behind a legacy, a skeleton in the X-ray laser closet that serves to remind other researchers that results are not always what they seem.

The Utah results were not the only ones reported in the early 1970s that were received skeptically. In 1974 P. Jaegle and colleagues at the University of Paris South in Orsay reported observing a small laser gain in an expanding aluminum plasma produced by an intense laser pulse.[24] However, others questioned the interpretation and proposed effects other than stimulated emission and amplification that could account for the observations.[25]

Slow Progress Toward Shorter Wavelengths

The 1970s did see a slow progress toward shorter wavelengths. A plateau was reached in 1976 when John Reintjes and several co-workers at the Naval Research Laboratory in Washington generated coherent light at a wavelength of 380 Å.[26] Their light source, however, was not a laser. They relied on what are called "nonlinear" effects, which can cause the frequency of light to be doubled (or multiplied by another low integer) when passing through certain materials. Starting with an intense pulse from a crystalline laser at the infrared wavelength of 10,600 Å, the Navy group quadrupled the frequency in special crystals to generate a wavelength of 2660 Å in the ultraviolet. They then focused that light into helium to generate the seventh harmonic (a sevenfold multiplication of frequency) at the 380-Å wavelength. Overall efficiency of the process was very low, less than one part in a million.

Developing lasers to directly produce light in the region between 100 and 1000 Å (called either the extreme ultraviolet or part of the soft X-ray region by various observers) proved much more difficult. In June 1977 some tantalizing preliminary results were reported by two leading Soviet researchers, I. I. Sobel'man of the Lebedev Physics Institute in Moscow and V. S. Letokhov of the Soviet Institute of Spectroscopy, also in Moscow.[27] Sobel'man showed a darkened spot on photographic film that might have indicated the production of stimulated emission from plasmas of calcium and titanium from which 10 and 12 electrons, respectively, had been removed. The emission was in the region of 350 to 850 Å, although he could not precisely identify wavelength. Letokhov reported "intensity anomalies" at wavelengths between 580 and 780 Å produced by chlorine ions from which 7 electrons had been removed.

At the time neither Sobel'man nor Letokhov claimed that they had observed X-ray lasing, stressing that further measurements were needed before they could be sure. In the couple of years that followed a handful of papers on the topic appeared in the open Russian scientific literature, but none went much further in their claims. Probably the strongest claim was one mentioned earlier that Sobel'man made at a 1979 conference. He said that when a longer-wavelength laser was used to produce a plasma containing calcium atoms missing 12 electrons each, "in rare shots the lasing at 600 angstroms was observed, but shot-to-shot reproducibility has not been achieved because of the difficulty to provide *[sic]* the needed plasma parameters."[28] American researchers who have studied the published Soviet works remain unconvinced that they demonstrated X-ray laser action.

Indeed, skepticism seemed to be the watchword in the United States even before Sobel'man and Letokhov had announced their initial results. In late 1976 the Defense Advanced Research Projects Agency had essentially given up on prospects for building an X-ray laser. DARPA is the Department of Defense agency responsible for "high-risk, high-payoff" research, that is, projects that are far from certain of success but that could provide very valuable results if they do succeed. As such, DARPA had been paying the bills for most X-ray laser research in the United States, but not getting much in the way of X-ray lasers to show for it. Frustrated by the lack of progress, DARPA decided to shift its support to the free-electron laser (see Chapter 4), which at the time was a new technology offering long-term promise for high-power output.

To confirm its judgment, DARPA asked Physical Dynamics Inc., a consulting firm in La Jolla, California, to study the potential applications of X-ray lasers. The company's report[29] cited fields such as medical imaging, microscopy, X-ray holography, materials research, spectroscopy, and the production of electronic components, nominally civilian applications that might have military implications. A classified appendix (which I have not seen) dealt with "space-based weapon systems." The tabulation of potential uses of X-ray lasers was an interesting one, but DARPA decided that none of them were important enough to justify continued support in view of the likelihood that no usable X-ray laser would be produced.

Because DARPA money had been behind most of the X-ray laser research in the United States, the cutoff caused a sharp decline in research. The United States did not abandon the field altogether, but most of the efforts were small ones requiring minimal support. In early 1979 there was a brief report in *Laser Focus* magazine indicating that the Lawrence Livermore National Laboratory had begun pursuing two approaches to demonstrating an X-ray laser.[30] One was based on pumping the laser with a high-power, short-pulse laser. The other, in the magazine's original words, "requires a nuclear device, is considered more promising, and is classified." Later those words would appear prophetic, but even those of us who were on the magazine's staff at the time weren't expecting to hear much more about it. Even the Soviets seemed to be giving up on the field. One can't help but wonder if they might have been directly or indirectly influenced by DARPA's decision.

The X-Ray Laser Revival

For a few years very little happened, and some observers were ready to conclude that the search for the X-ray laser had been abandoned. Before any obituaries could be written, a group at the University of Hull in England led by Geoffrey Pert reported observing "laser gain" at 182 Å in a highly ionized carbon plasma.[31] They vaporized carbon fibers a few micrometers thick with powerful infrared laser pulses lasting only one hundred trillionths of a second. As the highly ionized plasma from the carbon expanded and cooled, free electrons recombined with the carbon ions, apparently producing a population inversion on the transition on which they observed gain.

Perhaps with the earlier false alarms in mind, the British researchers were very cautious in their claims, deliberately steering clear of the term "X-ray laser." They labeled the wavelength "extreme ultraviolet." They subjected the results to a detailed analysis, and more than two years after the work was first announced published a detailed 24-page analysis in which they concluded that "the only result consistent with the full set of data is that amplification is observed, and that this would be ascribed to gain by stimulated emission."[32] Long before the Hull group could publish the final interpretation of their experiments, however, a new X-ray laser controversy hit the headlines. In early 1981 *Aviation Week & Space Technology* published an article by military editor Clarence A. Robinson, Jr., describing highly classified experiments in which X-ray lasing was observed at a wavelength of 14 Å.[33]

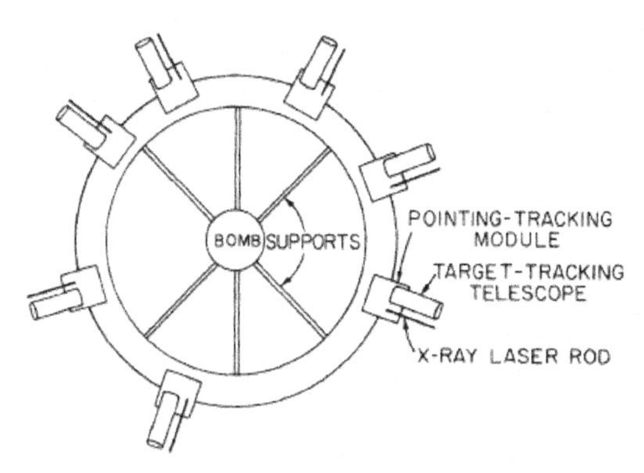

Basic idea of the X-ray laser battle station envisioned by military planners. An array of perhaps 50 X-ray lasers would surround a single nuclear bomb; only a few are shown here for simplicity. Each X-ray laser rod would require its own pointing-tracking module, probably equipped with a target-tracking telescope, which would locate the target and aim the X-ray laser so its output pulse would hit the target. In practice, the target-tracking telescope would probably be much larger than the X-ray laser rod because it would have to have large optics in order to track distant targets. However, it might be possible to mount tracking scopes on separate satellites–if extremely precise alignment could be maintained. (Drawing by Arthur Giordani.)

According to *Aviation Week*, researchers from the Lawrence Livermore National Laboratory used X rays from a small nuclear explosion to excite a small X-ray laser. The X-ray laser was housed in a vacuum chamber, and the experiment was performed at the Department of Energy's Nevada Nuclear Test Site, where nuclear weapons are tested in underground explosions. A pulse of several hundred trillion watts peak power, which an unnamed *Aviation Week* source said lasted "in the order of nanoseconds, one of the shortest pulses measured by Livermore," was produced.

The article went on to describe potential applications of such X-ray lasers in space-based weapon systems. One possibility cited involves a ring of about 50 laser rods surrounding a low-yield nuclear warhead. Each rod would be pointed at a target–a Soviet ballistic missile–then the bomb would be detonated to pump the lasers. The X rays from the explosion would stimulate each rod to direct a powerful pulse of X rays at its target. The X- ray pulse would be "so powerful that the beam [would evaporate] the target surface with radiation creating spallation," (nuclear reactions in which the nuclei of target atoms are shattered by energetic photons, such as X rays), according to a source identified only as a Pentagon official. "It is roughly akin to directing the energy in the nuclear pumping device in the laser beam." The article went on to say that ablative materials, which could provide some protection against longer-wavelength lasers (see Chapter 9) by evaporating from the laser-illuminated surface would not be an adequate defense against X-ray lasers. Twenty to thirty X-ray laser battle stations were said to be sufficient to handle a ballistic missile attack on the United States by destroying missiles in their "boost" phase (i.e. , as they rose out of the atmosphere) during a 30-min interval. Because detonation of the nuclear bomb would destroy the X-ray lasers, such battle stations would be strictly one-shot devices intended primarily to defend against a massive attack; a slower attack would require other measures.

The first reports said the laser systems would be stationed in orbit permanently, but there have been recent suggestions of a "pop-up" system that would be launched only on warning of an attack. The X-ray laser battle stations would be stationed on the ground or in submarines, on booster rockets that could launch them into orbit when needed. Advocates believe this approach could protect battle stations from enemy attack or sabotage, but it seems to have some limitations as well, which will be described later.

Although the *Aviation Week* article seemed to be a dramatic revelation at the time, it is possible to look back and see hints that something of the sort was in the works. The first was apparently the mention of Livermore's X-ray laser program in *Laser Focus* magazine,[34] but the magazine was unable to unearth more than that intriguing tidbit. The following year, a report in *Aviation Week* briefly mentioned "a nuclear explosive driven orbital laser ballistic missile defense system designed by Lawrence Livermore Laboratory." In that proposed concept excimer lasers producing ultraviolet beams would have been energized by X rays from a small nuclear explosion. Each laser would be independently aimed at a particular target. Parts of the description bear an unmistakable resemblance to the X-ray laser battle station proposal: "The lasers would be placed in a ring with the nuclear device in the center, and each device would pump 50 small rare-gas halide [excimer] lasers."[35]

Nonetheless, the early 1981 leak came as news to much of the military laser and X-ray laser community. Department of Energy officials were furious. Livermore management issued a blanket order that no one at the lab could say anything about the article or the experiments. Pentagon officials also would not comment, although strictly speaking the tests were the responsibility of the Department of Energy, which sponsored them as part of its program in nuclear weapons research. It is worth noting that secrecy is the watchword for the Department of Energy's nuclear test program, and standard regulations impose security restrictions on everything to do with nuclear tests–not just on X-ray laser experiments. In fact, the military branch of the Department of Energy (which is responsible for nuclear weapon development) seems even more obsessed with secrecy than the Pentagon. I have been told that the Department of Energy has been known to insist on new security investigations for new employees who already hold technically equivalent clearances from the Department of Defense.

Other sources were able to confirm that the experiments took place and to provide the name of one of the principal researchers, Livermore physicist George Chapline.[36] Chapline and Lowell Wood, another Livermore physicist, are generally credited with originating the idea of powering an X-ray laser with the intense X rays generated by a nuclear explosion. That concept is actually a variation on the earlier suggestion of using intense X rays to produce an X-ray population inversion.[37]

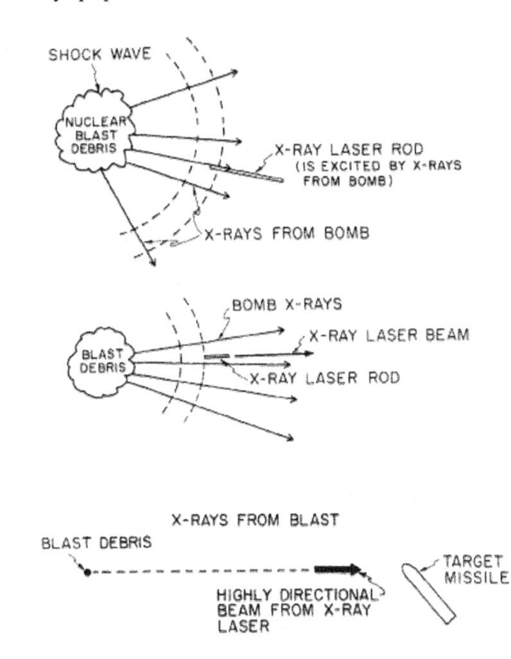

Operation of one X-ray laser rod attached to a battle station. Explosion of the nuclear bomb would generate X rays that would excite the X-ray laser (top). The laser would then emit a narrow beam of X rays, before the shock wave from the nuclear explosion hit it (middle). X rays produced by the blast would diffuse in space, but the energy in the X-ray laser beam would remain tightly focused far from the battle station, where it could disable its target (bottom). The X-ray pulse would be very short in duration. As shown in the bottom illustration, the X rays would be clumped together, not stretched out over a long straight line as an ordinary laser beam. It is possible that some X rays from the laser might go in the opposite direction from the main beam, but because the X rays from the blast are traveling along the length of the laser rod, most of the X rays will probably emerge from the outward-pointing end of the laser rod. (Drawing by Arthur Giordani.)

Aviation Week was uncharacteristically quiet after publishing its report. No major follow-up stories appeared. When *Omni* magazine asked for permission to reproduce one of the illustrations used with the original article to accompany a short report I had written on the X-ray laser, *Aviation Week* refused, although in other instances it had cooperated.[38] Even by the spring of 1983, two years after the original report, no real elaboration on the original article or its impact had appeared, although the X-ray laser concept had been mentioned in a couple of brief reports that referenced the original article.

Some observers have raised questions about details mentioned in the original report. The quoted pulse length of around one-billionth of a second is a thousand to a million times longer than would be expected from X-ray laser physics. Calling a pulse lasting one-billionth of a second "one of the shortest pulses measured by Livermore" also appears to reflect some misunderstanding. As far back as 1975, non-laser X-ray pulse lengths of about 20 trillionths of a second were measured with devices called streak cameras.[39] Such instruments are used in laser fusion experiments at Livermore by research groups with which Chapline and Wood are at least loosely affiliated. It is possible that the experimental conditions may have limited the time resolution or that researchers may not have wanted to risk their expensive streak camera by leaving it anywhere near a nuclear explosion. It is also possible that the reference to one-billionth of a second being "short" was intended as a comparison to the effects of a nuclear explosion, which last considerably longer. However, that is not what the article said.

One thing missing from the *Aviation Week* report was any mention of a critical parameter for space-based weapons: the beam divergence or rate of spread. Collimation or narrowing of the beam is one form of evidence for X-ray laser action, and some form of collimation is essential if the energy in the beam is to be delivered to a distant target. Given the problems of X-ray physics, however, it is possible that the Livermore group wasn't able to make the required measurements.

If the X-ray laser beam divergence is not adequate for some weapon applications, it may be possible to use other techniques to collimate the beam. Shortly after *Aviation Week* disclosed the Livermore experiments, Kamal Das Gupta of Texas Tech University in Lubbock reported intriguing experiments with crystals of germanium and gallium arsenide that, to his "utter surprise," produced X-ray beams with divergence as small as 10 millionths of a radian (10 μrad). His original paper[40] does not identify the reason for the collimation, but it appears to have something to do with crystal structure. His experiments did not involve X-ray lasers, but if the phenomenon could be used with a high-power beam, it is conceivable it could help narrow the output beam of an X-ray laser weapon system.

The weapon system concepts described in the *Aviation Week* article helped gain it widespread attention. They generally received a skeptical reception in the laser weapon community. One of the more charitable terms I heard used to describe them was "premature." The general consensus of sources I consider knowledgeable seems to be that the weaponization of X- ray lasers is a long way off, if it is possible at all.

The *Aviation Week* article stimulated widespread interest in the press and curiosity in the laser community. There was a bit too much interest for government security officials, who told some researchers that it was not a subject to be speculated about, particularly over the telephone. There was a later report in *Aviation Week* that the Los Alamos National Laboratory in New Mexico, which traditionally competes with Livermore for government support, was trying to start its own nuclear-pumped X-ray laser program. [41] Knowledgeable sources have told me that report is unfounded. It is possible that *Aviation Week* reporters were confused by Los Alamos interest in other unclassified approaches to demonstrating an X-ray laser that would not require a nuclear bomb.

The scientific community traditionally takes a jaundiced view of experiments reported by the press rather than described in proper detail in a scientific journal. This is more than simply a way of rationalizing a "no comment," although at times it may also serve that function. The function of the scientific journals is to lay details out for critical examination by other scientists. Without a detailed report of conditions under which an experiment was performed, other scientists cannot evaluate the results. X-ray laser researchers have seen enough false alarms to make them particularly wary.

Perhaps the best comment on the Livermore affair was made by Ronald W. Waynant of the Naval Research Laboratory while presenting an unclassified review of short-wavelength laser research at the 1982 Conference on Lasers and Electro-Optics. After pointing out that Kepros's spurious results were often attributed to the "hot dry climate" of Utah, he noted that the same conditions prevailed at the Nevada test site. He had to pause for the laughter to subside.[42]

Trying to Sell the X-ray Laser

The X-ray laser battle station concept returned to the public eye in September 1982 when *Aviation Week* reported that Edward Teller was trying to convince the Pentagon to support the program. [43] Teller evidently had been working quietly for a while. The previous October he had sent a classified letter to Congress that, reading between the lines in Congressional transcripts, evidently recommended stepping up X-ray laser research. The original program had cost $10 million. Teller asked President Reagan for $200 million a year to work on what he described as a "third-generation nuclear bomb," a term which the government has tried to classify.[44]

Understanding what this means requires a brief explanation of budgetary politics and military research. The Department of Energy's Division of Military Applications develops nuclear weapons and has traditionally assigned the main responsibility for bomb design and development to Livermore and Los Alamos, although both labs also do other work. Most funding for the X-ray laser program has been coming from the Department of Energy, *not* from the Department of Defense directed-energy weapon program.[45] In 1982 Livermore was said to be spending $15 million a year and having 70 people working on the X-ray laser program, with plans to increase funding to $37 million a year and staff to 185 people by 1987.[46] The Department of Defense is funding research on other types of short-wavelength lasers.

Other sources confirm that Teller indeed did discuss X-ray lasers with President Reagan and propose a massive development program. However, by going directly to the president, Teller is said to have offended Secretary of Defense Caspar Weinberger and other members of the Pentagon hierarchy.[47] And though Teller evidently was influential in convincing

President Reagan to push for missile defense in general, he has yet to convince the president, the Department of Defense, or Congress to start a crash X-ray laser program.

Problems with X-Ray Lasers

The hesitation to invest in such a massive program is understandable. Even if the X-ray laser works as advertised, something that is by no means certain, making it part of an effective weapon system will be a formidable task.

There are several levels of hardware problems. Although notable progress has been made recently,[48] X-ray optics are at best primitive and probably will never offer capabilities comparable to visible-light optics. This makes beam control a matter of pointing the laser at the target and blasting away. The laser itself will have to be a long, thin fiber or rod to produce a narrow beam. The need for the laser to be rigid (so it can be pointed accurately) will limit how long and thin it can be made. Outside of strength considerations, there is another inherent trade-off in design of the laser rods: the longer and narrower the rod, the more strongly the emitted X rays will be concentrated into a tight beam. The more concentrated the beam power, the deadlier the weapon to any target in the way; but the narrower the beam, the more likely it is to miss its target.

If it was to kill missiles thousands of kilometers away, an X-ray laser would have to have some impressive targeting and tracking equipment. Actually there would have to be one tracking system per laser rod to keep each rod directed at its separate target right up to the moment the bomb was detonated. That would mean that a battle station with 50 laser rods would have to have 50 pointing and tracking systems.

The need for so many pointing and tracking systems could turn out to be a fatal practical problem. That hardware is not cheap. In fiscal 1983 DARPA is investing $35 million in the Talon Gold pointing-tracking experiment for space-based laser weapons compared with only $22 million for the ALPHA chemical laser, a ground-based demonstration of the laser technology needed for space weaponry. In an operational chemical laser weapon the fire-control system could be used over and over again. In an X-ray laser weapon many of them would be blown up at once. That gets expensive.

Simultaneously aiming all the lasers so that they're pointed in exactly the right directions when the bomb goes off also presents many serious difficulties. The problems center on the need to maintain the required accuracy continually:

- For an X-ray laser rod to be pointed with an accuracy of one-thousandth of a radian, the relative positions of its two ends would have to be controlled within one part in a thousand. Even tighter tolerance would be needed if the beam had to be aimed more accurately. Yet long, thin rods are bendable and would be subject to small forces that could subtly misdirect them. The accumulation of mechanical motion required to point the rods could lead to a net rotation of the satellite itself, complicating the problem of aiming the lasers. Slight mechanical vibrations could knock some of the lasers off target. There are even slight residual gravitational forces on satellites[49] that might have to be considered.
- Ideally, the alignment of the laser rods should be adjusted continually to keep them pointed at their targets all the time. At the very least all laser rods must be aimed at their targets when the bomb goes off.
- Detonation of the bomb could trigger vibrations that would knock the lasers out of alignment with their targets. The problem would not be the nuclear explosion itself because the X rays carrying its energy to the lasers would travel at the speed of light, much faster than the shock wave generated by the explosion. Instead, the problem lies with the conventional explosive generally used to trigger a nuclear blast by forcing two subcritical masses of fissionable material together. The vibrations from this conventional explosion would have time to travel through parts of the battle station *before* the X rays were produced. Exactly how far they would get and how much damage they would do would depend on details of battle station design, such as how far the lasers can be from the nuclear explosion and how the elements of the bomb are pushed together to make it go off. Such details, if they are known, are understandably classified.

There are other potential technical problems as well. X-ray laser battle stations cannot verify that they have killed their targets after their bombs have exploded and turned them into clouds of radioactive debris. Even the advocates of X-ray laser weapons admit they could be foiled by launching an attack slowly enough that there would be targets for only a few of the lasers on each battle station. (The original *Aviation Week* article noted "there is a synergism in having chemical high-energy laser battle stations to engage a limited number of ballistic missiles over an extended period."[50]) The intense burst of electromagnetic radiation produced by a nuclear explosion (known as "electromagnetic pulse" or "EMP")[51] could also present problems for neighboring battle stations, or at least require some form of shielding, although the bombs used with X-ray lasers would produce only small explosions by the standards of nuclear bombs. And, as was mentioned earlier, air makes a good shield against X rays, particularly those with wavelengths longer than about 10 Å. The "pop-up" concept mentioned earlier could avoid one potential concern, the vulnerability of orbiting X-ray lasers, but it would raise other technical issues. The system would have to be packaged into a compact form able to withstand launch into space, yet ready to zap enemy missiles almost as soon as it reached orbit. Hitting targets thousands of kilometers (or thou- sands of miles) away might be a formidable problem. Precise beam direction would require both stable motion (which takes time to achieve) and precise target-tracking optics (which would inevitably be larger than the laser itself).

The larger the optics, the more massive the satellite; and the harder it is to get into orbit. It might be possible to put the target-tracking optics on a separate satellite, but this would require precise alignment of the two satellites. Target-tracking requirements could be loosened if the beam spread out more rapidly than a longer-wavelength laser beam, but this would only work if exceedingly powerful beams could be produced, able to deliver a lethal dose of X-ray energy to a broad expanse of space to disable a single target some- where in that volume. Even the energy available from a nuclear bomb has its limits, and those requirements may go beyond them.

Another concern is response time: how long would it take to put the system into orbit and ready to defend against nuclear missiles. Although it takes only a matter of minutes for a rocket to reach space, it can take longer to detect an attack, ready the launch, and stabilize the satellite's motion in orbit. If the time required is more than the half hour it takes an intercontinental ballistic missile to reach its target, the satellite would have to be in orbit before the attack began. That assumes the lasers could attack missiles anywhere on their route. If the laser system was designed to attack missiles in their boost phase, it would have to be in orbit before the attack was triggered. The serious technical problems that would be encountered in trying to "weaponize" X-ray lasers have tempered some of the initial enthusiasm. Perhaps more serious are the political and arms-control implications discussed in later chapters, particularly Chapters 16 and 17. A serious effort to develop X-ray laser battle stations and put them into orbit would require the dismantling of two decades of arms-control agreements, as mentioned at the start of this chapter. Such a move would certainly arouse strong opposition in the United States and its allies, to say nothing of vehement protests from the Soviet Union.

The sensitivity of the issue seems to be well recognized in Washington. It is evident in the placement of [deleted]s in unclassified transcripts of classified testimony at Congressional hearings. A careful examination of public documents won't reveal any official mention of X-ray laser concepts. But after watching this field for long enough, it's possible to fill in some of the [deleted]s, and in a number of cases the missing words are evidently "X- ray laser." For example, in a written response to a question by former Senator Harrison Schmitt, Douglas Tanimoto, then head of DARPA's directed-energy office, mentioned "[deleted] which must be used at altitudes above 100 kilometers because of atmospheric absorption,"[52] a reference that can only be to X-ray lasers. In his prepared statement to the Senate Appropriations Committee on the fiscal 1983 budget, Tanimoto identified three principal efforts DARPA was planning on short-wavelength lasers – free-electron lasers, excimer lasers, and [deleted],[53] evidently X-ray lasers. The need to play "fill in the blanks" to learn about X-ray laser plans is another curious part of the X-ray laser story.

References

1. Texts of the treaties appear in Appendix 2 of Bhupendra Jasani, ed, *Outer Space-A New Dimension of the Arms Race* (Oelgeschlager, Gunn, & Hain, Cambridge, Massachusetts, 1982).
2. The House Armed Services Committee stated its rationale for backing short-wavelength lasers in general in its *Report on the Department of Defense Authorization Act 1983* (House Report 97-482, April 13, 1982) p. 132; for a description of some of the politics involved see Jeff Hecht, "House and Senate squabble over laser weapons budget," *Lasers & Applications* **1** (1), 50-54 (September 1982). Official sources are careful to avoid saying "X-ray laser" in so many words, but my conversations with reliable sources have indicated that Battista is the principal support of X-ray lasers, and my conversations with Battista have done nothing to contradict that impression.
3. "Pentagon spurns Teller's new toy," *New Scientist* 96, 728 (December 16, 1982).
4. George C. Baldwin, Johndale C. Solem, and Vitallii I. Gol'danskii, "Approaches to the development of gamma-ray lasers," *Reviews of Modern Physics* 53 (4), 687-744 (October 1981); this statement is on p. 689.
5. S. Joma, *X-Ray Laser Applications Study* (Physical Dynamics Inc., La Jolla, California, July 1977, report PD-U-77-159), p. 19.
6. *Ibid.,* p. 19.
7. *Ibid.,* p. 26.
8. *Ibid.,* p. 19.
9. D. Jacoby, G. J. Pert, S. A. Ramsden, L. D. Shorrock, and G. J. Tallents, "Observation of gain in a possible extreme ultraviolet lasing system," *Optics Communications* 37 (3), 193-196, (May 1, 1981); D. Jacoby, G. J. Pert, L. D. Shorrock, and G. J. Tallents, "Observations of gains in the extreme ultraviolet," *Journal of Physics B: Atomic and Molecular Physics* **15,** 3557-3580 (1982); the experiments were first described at technical conferences in mid-1980, although the papers were not written until later.
10. I. I. Sobel'man, "Atomic Collision processes and UV and X-ray lasers," pp. 75-80 in *Electronic and Atomic Collisions,* N. Oda and K. Takayamagi, eds (North Holland, Amsterdam, 1980).
11. These penetration depths are the distances X rays can travel through the material before their intensity is reduced to *lie* (where *e* is the base of the natural logarithms) or about 0.37 of their original intensity. The data come from Joma, *op. cit.,* p. 103.
12. These figures are from my own calculations based on a graph showing the penetration of various wavelengths of electromagnetic radiation into the atmosphere from space that appears on p. 73 .of Herbert L. Anderson, ed., *Physics Vade Mecum* (American Institute of Physics, New York, 1981) and on tables of atmospheric characteristics listed on p. F-166 of the *CRC Handbook of Chemistry and Physics,* 62nd ed., (CRC Press, Boca Raton, Florida, 1981). The figures given for how far X rays can travel are for air at sea level density. The atmosphere does not end abruptly, instead tails off to very low densities at altitudes above 100 km. This could present problems if an X-ray laser is supposed to shoot at missiles in their boost phase, when they are still in the upper reaches of the atmosphere. My very crude calculations show that if an X-ray laser battle station at an altitude of 1000 km was shooting at a missile at an altitude of 100 km that was 3000 km away, a beam of 10-Å X rays would pass through enough air to reduce its intensity to a couple of billionths of the original level.
13. Herbert L. Anderson, ed., *Physics Vade Mecum* (American Institute of Physics, New York, 1981), p. 73.
14. Johndale C. Solem and George C. Baldwin, "Microholography of living organisms," *Science* **218,** 229-235 (October 15, 1982).
15. Joma, *op. cit.*
16. George C. Baldwin, Johndale C. Solem, and Vitalii I. Gol'danskii, "Approaches to the development of gamma-ray lasers," *Reviews of Modern Physics* 53 (4), 687-744 (October 1981).
17. *Ibid.,* p. 701.

18. For a detailed account see: Ronald W. Waynant and Raymond C. Elton, "Review of short-wavelength laser research," *Proceedings of the IEEE* **64** (7), 1059-1092 (July 1976).
19. John G. Kepros, Edward M. Eyring, and F. William Cagle, Jr., "Experimental evidence of an X-ray laser," *Proceedings of the National Academy of Sciences USA* 69 (7), 1744-1745 (July 1972); for a journalistic account see: "Wavefronts," *Laser Focus* (September 1972), p. 4.
20. Ronald W. Waynant and Raymond C. Elton, "Review of short-wavelength laser research," *Proceedings of the IEEE* **64** (7), 1059-1092 (July 1976); the Utah experiments are described on p. 1078.
21. *Laser Focus,* February 1972, p. 10.
22. Ronald W. Waynant, "Vacuum ultraviolet and X-ray lasers: an overview of progress," paper ThTl at 1982 Conference on Lasers and Electro-Optics, April 14-16, Phoenix; the essence of Waynant' s lively talk is unfortunately not preserved on paper in the conference proceedings. I had the good fortune to be in the audience.
23. The preliminary conference program distributed by the Optical Society of America listed Kepros as presenting paper ThT4, but that paper does not appear in the 1981 OSA annual meeting program and was not presented at the conference.
24. P. Jaegle, G. Jamelot, A. Carillon, A. Sureau, and P. Dhez, "Super radiant line in the soft X-ray range," *Physical Review Letters* 33, 1070-1073 (October 1974).
25. Ronald W. Waynant and Raymond C. Elton, "Review of short-wavelength laser research," *Proceedings of the IEEE* **64** (7), 1059-1092 (July 1976); the Orsay results are described on pp. 1077-1078.
26. John Reintjes, C. Y. She, R. C. Eckardt, N. E. Karangelen, R. A. Andrews, and R. C. Elton, "Seventh harmonic conversion of mode-locked laser pulses to 38.0 nm," *Applied Physics Letters* **30,** 480-482 (1977).
27. "New lasers emphasized at CLEA," *Laser Focus* **13** (8), 12 (August 1977).
28. The quote is from I. I. Sobel'man, "Atomic collision processes and UV and X-ray laser," pp. 75-80 in *Electronic and Atomic Collisions,* N. Oda and K. Takayanagi, eds (North Holland, Amsterdam, 1980); other relevant papers include A. A. Ilyukhin, G. V. Peregudov, E. N. Ragozin, I. I. Sobel'man, and V. E. Chirkov, "Concerning the problem of lasers for the far ultraviolet 500-700 Å," *JETP Letters* 25 (12), 535-539 (June 20, 1977); and L. A. Vainshtein *et al.,* "Stimulated emission in far ultraviolet due to transitions in multiply charged neonlike ions," *Soviet Journal of Quantum Electronics* **8** (2), 239-242 (February 1978).
29. S. Joma, *X-Ray Laser Applications Study* (Physical Dynamics Inc., La Jolla, California, July 1977).
30. "Livermore picks up X-ray laser effort," *Laser Focus,* February 1979, p. 4.
31. "Laser gain around 18.2 run produced in carbon plasma at Hull University," *Laser Focus 18* (8), 24-28 (August 1980); D. Jacoby, G. J. Pert, S. A. Ramsden, L. D. Shorrock, and G. J. Tallents, "Observation of gain in a possible extreme ultraviolet lasing system," *Optics Communications* 37 (3), 193-196 (May 1981); D. Jacoby, G. J. Pert, L. D. Shorrock, and G. J. Tallents, "Observations of gains in the extreme ultraviolet," *Journal of Physics B: Atomic and Molecular Physics 15*, 3557-3580 (1982).
32. D. Jacoby, G. J. Pert, L. D. Shorrock, and G. J. Tallents, "Observations of gains in the extreme ultraviolet," *Journal of Physics B: Atomic and Molecular Physics 15*, 3557-3580 (1982).
33. Clarence A. Robinson, Jr., "Advance made on high-energy laser," *Aviation Week* & *Space Technology,* February 23, 1982, pp. 25-27.
34. "Postdeadline reports," *Laser Focus,* February 1979, p. 4.
35. "Technology eyed to defend ICBMs, spacecraft," *Aviation Week* & *Space Technology,* July 28, 1980, pp. 32-42.
36. Jeff Hecht, "The X-ray laser flap," *Laser Focus* 17 (5), 6 (May 1981); Jeff Hecht, "X-ray laser potential rises," New Scientist 92, 166 (October 15, 1981).
37. Michael A. Duguay, "Soft X-ray lasers pumped by photoionization," pp. 557-579 in Stephen F. Jacobs, Marlon O. Scully, Murray Sargent III, and Cyrus D. Cantrell III, eds., *Laser-induced Fusion and X-Ray Laser Studies* (Addison-Wesley, Reading, Massachusetts, 1976).

38. Dick Teresi, private communication. In the end, no illustration was used with the short piece I wrote for Omni's Continuum section: Jeff Hecht, "X-ray laser," *Omni,* June 1981, p. 50.
39. "Streak camera attains 20-ps resolution at wavelengths from X-ray into the uv," *Laser Focus* 11 (9), 29-30 (September 1975).
40. Kamal Das Gupta, "Observation of an X-ray beam of 10 microradian divergence without using any collimator," presented at Thirteenth Annual Denver X-ray Conference, August 5, 1981, and published in *Advances in X-Ray Analysis,* Vol. 25 (Plenum Press, New York, 1982).
41. "Industry observer," *Aviation Week & Space Technology,* September 7, 1981, p. 15.
42. Ronald W. Waynant, "Vacuum ultraviolet and X-ray laser: an overview of progress," paper ThT1 at 1982 Conference on Lasers and Electro-Optics, April 14-16, Phoenix. Unfortunately, the lively parts of Waynant's talk, such as the somewhat irreverent speculation about exactly what happened in the Livermore experiments, are not preserved on paper in the conference proceedings. I had the good fortune to be in the audience to hear the informal version of his recollections.
43. "Washington roundup," *Aviation Week & Space Technology,* September 20, 1982.
44. Richard F. Harris, "Lab designing nuclear X-ray weapon?" *Livermore* (Calif.) *TriValley Herald,* September 24, 1982; "Teller says laser bomb is coming," *Los Angeles Times,* September 26, 1982.
45. The level of secrecy surrounding the X-ray laser program makes it impossible to be sure where all of its money originates, but it appears that a majority comes from the Department of Energy's Division of Military Applications (responsible for developing and testing nuclear weapons), with a smaller amount coming from the Department of Defense through the short-wavelength laser program of the Defense Advanced Research Projects Agency.
46. Harris, *op. cit.* (Ref. 44).
47. "Pentagon spurns Teller's new toy," *New Scientist* 96, 728 (December 16, 1982).
48. See, for example: D. Attwood and B. Henke, *Proceedings of Low-Energy X-Ray Diagnostics Conference* (American Institute of Physics, Conference Proceedings #75, New York, 1981).
49. The effects of residual gravity in spacecraft are discussed in "How to flatten spacetime," *New Scientist* 96, 563 (December 2, 1982).
50. Clarence A. Robinson, Jr., "Advance made on high-energy laser," *Aviation Week & Space Technology,* February 23, 1981, pp. 25-27.
51. Edward Teller, "Electromagnetic pulses from nuclear explosions," *IEEE Spectrum* 19 (10), 65 (October 1982).
52. Senate Committee on Appropriations, *Hearings on Department of Defense Appropriations Fiscal Year 1983,* part 4, p. 584.
53. *Ibid.,* p. 586. Tanimoto's testimony indicates that DARPA was planning to spend $2 million in fiscal 1983 on the [deleted], which evidently is X-ray lasers. That figure should not be taken as final because it reflects the Reagan Administration's original budget proposal for fiscal 1983, which was extensively modified by Congress. Changes included the addition of more money for "short-wavelength" laser research, but it is unclear how that money was divided among free-electron, excimer, and X-ray lasers.

7.
Particle-Beam Technology

You can get a rough idea of what is involved with particle-beam weaponry by taking a quick look at nature's version–lightning. Lightning occurs when natural processes in the atmosphere separate electrons from atoms and build up a high static voltage (an electric potential difference). That potential can be higher than 1 billion volts between clouds and the ground, enough to cause the "breakdown" of air, which is normally an insulator. The result is what we call lightning, the rapid flow of a powerful current between clouds and ground. The physics are really very different from those of man-made particle beams, but both can pack a big wallop.

Lightning leaves vivid evidence of its power wherever it strikes. An understanding of that power seems to be lurking in our instincts, the force that drives small children to hide or cling to their parents at the first thunderbolt. Most of us know enough about lightning to avoid making ourselves prime targets by standing in open fields or atop high buildings in thunderstorms. But even someone who does make such a mistake can live to tell the tale, for lightning is not exactly "directed energy." The powerful electric discharge follows a zigzag path on its way between cloud and ground, a path dependent on subtle atmospheric effects. Tall trees and buildings are tempting targets and may be hit often, but by no means always. Nature's beam weapon packs a mighty punch, but it often misses the mark.

Military researchers are trying to tame physical processes similar to those that produce lightning and use them as the basis of a directed-energy weapon. As in lightning, high voltages would give high energy to the particles: electrons, protons, or atoms carrying a positive or negative electric charge. (Electric and magnetic fields do not affect normal neutral atoms or other particles that have no net electric charge.) And, as with lightning, it is hard to make sure that the beam of energy hits precisely the right spot.

The roles envisioned for particle-beam weapons are similar to some of those for which high-energy laser weapons are being developed. Recognizing the similarities, the Pentagon lumps particle-beam weapons with lasers and microwaves under the heading "directed-energy" technology. There are indeed some similarities that extend beyond their potential missions. Lasers and particle-beam weapons could use similar fire-control equipment to spot targets. Particle-beam weaponry draws upon high-power electrical technology used in certain types of high-energy lasers. Yet there are also fundamental, and very important, differences.

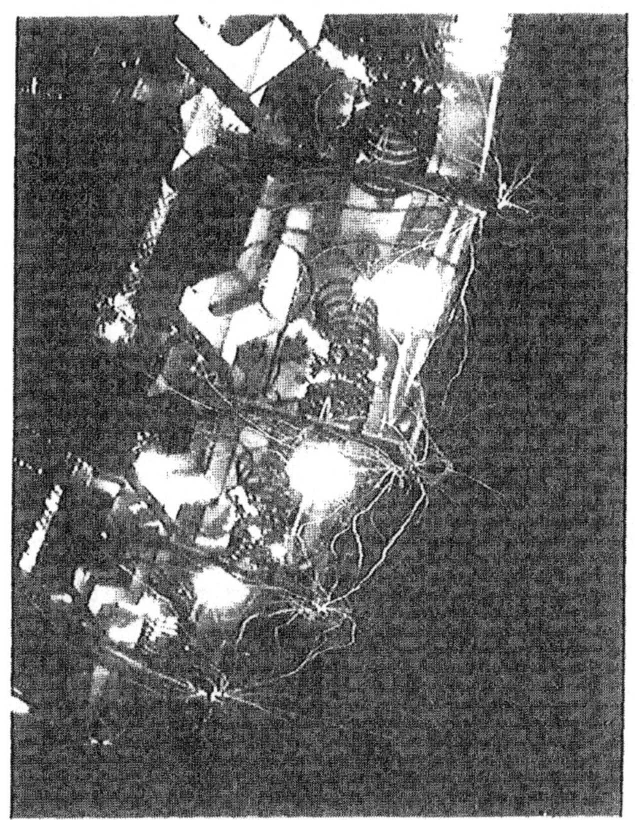

Sparks fly as a network of high-voltage switches operate during an experiment at Sandia National Laboratories in Albuquerque. These switches are handling a power of 30 trillion watts in an extremely short pulse; they are part of the Particle Beam Fusion Accelerator (PBFA-I) at Sandia. PBFA-I is used for fusion experiments, as described later in this chapter, but similar technology would probably be used in a particle-beam weapon. Note that this photo shows only energy lost while the particle-beam pulse is being switched; the beam itself is much more powerful. (Courtesy of Sandia National Laboratories, from *Sandia Technology*, October 1982.)

Because of these differences, many of the problems that must be solved to build particle-beam weapons are quite distinct from those facing laser weapon developers. Particles, unlike the photons in a laser beam, have mass and hence tend to penetrate farther into the target, where they are more likely to cause lethal damage than a laser beam. Yet, despite the problems described in Chapter 5, laser beams propagate through the atmosphere much better than beams of charged particles; that is, they follow a much more predictable path. Thus, the critical problem with particle-beam weapons is getting the beam to the target; once the particles hit they are virtually certain to damage it unless they have somehow lost all their energy. The differences add up, and though laser and particle-beam weapon programs are complementary in some ways, the two rarely cross each other's paths.

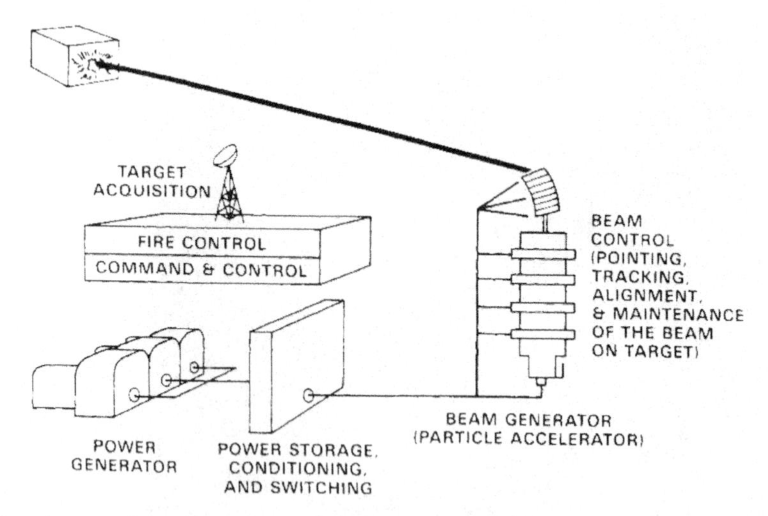

Key elements of a particle-beam weapon system, as envisioned by military planners. Energy to drive the beam generator comes from the components at lower left; the particles are then accelerated and pointed at the target. The beam-control system, which directs the beam, operates under the direction of the target-acquisition and fire-control equipment shown at left center. (Courtesy of Department of Defense,)

The particle-beam weapon program itself is split into two basic segments aimed at meeting different sets of military goals. Systems that generate beams of charged particles (such as electrons and protons) are being developed to go through the air. The interaction between the electric current in the beam of charged particles and the atmosphere holds the mutually repulsive particles together, but once the beam reaches the vacuum of outer space it rapidly disperses. In addition, charged beams would be bent in complex and hard-to-predict ways by the earth's magnetic field when traveling long distances and a build-up of charge in space would rob the beam of energy and eventually stop it all together. In contrast, beams of high-energy neutral particles would hold together in space without being bent by magnetic fields, a requirement for long-distance space applications such as ballistic missile defense. However, they would break up in air, or simply have electrons stripped from them to make charged particle beams, which are easier to produce

There are clearly many obstacles to be overcome. The Pentagon's position is that, "Particle-beam technology is in the very early research and exploratory development phases with fundamental issues of feasibility to be resolved. There is an enormous gulf between the technology required for fulfillment of the conceptual payoffs and the state of the art."[1] In short, military planners aren't yet convinced that the whole idea is feasible, and even if it is, there remains a long way to go before the concept can be translated into an operating weapon system. Some advocates of particle-beam weapons are much more optimistic, but there are also pessimists who think the whole idea is ridiculous. Caution seems to be the watchword in the Pentagon; for fiscal 1983 particle-beam technology was allocated a little over one-tenth of the directed-energy budget.

Understanding the problems and promise of particle-beam weaponry requires a brief overview of the technology. The description that follows is of necessity oversimplified because the details of the physics involved are extremely complex.[2]

Beam Generation

Generation of a beam of energetic particles starts with the production of the particles themselves, which may be electrons, protons, positive ions (atoms that have lost one or more electrons), or negative ions (atoms with one or more extra electrons). These charged particles are then accelerated by electrical and/or magnetic fields and aimed at a target. The process is nowhere near as simple as this description makes it sound, however. Production of high-energy beams requires electrical potentials of millions of volts, massive amounts of electrical power, and the ability to switch those high voltages and currents in intervals measured in billionths of a second.

Particle accelerator technology has been marching forward over the last 50 years or so. The main stimulus has been strong financial support for research in subatomic physics, which only recently has begun to slack off. New uses for accelerator technology have begun to emerge over the past couple of decades, stimulated both by its availability and by the interest of accelerator designers in keeping busy. The last two decades have also seen impressive developments in "pulsed power," the generation and manipulation of electrical voltages in the "megavolt" (million-volt) range. These developments have helped stimulate military interest in particle-beam weapons.

The first stage of a particle-beam generator is a source of the particles. In an electron accelerator the source is a "spark" formed by applying a high-voltage electrical pulse across a pair of conducting electrodes in a vacuum, called a diode in electrical terminology. The negatively charged electrons are emitted from the negatively charged electrode (the cathode) toward the positive anode. Instead of being collected by the anode, the electrons pass by or through it (depending on the design) and into the accelerator.

A pulse of ions starts in the same way, with a short, high-voltage pulse applied to a pair of electrodes. Positively charged ions (atoms that have lost one or more electrons) come from the positive electrode, produced either from a solid material or a discharge in a gas. If the goal is negatively charged ions, electrons are injected into the positive electrode so they are likely to attach themselves to the gas atoms near the electrode, producing negatively charged ions that can be drawn into the accelerator. By adjusting electrical voltages, it is possible to inject into the accelerator only certain ions, such as those that have lost only one electron. The same basic approach can be used to generate beams of protons by stripping the single electron from hydrogen atoms, or beams of ionized uranium, the heaviest naturally occurring element.

Electron or particle beams powerful enough to be weapons are not produced continuously but rather in short bursts or pulses (sometimes called "bolts" in analogy to lightning). The particle generator and accelerator, thus, are powered by short pulses of electric current at high voltages. The simplest way to provide this power is to store electrical charge in a bank of devices called "capacitors," then rapidly discharge the devices when the energy is needed. In its simplest form a capacitor is a pair of metal plates separated by an insulator, which builds up an electric charge when a voltage is applied across it. This stored electric charge represents energy that can be released by connecting the leads from the two plates so a current flows between them for a brief time until the stored energy is used up (or "discharged"). Complex capacitors can store enough energy to deliver a high current in a high-voltage pulse. The problem is that capacitors are bulky, expensive, and comparatively inefficient for energy storage. They are fine for demonstration systems or laboratory accelerators but not for a practical weapon system.

Other types of power sources are being studied in the search for a compact, lightweight source of high-power pulses. One relies on explosives to force an electrically conductive plasma (ionized gas) into a generator coil housed inside a superconducting magnet. The plasma traps and compresses the magnetic field inside the coil, thus producing an electrical pulse in the generator.[3] Published reports indicate that the Soviet Union may be working on a similar approach.[4] Also under investigation is the possibility of storing and extracting energy in the form of mechanical rotation.[5]

There are other complexities with pulsed power. Simply transporting high-power pulses can be a problem. Some "conditioning" is generally needed to convert the electrical pulse generated by the source into the form needed by the particle-beam generator, a task that may require multiplying the already high voltages in the pulse. The seemingly mundane operation of switching is not exactly mundane when hundreds of thousands or millions of volts are involved, and in weapon systems such high voltages will have to be switched many times per second. Impressive work has been done in the last couple of decades,[6] but there is still a long way to go.

Particle Accelerators

Once the beam of electrons or ions has been generated, it must be accelerated to high velocities, and hence to high energies as well. This requires passing the pulse through a strong electrical field, which transfers energy to the particle. (Magnetic-field acceleration is also possible, but it is used mainly in areas other than weaponry). A particle's energy depends on its charge and the intensity of the field it has passed through, so physicists often measure energy in "electron volts," that is, the amount of energy an electron picks up when passing through a 1-V electric field. Energies of particles used in weapon research are many millions of electron volts.

The power in a particle beam is the energy per particle times the number of particles in the pulse. The number of particles in the beam is called the current in the same way that the number of electrons passing through a wire determines the amount of electrical current. Although research accelerators have demonstrated high energy per particle, they have not achieved the high currents that would be needed in a weapon system. Military researchers are working on special high-current accelerators, such as the Advanced Test Accelerator described later in this chapter, in order to see what happens at high current levels.

There is no general consensus on the type of accelerator that would be used in a weapon system, and several different concepts are being studied. Some are intended specifically for weaponry, while others are primarily intended for other uses. There are enough common elements of the technology that the Pentagon expects to learn valuable information from a variety of tests that aren't directly supported by the directed-energy program.

For the past few years, much of the Pentagon's particle-beam program was aimed at the development of accelerators. The "mainstream" of that effort was development of two linear electron accelerators at the Lawrence Livermore National Laboratory.[7] The first was the Experimental Test Accelerator (ETA), which began operation in June 1979. This system was the outgrowth of a Navy program originally called "Chair Heritage," which was transferred to the Defense Advanced Research Projects Agency in 1978. It delivered 10,000-amp pulses of electrons carrying 5 million electron volts each and lasting 50 billionths of a second. It could produce 1000 shots/sec in short bursts, although on the average it could deliver only five pulses per second.

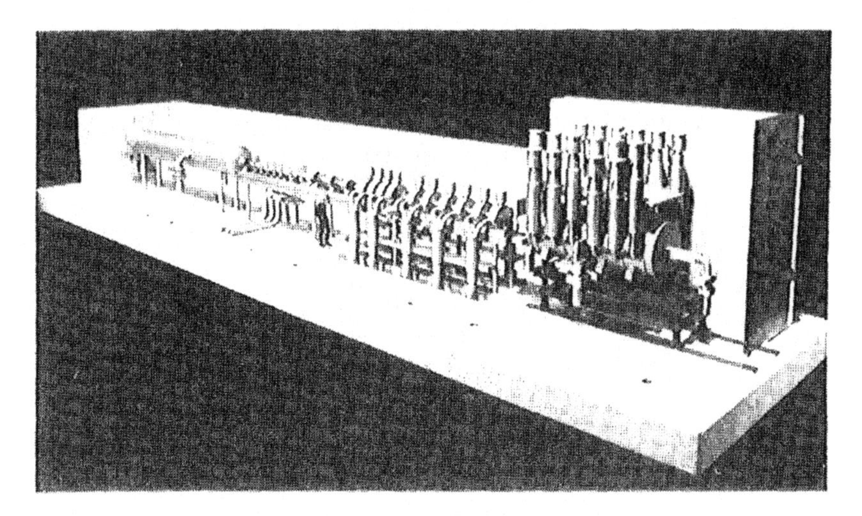

Model of the Experimental Test Accelerator built at the Lawrence Livermore National Laboratory, with a model workman shown just left of center to indicate the scale. (Courtesy of Department of Defense.)

The Experimental Test Accelerator was intended only as a test bed for the larger Advanced Test Accelerator (ATA), which is designed to generate pulses of 50-million-electron-volt electrons at similar rates. Livermore developers were encouraged by tests of the smaller system. During its first year of operation ETA "achieved 80% of all design goals, and has produced over two million pulses despite a more than two-month shutdown for repair of earthquake-induced damage."[8] The Advanced Test Accelerator was nearing completion as this book was being finished. DARPA says that when the system is completed it will be "the free world's most powerful accelerator."[9] In late 1982 the Livermore public affairs office was talking about plans to show the accelerator off to the press in early 1983, a rather unusual move for the directed-energy weapon program, but one that as of May 1983 had yet to materialize.

Artist's conception of the Advanced Test Accelerator at the Lawrence Livermore National Laboratory, with cutaways showing equipment inside the building. The small secondary building at the right is about 10 ft tall; the tunnel underneath the two buildings is about the height of a man. (Courtesy of Department of Defense.)

Both the Experimental Test Accelerator and the Advanced Test Accelerator are linear induction accelerators in which a series of accelerating elements are arranged in a straight line. In operation a pulsed electrical voltage swiftly changes the electric field applied to the beam, thus accelerating the particles. This process is repeated through the series of accelerating elements, raising the particle energy to high levels.

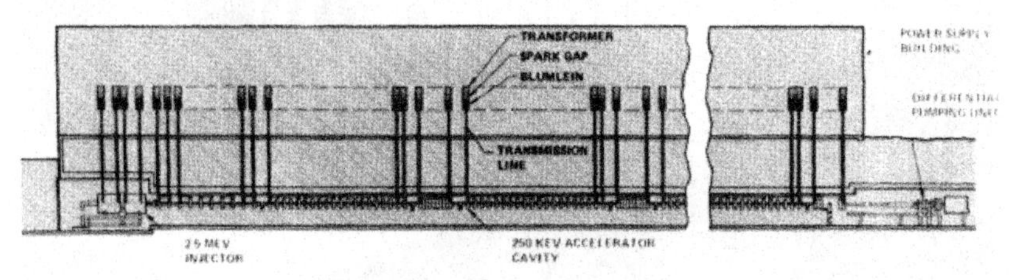

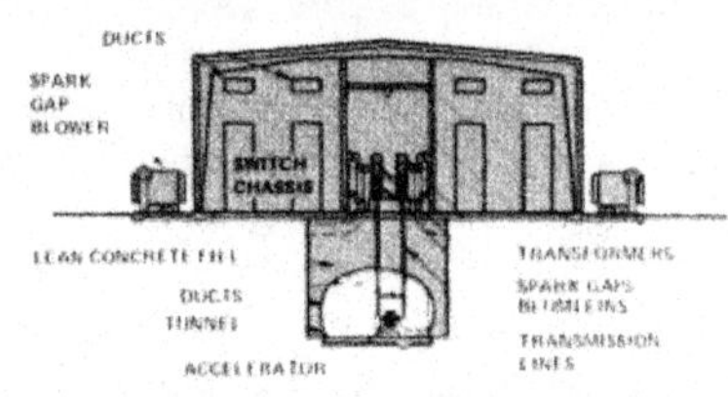

Major components of the Advanced Test Accelerator at the Lawrence Livermore National Laboratory. The beam injector is at the left end. Most of the length of the system is made up of 190 accelerator cells, totaling 256 ft, which each add 0.25 million volts of energy to the electrons passing through them. At right, past the end of the power supply building, are the beam-transport system and, at the right edge, the experimental target area. The lower drawing shows the cross section of the accelerator tunnel. (Courtesy of Lawrence Livermore National Laboratory.)

A somewhat similar approach is used in the Radial Line Accelerator (RADLAC-I) at Sandia National Laboratories in Albuquerque, New Mexico. The main difference is in the drive circuitry. This system can accelerate electrons to an energy of 9 million electron volts and generate a current of 25,000 amp. Funded by the Department of Energy and the Air Force Weapons Laboratory, RADLAC-1 zaps targets to produce bursts of X rays that simulate the effects of nuclear explosions. A larger, more powerful version, RADLAC-11, was nearing completion as this book was being readied for publication.[10]

DARPA's other major accelerator program, code-named "White Horse," is at the Los Alamos National Laboratory in New Mexico. The accelerator will be driven with a radio frequency field oscillating 80 million times per second (a frequency of 80 million hertz or 80 MHz), which forms an electric field with four poles rather than the standard two. This approach, originally developed in the Soviet Union and refined at Los Alamos, focuses the beam at the same time it accelerates it. The system will be used to accelerate negatively charged hydrogen ions; at the end of the accelerator the extra electrons will be removed from the hydrogen atoms to produce a neutral beam of the type needed for use in space.[12]

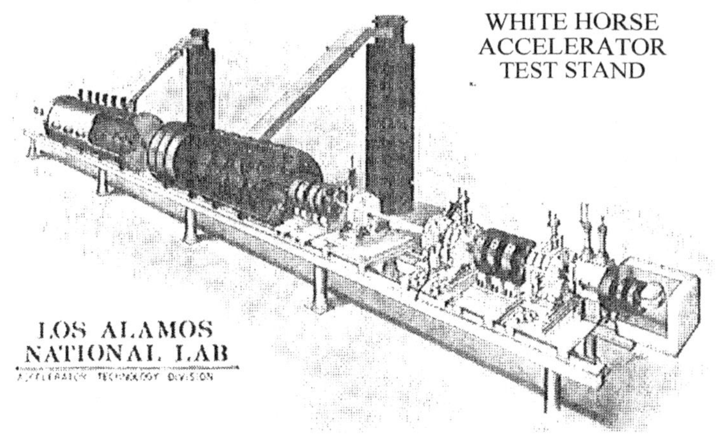

The White Horse Accelerator at the Los Alamos National Laboratory is shown in an artist's conception. This system will be used to test the feasibility of producing beams of neutral particles in experiments sponsored by the Defense Advanced Research Projects Agency. (Courtesy of the Department of Defense.)

A number of other large accelerators are operating, under development, or in conceptual design for other military-related programs, notably inertial confinement fusion and nuclear effects simulation. The idea of inertial fusion is to focus an intense particle-beam pulse onto a small target containing hydrogen isotopes, producing a hot, dense plasma in which the hydrogen nuclei would fuse together and release energy.[13] In the long term this approach might be used to generate power in a nuclear fusion reactor. For now, however, the main interest is in simulating, on a microscopic scale, the explosions of hydrogen bombs. The physics involved in these microexplosions turns out to be quite similar to that involved in the blast of a hydrogen bomb, and hence the results of inertial fusion experiments are of particular interest to nuclear weapon designers. (Such experiments and their results are classified.) The radiation produced by an inertial fusion microexplosion is also similar to that produced by a hydrogen bomb and can be used to study the effects that fusion blasts might have on military hardware. So far these military applications have justified essentially all of the money the United States government has spent on inertial fusion research. Although laser fusion has accounted for the bulk of the U.S. effort, there is also a major program in substituting particle beams for the lasers. Work on beams of light ions is centered at Sandia National Laboratories, which first started looking at electron beams and shifted in 1979 to protons. Sandia's Particle Beam Fusion Accelerator (PBFA-I) can produce an electron pulse of 30 trillion watts, and researchers at Sandia are working on ion diodes to generate and direct this power in the form of ion beams. A larger facility, called PBFA-II, is planned for completion about 1986.[14] The Department of Energy also sponsors a smaller program on heavy-ion beam fusion, conducted at the Los Alamos National Laboratory and elsewhere.[15]

Particle beams are also used in a different type of nuclear effects testing. In this approach intense electron beams from generators like the RADLAC system mentioned earlier strike targets that generate "hard radiation" such as X and gamma rays. Military equipment can be exposed to this radiation to see what would happen in a nuclear war. Simulation of nuclear effects has become a large program because it is far less expensive than testing nuclear weapons underground and offers a convenient way to identify and fix vulnerabilities.

Although these test accelerators were originally intended for other purposes, they can help the Pentagon resolve some key concerns about accelerator technology: Is it possible to generate the high-energy particles required for weaponry in a reasonably efficient and compact system? Can high enough particle fluxes be produced? Can particle generators produce pulses fast enough to zap a barrage of attacking craft? And how efficiently can different types of beams be generated?

Beam Propagation

DARPA's two major programs, at Livermore and Los Alamos, are looking at questions of accelerator technology. There is particular emphasis on assessing the potential for generating high-current pulses at high firing rates, a capability not needed in research accelerators. But these systems are also test beds for answering a more difficult question: Can particle beams propagate through the air and/or space well enough to accurately hit military targets? Livermore's Advanced Test Accelerator is designed to look at the prospects for propagation of charged particle beams in the atmosphere. Los Alamos's White Horse program is intended to look at the prospects for producing a beam of neutral particles that could travel long distances through outer space.

A charged particle beam is subject to two conflicting effects caused by the electrical charge carried by the particles. Like charges repel each other, but the flow of many like charges in the same direction (in essence an electric current) generates a magnetic field surrounding the beam that tends to pinch it together. This self-pinching effect is only expected to occur when a high- current beam is traveling through the air, in which case it would typically keep the beam diameter down to a centimeter or so.[16]

Air tends to attenuate particle beams and absorb some of the energy they carry. Under normal conditions an electron beam could make its way through about 200 m of air before half of its energy had been absorbed by the atmosphere. For single pulses this would tend to limit the range of a weapon. However, theorists predict that nature may be kind to the weapon system developer in this case. In a few millionths of a second the air heated by the first pulse will have expanded and cleared out a path for the second beam. In theory, this "hole-punching" effect should make it possible to transmit beams for a few kilometers.[17]

Livermore's Advanced Test Accelerator is intended to provide the high beam currents needed to test these predictions experimentally. According to DARPA,

> ATA performance parameters are greater than the minimum expected to be required for stable propagation of electron beams at full atmospheric densities. Initial experiments will evaluate a propagation mode suitable for experimental ranges. The beam parameters will be modified in subsequent tests to provide a preliminary assessment of propagation modes potentially capable of operationally acceptable ranges.[18]

Translated from the Pentagon bureaucratese into English, this means that the experiments will see if electron beams can make their way through a few kilometers of air to zap military targets.

Neutral Beams for Missile Defense

Charged particle beams are manifestly unsuitable for sending thousands of kilometers through space to destroy enemy missiles. Not only would the mutual repulsion of the particles cause the beam to break up, but the earth's magnetic field would bend the beam in complex and unpredictable ways, and the one-way transfer of change in space would build up powerful forces that would eventually halt the beam. These problems can be avoided by using beams of neutral particles, a concept being tested by DARPA's White Horse program at Los Alamos.

Accelerators only work on charged particles, thus, the first difficulty is producing an energetic beam of neutral particles. The approach being used at Los Alamos is to start with negatively charged hydrogen atoms, protons surrounded by two electrons, rather than the one normally found in a hydrogen atom. After being accelerated to high energies, these negative ions would pass through a gas or some other medium that would strip off the extra electrons to leave a beam of neutral particles.

It doesn't take much air to break up a beam of neutral particles because it is not subject to the same cohesive effects as a beam of charged particles. Fortunately, in the vacuum of space there is nothing to get in the way, and uncharged particles would not be affected by the earth's magnetic field. Once the beam path was defined, the particles should travel in a straight line until something (hopefully the target) gets in the way. Unlike massless photons, massive neutral particles would be subject to the earth's gravitational field; but the effects on particles traveling near the speed of light should be small and fairly easy to predict.

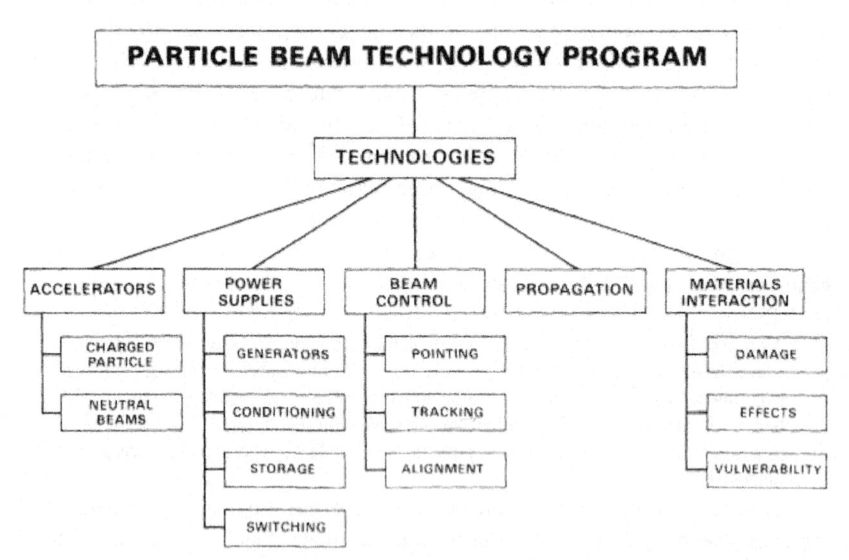

Major elements of the particle-beam technology program, as identified by the Department of Defense. Accelerators boost the particles to high speeds and high energies, power supplies provide the energy needed by the accelerators, beam control covers aiming the beam at the target, propagation is getting the beam through the intervening air or space to the target, and materials interaction involves what happens once the beam hits the target. (Courtesy of Department of Defense.)

The Los Alamos program is addressing questions that affect both beam production and propagation. In DARPA's words, "The critical issue for neutral particle beam system concepts is successful achievement of the system elements that contribute to the final beam divergence, and thus to the maximum effective range."[19] Obtaining the narrow beam divergence needed over long distances is possible only if a way can be found to strip the extra electron from each hydrogen ion without harming beam quality–something which is now a problem. Beam quality can be affected in various ways. Even a slight scattering of the negative ions as the extra electrons are being removed from them could deflect the resulting neutral particles just enough to cause them to miss their target thousands of kilometers away. Such scattering could also disperse the beam, spreading it out so much that its intensity would be too low to do any damage once it reached the target. Scattering also could spread particle energies over a wider range than desirable, an effect which would lengthen the bunch of particles as it travels because the velocity of a particle is proportional to its energy. The more the spread in velocity, the more the beam will lengthen as it travels, thus dissipating the high peak power pulse that makes particle beams attractive for weaponry.

In addition to its two main programs, DARPA is also sponsoring work on what it calls its "technology base." The goal of that effort is to gain an improved knowledge of particle-beam technology that would help in adapting it for weaponry (or in being sure that it couldn't be used in weapons).

The Soviet Effort

The Soviet Union generally is considered to have a large program in particle-beam technology, but its precise magnitude and goals remain a subject of debate. Because the major applications of particle beams have traditionally been in the study of subatomic physics, generally considered to lie in the unthreatening realm of "pure research," the United States and Soviet Union have openly exchanged ideas. Many concepts have been published in the open scientific literature, and interestingly, a number of ideas now being pursued in the United States originated in the Soviet Union.

Pentagon officials seem to have mixed feelings on the nature of the Soviet program. A "fact sheet" on the particle beam program issued in early 1982 stated: "The Soviet effort is judged to be larger than that of the United States (particularly in the area of accelerators for fusion applications), and to have been in progress longer. However, no direct correlation between Soviet particle beam work and weapons-related work has been established."[20] Yet the booklet *Soviet Military Power,* issued by the Department of Defense in late 1981, said: "The Soviets have been interested in particle beam weapons (PBW) concepts since the early 1950s. There is considerable work within the USSR in areas of technology relevant to such weapons."[21]

Much more alarmist views have been aired in the pages of *Aviation Week & Space Technology.* Military editor Clarence A. Robinson, Jr., opened one of his first major articles on the subject by flatly stating: "[The] Soviet Union is developing a charged particle beam device designed to destroy U.S. intercontinental and submarine-launched ballistic missile nuclear warheads. Development tests are being conducted at a facility in Soviet Central Asia."[22] In mid-1980 the magazine published another article warning that a

> directed-energy weapon that could be the first step in a revolutionary concept of warfare is being constructed by the Soviet Union at Sary Shagan, a ballistic missile range near the Sino-Soviet border in southern Russia. . . . Many U.S. intelligence analysts believe the weapon is an early prototype of a new-design charged particle beam device, and that it may be used within a year or so in tests against ballistic missile targets.[23]

At this writing no such tests have come to the public eye, in *Aviation Week* or any other medium. That could reflect exaggeration or inaccurate assessment on the part of *Aviation Week* or its sources, tightened security in the United States or Soviet Union preventing leaks of information on such tests, or a failure of the Soviet system to work as planned. Without access to classified Soviet sources, I cannot be sure which is the proper interpretation. Although the Soviet military program is out of sight to Westerners, a study of the open Russian scientific literature leaves little doubt that there is a large program in pulsed-power and particle-beam generation technology. A 1978 report by the Rand Corporation, based on a survey of Soviet scientific journals published from 1962 through 1976, concluded that "the scope of Soviet pulsed-power research and development appears to be much larger than that required for the explicitly stated purposes of . . . the conversion of energy from controlled thermonuclear reactions." The report notes that public descriptions of the intended applications of pulsed-power research are "un- usually vague, even by Soviet standards." Although refusing to state flatly that the large Soviet effort included work on particle-beam weaponry, the report concludes, "it is highly probable that it does."[24]

Problems with Beam Control

As is the case with high-energy lasers, it seems much easier to build large particle-beam generators than it is to do anything militarily useful with them, although one researcher noted that it is even easier *not* to build them. At present the Pentagon is concentrating on assessing prospects for getting the particle beam through the atmosphere or space to its target. However, there is also an even more serious uncertainty–beam control, the job of pointing a particle beam at a target and making it track the target. Conceptually what is needed seems to be some kind of electrical and/or magnetic field capable of deflecting the beam as it emerges from the accelerator, somewhat like an electron gun deflects the electron beam that writes on the inside of a television picture tube. Moving from that vague concept to practical hardware will be no easy task.

The Pentagon's fact sheet on particle-beam technology lays out the problem in matter-of-fact bureaucratese: "Beam-control subsystems for charged and neutral beams present critical technology requirements that are beyond the present state of the art in all cases and without a technology basis in some cases."[25] Although clearly having some reservations about the whole idea, military researchers are attacking the beam-control problem with a "nominal" development effort, including work on a system for beam sensing and internal alignment.

Exactly how critical a problem beam direction will pose is a function of the nature of the mission. Four different concepts for particle-beam weapons emerged from a study conducted for the Pentagon in the late 1970s:

- Short-range weapons that would use an intense charged particle beam to produce a shower of high-energy secondary radiation when the particles hit atoms in the atmosphere. This cone of secondary radiation, energetic charged particles and perhaps X rays and gamma rays, would destroy military electronics and possibly kill soldiers. With a range of around 1 km and a broad swath of lethal output, such a system would not require precise pointing and tracking.

- Medium-range weapons for use in the atmosphere that would direct a tightly focused particle beam over distances to 5 km for destruction of "hardened" targets such as nuclear missile warheads. This would require precise pointing and tracking.
- Longer-range atmospheric weapons, which would use a tightly focused beam of charged particles to destroy targets up to 10 km away, but would require higher pulse energies and more precise pointing and tracking.
- Space-based (i.e., orbiting) neutral particle-beam weapons with ranges of hundreds or perhaps thousands of kilometers for defense against ballistic missile attack. Hitting the missiles at such distances would require extremely precise pointing and tracking.[26]

Some critics of particle-beam weapons consider the beam-direction problem fatal. For example, a study group working at the Massachusetts Institute of Technology concluded that "there is serious doubt whether the necessary fire control and beam control systems needed for a weapon are technically feasible."[27] Although there are some optimists, most observers appear to agree that the problem is serious.

Target Interactions

If a tightly focused particle beam can hit its target, it can do great damage. An ion beam could deposit around 1 million joules of energy in a long, narrow cone within the target, causing extensive damage. High-energy electrons can make their way through several feet of aluminum, far more protection than would be found on any mobile target. A hit from a particle beam would almost instantaneously detonate chemical explosives such as rocket fuels, cause structural damage, and disrupt electronic equipment.[28] More disruption would come from the high-energy secondary radiation generated by the interaction between beam and target. Most military analysts believe that the deep penetration of particle beams would make it very hard to protect potential targets from destruction.

Secondary radiation and diffuse particle beams would present rather different threats, focusing on electronic components that are vital to the function of weapons, and–because of the lower beam intensity–causing minimal or no structural damage. Like other quiet-kill mechanisms that cause no damage visible at a distance, they could leave the user of the weapon uncertain if a target has been disabled.

Questions of Feasibility

There are also other important questions that remain to be answered in assessing the feasibility of particle-beam weapons. Many center on the inherent bulk of particle-beam generators and the power supplies they require. One cynic once cracked that the only way to kill someone with a laser was to drop it on him. That comment would have been more appropriately applied to particle-beam weapons, which with current technology tend to weigh in much heavier than chemically fueled lasers.

There are undoubtedly attractions to particle-beam weaponry. At least in theory, a particle-beam generator could fire off bursts of shots rapidly and kill targets with a single shot (although an initial shot might be needed to punch a hole in the atmosphere first). Depending on beam- and fire-control capabilities, a particle-beam weapon might be able to shoot at tens of targets per second, a capability that would be the envy of the designers of the Death Star in *Star Wars.*

Barring bad news from the Livermore and Los Alamos tests, the debate over the feasibility of particle-beam weaponry will probably continue unresolved over the next few years. The Pentagon's $50-million a year effort, roughly 10% of the directed-energy budget in 1983, reflects a cautious approach that tries to balance the promises of particle-beam technology with the problems that first have to be overcome. In the minds of the Pentagon planners who set priorities, particle beams are a much less mature technology than high-energy lasers. Time will tell how right they are.

References

1. "Fact Sheet: Particle Beam Technology," Department of Defense, February 1982, p. 3.
2. For an in-depth treatment see: R. B. Miller, *An Introduction to the Physics of Intense Charged Particle Beams* (Plenum Press, New York, 1982); for a briefer and more elementary description of electron and accelerator physics, see an introductory-level college physics text. I found useful parts of Chapter 5 in D. Elwell and A. J. Pointon, *Physics for Engineers and Scientists,* 2nd ed. (Ellis Horwood Ltd., Chichester, England, 1978).
3. *Pulsed Power Research and Development at Sandia National Laboratories* (Sandia National Laboratories, Albuquerque, New Mexico, 1981) p. 24.
4. "Soviets build directed-energy weapon," *Aviation Week & Space Technology,* July 28, 1980, pp. 47-50.
5. Richard J. Foley and William F. Weldon, *Compensated Pulsed Alternator* (Lawrence Livermore National Laboratory, Livermore, California, October 1980).
6. See, for example, R. B. Miller, *An Introduction to the Physics of Intense Charged Particle Beams* (Plenum Press, New York, 1982), pp. 2-19; and *Pulsed Power Research and Technology at Sandia National Laboratories* (Sandia National Laboratories, Albuquerque, New Mexico, 1981), pp. 18-26.
7. "Fact Sheet: Particle Beam Technology," Department of Defense, February 1982, p. 5.
8. William A. Barletta, "The Advanced Test Accelerator," *Military Electronics/Countermeasures* 7 (8), 21-26 (August 1981).
9. Defense Advanced Research Projects Agency, *Fiscal Year 1983 Research and Development Program,* March 30, 1982, p. 111-38.
10. *Pulsed Power Research and Technology at Sandia National Laboratories* (Sandia National Laboratories, Albuquerque, New Mexico, 1981).
11. The program was originally called "Sipapu," a Native American word meaning sacred fire, but the name was changed because officials were concerned its use might offend the large Native American population in the Los Alamos area.
12. *Los Alamos National Laboratory Annual Report 1980* (Los Alamos National Laboratory, Los Alamos, New Mexico) pp. 34-35.
13. For descriptions of the physics of inertial-confinement fusion with particle beams, see: Chapter 8 in R. B. Miller, *An Introduction to the Physics of Intense Charged Particle Beams*

(Plenum Press, New York, 1982), and "Drivers for pulsed power fusion," *Sandia Technology,* October 1982, pp. 1-15.

14. For descriptions of the Sandia program, see: *Pulsed Power Research and Technology at Sandia National Laboratories* (Sandia National Laboratories, Albuquerque, New Mexico, 1981); "Drivers for pulsed power fusion," *Sandia Technology,* October 1982, pp. 1-15; and Gerold Yonas and J. Pace VanDevender, "Advances in ICF using light ion beams," paper presented at 6th International Workshop on Laser Interaction and Related Plasma Phenomena, October 27, 1982, Monterey, California. Target experiments with PBFA-I began in 1982.
15. The physics of fusion with heavy ions is covered in R. B. Miller, *op. cit.;* see also: John Lawson and Derek Beynon, "Heavy ions beam in on fusion," *New Scientist* 95, 565-568 (August 26, 1982).
16. William A. Barletta, "The Advanced Test Accelerator," *Military Electronics/Countermeasures* 7 (8), 21-26 (August 1981).
17. *Ibid.,* p. 22.
18. Defense Advanced Research Projects Agency, *Fiscal Year 1983 Research and Development Program,* March 30, 1982, p. III-38.
19. *Ibid.,* p. III-39.
20. "Fact Sheet: Particle Beam Technology," Department of Defense, February 1982, p. 7; the same statement appears in the February 1983 edition.
21. Department of Defense, *Soviet Military Power* (U.S. Government Printing Office, Washington, D.C., October 1981), p. 75. A similar but not identical statement appears on p. 75 of the March 1983 edition.
22. Clarence A. Robinson, Jr., "Soviets push for beam weapon," *Aviation Week & Space Technology,* May 2, 1977, pp. 16-23.
23. "Soviets build directed-energy weapon," *Aviation Week & Space Technology,* July 28, 1980, pp. 47-50.
24. Simon Kassel, *Pulsed Power Research and Development in the USSR* (Rand Corporation, Santa Monica: California, May 1978), pp. 114, 118. The report as a whole summarizes one and a half decades of Soviet research published in the open Russian literature.
25. "Fact Sheet: Particle Beam Technology," Department of Defense, February 1982, p. 6.
26. Clarence A. Robinson, Jr., "Beam weapons effort to grow," *Aviation Week & Space Technology*, April 2, 1979, pp. 12-16. This account of the report that studied particle beam technology for the Department of Defense mentions only ranges of hundreds of kilometers for space-based particle beam weapons. Other accounts generally assume that the range would have to be a few thousand kilometers. Although the longer range would put much more stringent demands on particle-beam technology, the shorter range would require unrealistically large numbers of particle-beam battle stations. My rough estimate puts the figure at somewhere around one thousand.
27. G. Bekefi, B. T. Feld, J. Parmentola, and Kosta Tsipis, *Particle Beam Weapons* (Program in Science and Technology for International Security, Department of Physics, Massachusetts Institute of Technology, Cambridge, Massachusetts, December 1978), p. 59; much the same conclusions are contained in a Jess-detailed but more accessible article, John Parmentola and Kosta Tsipis, "Particle-beam weapons," *Scientific American* 240 (4), 54-65 (April 1979).
28. William A. Barletta, "The Advanced Test Accelerator," *Military Electronics/Countermeasures* 7 (8), 21-26 (August 1981).

8. High-Power Microwaves The Zapping of Hardware

The Reagan Administration and the Pentagon almost invariably list microwaves as the third and last type of "directed-energy" weapon under development. Last, in this case, is indeed least in terms of program size. Of the roughly $500 million a year budgeted for directed-energy weapons, less than 1% goes for microwaves. In fact, the microwave budget is small enough that the Pentagon's directed-energy office doesn't break it out separately. When I asked for an estimate of the total set aside for microwaves, I was told it was in the range of $1.5 to $5 million, but that no definite figure was readily available.[1] The most immediate goal of that modest program is to learn if there are ways that microwaves could cause militarily useful damage to potential targets.

Microwave hazards have been in the headlines before, but the general press seemed to keep the issue in unusually good perspective in its coverage of President Reagan's March 23, 1983, speech proposing development of beam weapons for ballistic missile defense. It may have been that lasers have become such a symbol of space-age technology that they stole all the headlines. Or the press may have remembered the great microwave scare of the 1970s and decided that there was nothing new to report.

Looking back, it seems that the microwave scare of a decade ago probably reflected people's reactions to the then-new microwave oven. This new oven convincingly demonstrated that it could cook food very rapidly, but in a somewhat unusual or even mysterious way. I recall a friend who was mystified when he broke open a cookie that he had put in a microwave oven and discovered that he had burnt the center but not the outside. If microwaves could cook food inside an oven so effectively, people began to wonder, what were they doing to the people standing outside? These fears were magnified when *Consumer Reports* criticized makers of some microwave ovens for letting some of the microwaves leak out and complained that the industry as a whole had not developed an adequate safety standard for emission from ovens. In 1973 sales of microwave ovens slumped, and some makers began to counterattack. One noted scientist, James Van Allen of the University of Iowa, offered to back up industry claims by sitting "on top of my Amana Radarange for a solid year while it is in full operation."[2]

The controversy simmered for a while but soon began heating up. In early 1976 the State Department revealed that the United States embassy in Moscow had been irradiated by low levels of microwaves for over 20 years, stimulating speculation about dastardly communist plots. Soon afterwards, the American public was warned of the hazards of exposure to microwaves in a book titled *The Zapping of America: Microwaves, Their Deadly Risk, and the Cover Up.*[3] Written in an "expose" style, it argued that much of the public is endangered by microwaves from communication and radar systems, microwave ovens, and a variety of industrial equipment. The issue was further heated by public confusion about the nature of microwave radiation. Concern over the safety of nuclear power plants also was growing, making "radiation" a bad word to many people who didn't realize the term covered the entire electromagnetic spectrum.[4]

Scientists who believed that microwave hazards were modest at low exposure levels counterattacked, charging that *The Zapping of America* was riddled with errors. They backed up their assertions by saying that studies showing dangers from low exposure levels were not performed properly. They bolstered their arguments with a government study that showed no detectable long-term health effects in former workers in the microwave-irradiated Moscow embassy.[5]

The public controversy mirrored a debate among scientists over what levels of microwave exposure were safe for humans. At the time the Soviet Union had standards limiting microwave exposure to only one-thousandth to one ten-thousandth the levels permitted under the informal standards prevalent in the United States (although those limits apparently did not apply to the U.S. embassy). American researchers generally were skeptical about the validity of the studies the Soviets used to determine their low exposure limits, but eventually the United States did establish formal safety rules. The current exposure limits in the United States are somewhat lower than those used earlier but are still much higher than Soviet standards.

Today, much uncertainty remains about health effects and biological interactions involving microwaves, with the largest questions concerned with long-term exposure to low levels. However, there appears to be no real indication that many people were ever exposed to serious microwave hazards, except engineers and technicians working directly with microwave equipment used for such applications as radar and communications.[6] Users of microwave ovens are in no serious danger except if they do something stupid such as stick their hands into a working oven[7] or unless they rely on heart pacemakers, which are extremely sensitive to electromagnetic interference generated by the oven. Current microwave ovens comply with federal safety standards and are finding their way into many kitchens; the Association of Home Appliance Manufacturers reported that over 350,000 microwave ovens were shipped in March 1983.[8]

It sometimes seems that whatever scares the public attracts the Pentagon, and that may have happened with microwaves. During the great microwave scare, there was wild speculation that low levels of microwave irradiation could scramble people's thoughts, but no such effects have ever been demonstrated. Instead, the Pentagon is interested in using microwaves to cook or scramble the electronic thoughts of military hardware.

What stimulated military interest in the possibility of microwave weapons is the availability of extremely high powers. Pentagon officials have talked about prospects for generating pulses of 100 billion watts at a wavelength of 10 cm.[9] With prospects appearing good for such high powers, military planners are trying to find ways to put them to use. In theory, possibilities include jamming of communications links, zapping electronic circuits with signal-scrambling electromagnetic interference, and heating targets with microwave intensities comparable to those in an oven. What is practical remains unclear; microwave effects on hardware are not well understood. So far the Pentagon's microwave weapon program is a small one directed at understanding microwave effects that might be of some military use.

High-Power Microwaves

The original impetus for development of microwave generators came from the development of military radar and communication systems. Microwaves offered important advantages over lower-frequency radio waves, including finer radar resolution, because of their shorter wavelength, and higher information-handling capacity for communications, because of their higher frequency. Higher powers increased the operating range and signal quality in both radar and communication systems.

The engineers who developed microwave generators for World War II radars soon learned that powerful microwave beams could heat objects in the way. Legend holds that the discovery was made inadvertently when an engineer cooked his lunch by leaving in next to a microwave tube. That discovery would eventually lead to the microwave oven in which powers of about one hundred to several hundred watts are confined in a small cavity to heat or cook foods. (There are also industrial microwave ovens that may use powers of a few thousand watts.) Although the military sponsored most microwave development, it could see little prospect of putting the heating phenomenon to practical use. By the mid-1960s, the technology had reached a point where individual microwave sources could generate average powers of hundreds of thousands of watts at wavelengths around 10 cm (corresponding to a frequency of a few gigahertz, or billions of cycles per second). Today that technology is used in military radars for early warning of enemy attack and other purposes and in a variety of communication systems.[10]

Many radar systems now are limited in range not by the microwave power available but by the curvature of the earth. However, microwave technology has not stood still, pushing forward both to higher powers and to higher frequencies. At frequencies between 100 MHz (megahertz) and 1000 GHz (gigahertz), where microwave and related technologies are applicable, the output power drops as operating frequency increases. Thus, the highest powers are available only at the lowest frequencies (or, equivalently, at the longest wavelengths). Over the years engineers have steadily pushed the power and frequency limits up, making powers previously attainable only at low frequencies also available at higher frequencies. They have also increased the efficiency of microwave sources and thus reduced the amount of waste heat they generate.

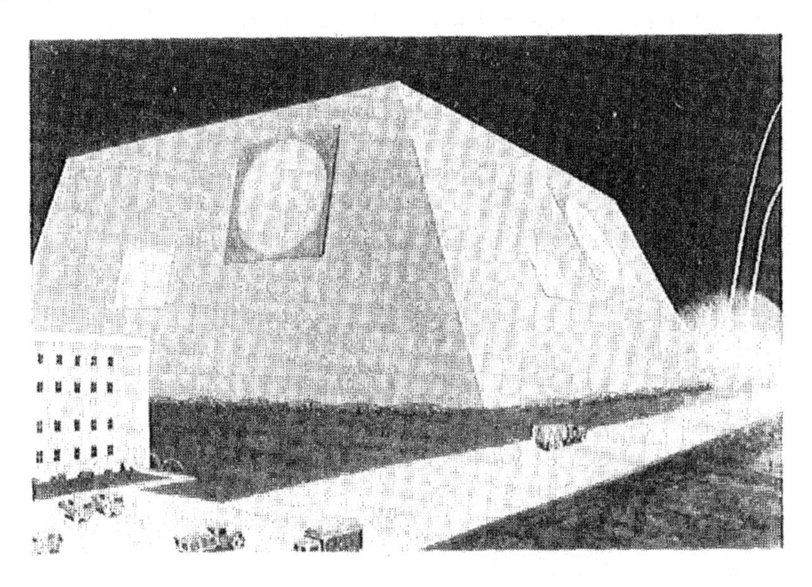

A representative monster-sized radar. In this case an artist's conception of a new radar the Soviet Union has developed for use as part of its antiballistic missile system to defend Moscow. The four-sided structure is roughly 35 m (120 ft) high and 150 m (500 ft) wide. (Courtesy of Department of Defense.)

Past and expected improvements in the technology were chronicled in a 1978 paper by Ervin J. Nalos, an engineering supervisor at the Boeing Aerospace Corp. In 1960 it was possible to generate average or continuous powers (*not* peak power in short pulses) of about 1000 W at a frequency of 10 GHz (3-cm wavelength) with roughly 10% efficiency. By 1980, he said, efficiency at that frequency would reach 40% and output power 100,000 W. He expected that by the year 2000, power would reach 1 million watts and efficiency 85%.[11]

New technology involving the coupling of electron beams to microwave equipment is pointing to even higher powers (particularly in short pulses) and higher frequencies. There is essentially a continuum of devices, reaching up to the upper limits of microwave frequencies, sometimes called millimeter waves because their wavelengths are a few millimeters. These include devices that rely on a beam of electrons passing through a magnetic field that varies in space, which are in a sense the microwave equivalent of the free-electron laser described in Chapter 4, although the physical details are different.

High-power microwave technology has advanced enough beyond the military's immediate needs that it might almost deserve the same label that was attached to the laser in the 1960s: "A solution looking for a problem." One concept that has gotten much attention from proposers of "blue-sky" systems for use in the next century is power beaming, that is, the transmission of large quantities of energy from one point to another. The concept has taken its most concrete form in the idea of the solar power satellite proposed by Peter E. Glaser of Arthur D. Little Inc.[12] and has been adopted enthusiastically by advocates of the industrialization and colonization of space.

Solar power satellites would collect solar energy in space, then beam it down to earth where it would be used. Advocates say the concept would need much less land than ground-based solar power. Also, the satellite would be in a high enough orbit that it could generate power nearly continuously, without interruption by night or bad weather. Glaser's conceptual design eventually settled on microwaves to beam the power down to earth, although laser transmission has also been considered.[13]

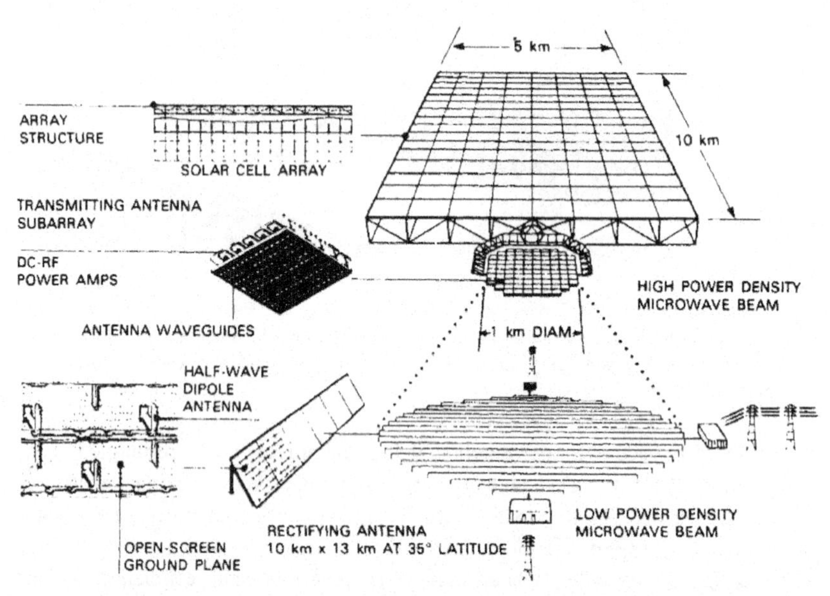

Basic components of the reference solar power satellite system (using microwave transmission) evaluated by the Department of Energy and the National Aeronautics and Space Administration. The large array of solar cells at upper right would convert solar energy into electricity that would drive the 1-km (0.6-mile) microwave antenna on the satellite. The microwaves would be beamed down to earth, where a rectifying antenna covering as much territory as a moderate- sized city would convert them back into electricity. Concern over possible heating of the ionosphere would limit power density on the ground. (Courtesy of Department of Energy. From Ref. 14, p. 61.)

The proposed power satellites would be truly monstrous. A "baseline" design that was evaluated by the Department of Energy and the National Aeronautics and Space Administration called for a transmitting antenna 1 km square that would produce about 7 billion watts at a frequency of 2.45 GHz (a wavelength of about 12 cm). The satellite would be in an orbit nearly 36,000 km (22,000 miles) above the earth, which is called a geosynchronous orbit because the satellite would circle the earth once each day, appearing to remain stationary above the same point on the planet. To avoid heating of the ionosphere, power density in the atmosphere and on the earth would be limited to 0.023 W/cm^2. The receiving antenna would be an oval 10 by 13 km across that would generate about 5 billion watts of electricity from the received microwave power. Laser transmission would allow the use of a much smaller antenna because of the better focusing inherent at the shorter wavelength and the absence of ionospheric heating. It would also avoid concern over health effects of microwaves, although the higher power density of the laser beam would be a serious hazard in itself. However, NASA and the Department of Energy settled on the microwave approach for their study because the technology seemed nearer at hand and because microwaves, unlike lasers, can penetrate cloud cover.[14] An additional concern, implicit but rarely stated, is that a high-energy laser might be mistaken for a weapon system, although laser advocates have suggested ways in which the transmitter could be designed so it couldn't function as a weapon.

Space development enthusiasts are excited about solar power satellites, but the concept is dormant on an official level. The small government-funded research project became one of Congress's proverbial "political footballs." Aided by at best a lukewarm endorsement by the DOE/NASA study committee, Congressional opponents were able to drop back and punt the program clear out of the budget.

Weapon Applications

High-power microwave technology remains available for other power-beaming applications, such as directed-energy weapons and delivering power to remote sensors without having to lay a power cable. Weapon-like effects of microwave heating have been demonstrated. Steel wool has been ignited by a microwave transmitter 14 m (45 ft) away; a mixture of aluminum chips, gas vapor, and air ignited at 76 m (250 ft); and photoflash bulbs set off at a distance of 260 m (850 ft).[15] Nonetheless, as in the case of laser weapons, simple demonstrations do not represent an effective weapon system. The effects of microwaves on potential targets are not well understood, and learning more about microwave effects is the main goal of the Pentagon's program.

Microwaves can produce various types of militarily useful effects:

- Jamming microwave radar and communication systems by overpowering the signals that their antennas are trying to receive. The effect is similar to that of a powerful static source on an AM radio or of the noise from a low-flying jet on a conversation. This is called the "negation of information" by military engineers and requires power levels of 10^{-8} to 10^{-6} W/cm^2, well within even the Soviet Union's strict standards limiting the continuous exposure to microwaves.
- Beaming power to remote sensors and other equipment instead of relying on wires or internal batteries. The remote equipment thus gets its energy by absorbing the transmitted microwaves. This requires powers of 0.01-0.1 W/cm^2, above U.S. standards for continuous exposure of humans to microwaves.
- Disrupting the operation of sensors and electronic equipment by inducing

microwave currents that drown out the signals in the equipment. This is distinct from jamming in that it is aimed at equipment that is not supposed to be picking up microwave signals. It would probably require powers of 10-100 W/cm^2.

- Heating of targets in "reasonably short" illumination times by effects similar to those in a microwave oven, which would require powers around 1000-10,000 W/cm^2.[16]

The strength of all these effects, particularly heating, will depend on the type of target. Metals, which make up the outer surfaces of the commonest military targets, could provide a shield because they tend to reflect much of the incident microwave power. In contrast, microwave effects would be enhanced by materials that tend to absorb microwaves, such as those that play a role in the "Stealth" aircraft designed to be hard to spot with radar.

Despite some major studies triggered by the microwave safety debate of the last decade, the precise nature of microwave effects on humans remains uncertain. Thermal effects can occur at high microwave powers because body tissue is a fairly good absorber of microwaves. (Chemically speaking, the human body is rather similar to the food cooked in a microwave oven.) The effects of long-term exposure to low levels of microwaves are harder to quantify. Researchers have found that long-term exposures to microwave powers of 0.1 W/cm^2 (well below that *inside* a microwave oven) can cause cataracts and extensive thermal damage to body tissue. They say that less obvious long-term effects could show up at power densities as low as 0.001 W/cm^2.[11]

Microwave weaponry is, however, rather less threatening to people than those numbers might indicate. Long-term effects are of little use on a battle- field, where the goal is to disable enemy soldiers immediately. In addition, fairly simple metal shielding such as an aluminum-foil suit can reduce the intensity significantly, a technique already used at times to protect microwave technicians.

The exact nature of the missions envisioned for microwave weaponry is at best vaguely defined, as would be expected from a program that is only in the early stages of defining itself. In space the Pentagon's main interest lies in "the potential for either destruction of space vehicles or bum out of their electronic components at very long range," according to the *High Frontier* report. High power microwave pulses are said to be "potentially lethal" when directed against targets including cruise missiles, aircraft, and what the Pentagon calls "remotely piloted vehicles," small aircraft that are remotely controlled. In addition, satellites might be vulnerable to heating effects or to microwave-induced currents that could damage their electronic circuitry.[18]

The same report asserts that technical challenges involved in developing microwave weapons "appear lower than for any other directed-energy weapon," a conclusion many other observers would dispute. Nonetheless, the study warns that "there is uncertainty and lack of agreement on the lethality of microwave energy at the power levels studied. Furthermore, the uncertainty is associated with details of the target design since a major kill mechanism is leakage energy getting into the electronics."[19] That makes the problem a particularly hard one to solve because potential adversaries jealously guard the design details of their military systems.

Size and Cost Tradeoffs

There are some clear attractions to microwave weapon concepts. Microwave sources are much cheaper than lasers. One-thousand-watt microwave tubes can be manufactured in quantities of several hundred thousand for well under $50 each. Power from a large infrared carbon dioxide laser costs about as much per watt. Thus, a dollar buys roughly 1000 times more microwave power than laser power,[20] although that ratio might come down if lasers were manufactured in as large quantity.

The most obvious problem with microwaves is the size of the equipment needed to direct and focus the beam over long distances. Microwaves are subject to the same diffraction limit that sets a lower limit on how small a spot a laser beam can be focused. This means that the angular spot size (measured in radians) is roughly proportional to the wavelength divided by the diameter of the output aperture, an antenna, in the case of microwaves. Because microwave wavelengths are about 1000 times longer than those produced by infrared lasers, the antenna diameter must be 1000 times that of the laser mirror to get the same size focal spot. To produce a focal spot of about one millionth of a radian, as desired for some long-distance laser weapons, using 30-GHz (1-cm) microwaves, the antenna would have to be about *10 km* across. The power requirements may not be out of this world, but the antenna diameter is.

The picture is not necessarily that bad because microwaves need not be focused so tightly if enough power is available. However, a spread-out beam could inadvertently destroy friendly forces. One of the major attractions of laser and particle-beam weapons is their potential to pick out one enemy target from among many objects in view, including friendly forces. A broad microwave beam would simply zap everything in its way.

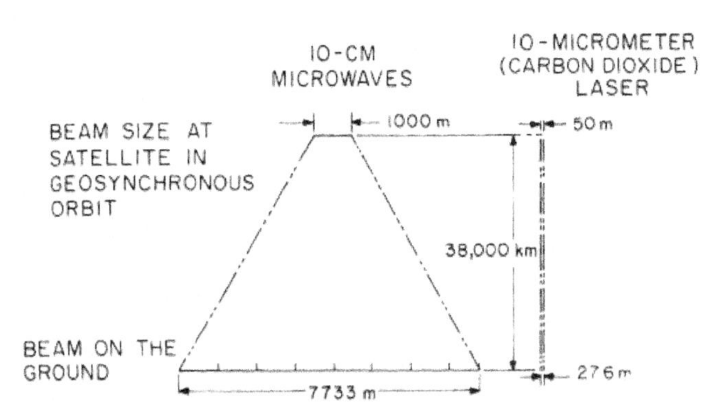

The fact that microwaves are much longer than light waves makes dramatic differences in the "optics" of laser and microwave systems. This drawing shows the different dimensions needed for microwave and laser transmission of energy from a solar power satellite to the ground. In this example the microwave wavelength is taken as 10 cm and the laser wavelength as 10 μm (the output of a carbon dioxide laser), a factor of 10,000 shorter. Diameters of the transmitting antenna and mirror are shown at top in meters (1 m is about 3.3 ft); those of the receivers on the ground are shown at the bottom. (Drawing by Arthur Giordani based on information in Ref. 13.)

It is not as important to limit the spreading of the beam in a short-range weapon because the beam doesn't get that much chance to spread out. For example, if a theoretically perfect 1-m antenna emits a diffraction-limited (i.e., as narrow as possible) beam of 1-cm (30 GHz) microwaves, the beam would spread out to about 12 m (40 ft) in diameter after traveling 1 km (0.6 mile). That is roughly the same size as a fighter plane such as the F/SA or the MiG-21. At 2 km the beam diameter would double, and the power density would drop by a factor of 4.

Pentagon researchers have pointed out some potential problems with microwave weapons. High-power microwaves ionize air molecules by stripping electrons from them much more readily than do laser beams; this presents problems because ionized air can block the beam. Another problem is that many potential military targets are sheathed in metal that strongly reflects microwaves, making it harder to transfer microwave energy to them.[21]

In some areas microwave weapons could draw upon more established technologies. Beam-control methods for microwaves are well advanced, thanks in large part to radar system developers. Large "phased-array" antennas have been developed in which many separate emitting elements are adjusted electronically to change the direction of the beam.[22] The effect is the same as tilting the antenna, but it is much easier than physically moving a large antenna or even the smaller components. Fire-control techniques would probably be similar to those developed for other types of directed-energy weapons, although there might be a problem in seeing internal microwave damage to a target.

Questions of Feasibility

Despite the availability of high microwave powers, some big question marks remain for microwave weapons. The most likely damage mechanism, which disables internal electronics, is unlikely to occur at a precisely predictable threshold. Although shielding against microwaves cannot be perfect, techniques are improving. Witness the use of microprocessors in home microwave ovens. Developers of military systems can spend even more money on the problem and are already being pressed to shield equipment against electromagnetic pulse effects of nuclear bombs, something which is likely to help protect against microwave attack as well. Defense against microwave attack might be easy enough to render the whole idea impractical. Compounding the problem will be the difficulty in telling from afar if a critical electronic system has been disabled inside the target even if no effects are visible outside.

Higher microwave powers could bake their targets, but only if they came close enough or used very large antennas. The problems are most obvious when long distances are involved. For example, if the 1-km (0.6 mile) microwave antenna considered for a solar power satellite was focused as tightly as possible onto the earth's surface from its orbit 36,000 km (22,000 miles) above, it would produce a spot 3 km (two miles) in diameter. The average power density in that area would be an unhealthy level slightly under 0.1 W/cm^2, dangerous to someone who stayed there but far, far below the level needed to cook anything quickly. To reach the 1000-W/cm^2 threshold for rapid heating, the focal spot would have to be shrunk to about 30 m (100 ft) across, which would make it necessary to move the satellite down to an altitude of about 360 km (220 miles). Such a huge antenna would be unlikely to last long at such an altitude. Even without that problem, the notion of an orbiting antenna 1 km across makes a chemical laser battle satellite seem a veritable model of miniaturization.

Although microwave weapons do not seem to be right around the corner, the Department of Defense sees enough promise in the technology that it has supported a modest effort since the late 1970s. The microwave program is the newest thrust in the directed-energy effort, and military researchers are concentrating on understanding microwave effects and evaluating their potential uses. According to Major General Donald L. Lamberson, then assistant for directed-energy weapons to the undersecretary of Defense for Research and Engineering, "We have concluded that if experiments verify that appropriate damage mechanisms exist, then the technology has promise as a weapon system."[23] It is not clear what the Soviet Union is doing, although the Pentagon's booklet *Soviet Military Power* states: "Indications of Soviet interest in radio frequency technologies, particularly the capability to develop very high peak-power microwave generators, indicate that the Soviets intend to develop such a [microwave directed-energy] weapon."[24]

The small size of the microwave weapon program reflects the state of the art and the serious technical uncertainties that must be resolved. Military roles for microwave weapons need to be defined and matched to the available technology. The concept of powerful microwave weapons seems a long way from reality, but some microwave engineers and military planners believe that some bright ideas could make it possible.

References

1. Jeff Hecht, "Arms race spreads to microwaves," *New Scientist* 96, 728 (December 16, 1982).
2. "Looking back," *Microwaves* & *RF* 22(5), 192 (May 1983); this retrospective column looks back to events reported in the magazine's May 1973 issue.
3. Paul Brodeur, *The Zapping of America: Microwaves, Their Deadly Risk, and the Cover Up* (W.W. Norton, New York, 1977); portions of this book were published earlier in *The New Yorker*.
4. The term "electromagnetic radiation" is a generic one for light, microwaves, X rays, radio waves, and so on, as described in Chapter 4. Many forms are harmless (except at extremely high levels), but people often are misled by the term radiation, thinking it means the type of harmful radiation produced by radioactive materials and nuclear reactions or even confusing it with radioactive material. The proper term for such intrinsically dangerous

radiation is "ionizing radiation" (also sometimes called "hard" radiation), which comes from the fact that the photons carry enough energy to ionize normal atoms. X rays and gamma rays are ionizing radiation, but microwaves are not. Early physicists thought that some particles, such as electrons and neutrons, emitted during nuclear reactions were really forms of radiation, and sometimes these energetic particles are also called "radiation."

5. One such argument is Herbert Pollack, "Microwave anxieties: The U.S. embassy in Moscow, a case in point," paper 31.6 in *Conference Record, International Conference on Communications,* June 10-14, 1979, Boston, published by Institute of Electrical and Electronics Engineers, New York.
6. Eric J. Lerner, "RF radiation: biological effects," *IEEE Spectrum* **17** (12), 51-59 (December 1980).
7. "Ovens," *Microwave News* 3 (4), 7 (May 1983).
8. *Ibid.*
9. Douglas Tanimoto, talk at Electro-Optical Systems & Technology meeting, November 1980, Boston, sponsored by American Institute of Aeronautics and Astronautics.
10. Ervin J. Nalos, "New developments in electromagnetic energy beaming," *Proceedings of the IEEE* **66** (3), 276-289 (March 1978); see also reports on Microwave Power Tube conferences in the July 1980 and July 1982 issues of *Microwave Journal.*
11. *Ibid.* p. 277; note that these powers are at higher frequencies than those mentioned earlier and also represent continuous emission or an average over many pulses, not the peak instantaneous power in a short pulse.
12. His idea originally appeared in Peter E. Glaser, "Power from the sun: its future," *Science* **162,** 857-886 (November 1968); for a more-current outline of his ideas see: Peter E. Glaser, "The earth benefits of solar power satellites," *Space Solar Power Review* **1** (1&2), 9-38 (1980). The continued existence of a journal on a topic for which no federal funding is available illustrates the appeal of the idea.
13. Claud N. Bain, *Potential of Laser for SPS Power Transmission,* (PRC Energy Analysis Co., McLean, Virginia, September 1978; report to Department of Energy).
14. Department of Energy, *Program Assessment Report Statement of Findings Satellite Power Systems Concept Development and Evaluation Program,* November 1980.
15. Edward L. Safford, Jr., *Modern Radar: Theory, Operation & Maintenance,* 2nd ed (TAB Books, Blue Ridge Summit, Pennsylvania, 1981), p. 23.
16. Nalos, *op. cit.,* p. 283.
17. Samuel Koslov, "Radio-frequency radiation: the buildup of knowledge against the background of concern," paper 31.2 in *Conference Record International Conference on Communications* June 10-14, 1979, Boston, published by Institute of Electrical and Electronics Engineers, New York.
18. Daniel O. Graham, *High Frontier A New National Strategy* (High Frontier, Washington, D.C., 1982), p. 142.
19. *Ibid.*
20. Nalos, *op. cit.,* p. 280.
21. Tanimoto, *op. cit.*
22. Safford, *op. cit.,* p. 232.
23. Donald L. Lamberson, plenary paper TuLI titled "Department of Defense Directed Energy Program," presented May 17, 1983 at the Conference on Lasers and Electro-Optics in Baltimore. This quote is taken from the written text supplied by General Lamberson at the conference; the technical digest contains only a scant five-line summary.
24. Department of Defense, *Soviet Military Power 1983* (U.S. Government Printing Office, Washington, D.C., March 1983), p. 75. A similar statement appears on p. 75 of the first edition of this booklet, published in October 1981).

9. Countermeasures, Counter-Countermeasures, *Ad Infinitum*

We often see the arms race as a continual program to build bigger and "better" bombs to add to stockpiles that already are more than adequate to blow humankind into oblivion many times over. However, there is a subtler and in some ways more technologically demanding sort of arms race that has attracted much less attention: the game of "countermeasures," which today occupies the talents of many of the best engineers on both sides of the Iron Curtain.

The term "countermeasures" comes from the idea of taking measures to counter an enemy attack. As the name implies, the goal is to make enemy weapons ineffective. For example, during World War II, Britain developed radio countermeasures which could confuse the radio-based navigation systems of German bombers, causing them to get lost and miss their targets. The other side can try to defeat countermeasures by developing "counter-countermeasures" to thwart the countermeasures, or by shifting to new and different weapons that wouldn't be affected. In theory, the game could go on to counter-counter-countermeasures and still further levels of sophistication.

To an engineer, developing countermeasures is an elegant chess game where moves are made by refining technologies. The field presents the sort of technological challenges that stimulate the engineering mind. The inherent competition between the United States and the Soviet Union also spurs effort. As a result, today's countermeasures are an incredibly sophisticated black art–"black" both in the traditional sense of being mysterious and in the modem engineering slang of being highly classified. Many of the most refined countermeasures are electronic, intended for purposes such as obscuring the vision of radars, confusing navigation and fire-control equipment, or otherwise messing up the machinery of war.

Although the term "countermeasure" is a recent creation, the concept of trying to blunt the power of enemy weaponry is almost as old as military engineering. The medieval knight's suit of armor or chain mail was a simple and reasonably effective countermeasure against the prevalent weapons, swords, and lances. Armor was made obsolete by a new generation of weapons that could launch arrows and bullets with enough force to pierce the thin layer of metal. A similar cycle is being played out today as engineers increase both the penetrating power of artillery shells and the thickness of tank armor.

Countermeasures are not glamorous enough to have gotten much attention in science fiction, but one concept does deserve mention: the force shield, capable of preventing any enemy projectile or beam from passing through. In many stories the shield is somehow imperfect, and the plot turns on the nature of the imperfection and the events it causes. A truly impenetrable barrier might seem to be an ideal defense system for a peace-loving people, but no one knows how to build one.

Beam-Weapon Countermeasures

Beam weapons are not omnipotent enough to require force fields as countermeasures. A simple reflective surface would offer some protection against high-energy laser attack. That, however, is something which laser weapon developers are already considering; no one is foolish enough to think that the enemy would be helpful enough to paint all its missiles with laser- light-absorbing black paint.

The potential effectiveness of beam weapon countermeasures is far from clear. To some observers outside the Pentagon, the potential of countermeasures seems so great that they recommend building large lasers only to test countermeasures. [1] Others doubt that countermeasures would be so effective. They also point out that even if useful countermeasures could be developed, it would take time to put them into use, in line with an axiom cited in an Air Force textbook on electronic warfare:

> We should not fail to develop a technique because it has a simple counter. For we benefit from the inevitable delay between the time that the enemy is certain that we are using the technique and the time that he can make the counter operational. In that time we can be working on the next step in the ladder to maintain our technological superiority.[2]

At least in theory, several basic types of countermeasures could reduce the effectiveness of a beam weapon. They could operate in various ways:

- Blocking the beam or shielding the target, perhaps by causing a beam-absorbing material to get in the way
- Preventing damaging interactions between the beam and target by reflecting most of the beam energy
- Spreading the beam out over a large enough area that it couldn't do much harm, such as by spinning a missile (some existing missiles are said to spin in flight)
- Confusing the weapon system with decoy or false targets or by firing targets at a rate much faster or slower than it can handle
- Taking evasive action once beam weapon attack is detected
- Disabling the beam weapon or its fire-control system, for example by blinding the sensors it uses to detect targets
- Switching to types of equipment less vulnerable to beam weapon attack (for example, by replacing solar-cell panels on satellites with internal power supplies).

Effectiveness of countermeasures will depend on their nature and those of the target and beam weapon. If the beam could do its damage in an essentially instantaneous pulse, for example, it wouldn't do any good to spin the target or detect the beam because the first pulse could destroy it almost instantly. Such countermeasures would be useful if the beam had to stay pointed at the same point on a target for a long "dwell time." However, in this case "long" would probably mean a matter of seconds, generally not long enough to get out of the way.

Another way to look at countermeasures that is probably more relevant for beam weapons is to see what part of the weapon system they act against. Some are directed at the beam itself. Others are intended to foil the fire- control and/or beam-direction subsystems. To keep the descriptions of countermeasures manageable and in a fairly logical sequence, I will start with countermeasures against the beam, then turn to those directed more generally at fire-control equipment.

Laser Countermeasures

The simplest way to deflect laser attack is with a mirror. No mirror is 100% reflective, and the real world is full of other complications, but the high natural reflectivity of some materials can make it hard for laser weapons to do their job. Near room temperature, metals typically reflect most of the infrared light that reaches them, and thus absorb only a small fraction of the energy from an infrared laser beam. That small fraction is enough to heat the metal, and typically the absorption increases (and reflection decreases) as the temperature rises.

Many metals are very strong reflectors. Aluminum, a very common aerospace material, when polished, reflects about 98% of the 10-μm light from a carbon dioxide laser, and over 95% of the light at the 3.5- to 4-μm wavelengths of deuterium fluoride chemical lasers. The difference sounds small, but it means that roughly twice as much energy is absorbed from the shorter-wavelength beam. Another common aerospace material, titanium, has about 90% reflectivity at 10 μm and a little under 80% in the 4-μm region.[3] The metal's absorption is large enough that carbon dioxide lasers have become a standard tool for cutting tough sheets of titanium. The reflectivity of clean, polished metals generally drops at shorter wavelengths; aluminum reflects a little over 90% of the light in the 0.4- to 0.7-μm visible region, while titanium has 50 to 60% reflectivity in that region. This higher absorption has helped stimulate military interest in short-wavelength lasers. In a sense aluminum's high reflectivity is a built-in countermeasure against infrared laser weapons. This is important because for many years aluminum has been the most common material for the skins of aircraft and missiles. Lately, military researchers have been looking at the possibility of replacing aluminum with other materials, including nonmetallic composites that are lighter and stronger, and "stealthy" materials that make planes and missiles hard to spot on radar. This could create a problem if lasers make it to the battlefield because these new materials typically absorb more infrared light than aluminum, and hence would be more vulnerable to laser attack.

It might seem tempting to use a mirror-like metal surface to reflect the laser beam back at the laser to knock out its sensors and perhaps cause other damage, but the idea is not practical. Aerodynamics demand flat or smoothly curved surfaces, which would have to be aligned at a precise angle to reflect a large fraction of the laser beam back toward the laser, and even then the beam would spread out on the return trip so the intensity at the laser would be low. There are special "retroreflective" mirrors that return a laser beam along the same path it traveled to the mirror–but they aren't flat. Flatness also turns out to be desirable to minimize vulnerability to laser attack, although in any aircraft exposed to the real world, perfect flatness is unattainable. Small surface irregularities, such as dents, scratches, and junctions of metal plates, can be starting points for laser damage, although they may not cause other mechanical weaknesses. These irregularities are unusually vulnerable because of effects such as the concentration of optical power and weakening of the metal's structural lattice. Even dirt might be dangerous because it could increase the absorption of laser light.

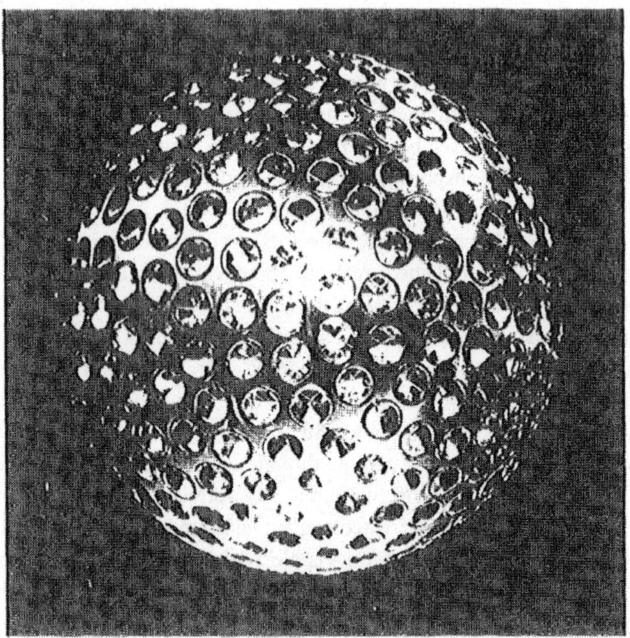

The Laser Geodynamic Satellite (LAGEOS I) is a satellite with built-in countermeasures that make it unlikely anyone will take pot shots at it with a laser weapon. Launched by NASA in 1976, the surface of the 2-ft diameter satellite is covered with 426 special mirrors called retroreflectors, which return incident laser beams right back where they came from. It is used with low-power laser ranging instruments for precisely measuring positions and altitudes on the earth; the surface is covered with retroreflectors to aid those measurements. That built-in defense did not come cheap–with each retroreflector costing about $600 in 1975, the total came to $250,000. The mirrors are not designed to withstand high powers or to concentrate the returned light onto a small area at the laser site. (Courtesy of Perkin-Elmer Corp.)

One way to counter this problem would be to polish the surface, an idea that conjures up images of soldiers polishing aircraft before going into battle. Another possibility is to apply a strongly reflective coating. Researchers have found that special coatings can enhance the infrared reflectivity of metals to as high as 99.1%, and theoretical predictions reach 99.8% reflectivity.[4] That does not mean such surfaces are invulnerable to laser attack. They can get dirty, and even when clean absorb a little light. Enough laser power can eventually heat them to temperatures where they absorb more light.

More complex countermeasures based on similar ideas have also been proposed. One suggestion is to enclose nuclear warheads in reflective balloons. Another is to spin missiles or warheads to disperse the laser energy over a larger surface area.[5] Both approaches have the disadvantages of requiring redesign of many potential targets. They might also complicate the task of guiding the ballistic missiles or nuclear warheads they are designed to protect. And both would be vulnerable to defeat by higher-power lasers.

The Pentagon is also looking at ways to block a laser beam from reaching its target. To a certain extent, this can occur naturally when an intense laser beam heats a surface enough to vaporize material from it. Some of the resulting vapor stays above the surface, where it absorbs more energy and is eventually ionized to form a plasma. This plasma, in turn, absorbs more energy from the laser beam. At high enough laser intensities, typically a few tens of millions of watts per square centimeter for pulsed carbon dioxide lasers, this plasma forms a wave that travels through the air back toward the laser. This "laser-supported absorption wave" keeps traveling along the beam until it reaches a point where there isn't enough power density in the laser beam to keep it going. Engineers using lasers in metalworking see the effect as a sharp noise and a bright flash of light, which may be repeated several times in a thousandth of a second. This phenomenon has a sharp threshold and can prevent the laser beam from reaching the surface, causing the amount of laser power delivered to the target to drop abruptly as the laser's output is increased.[6]

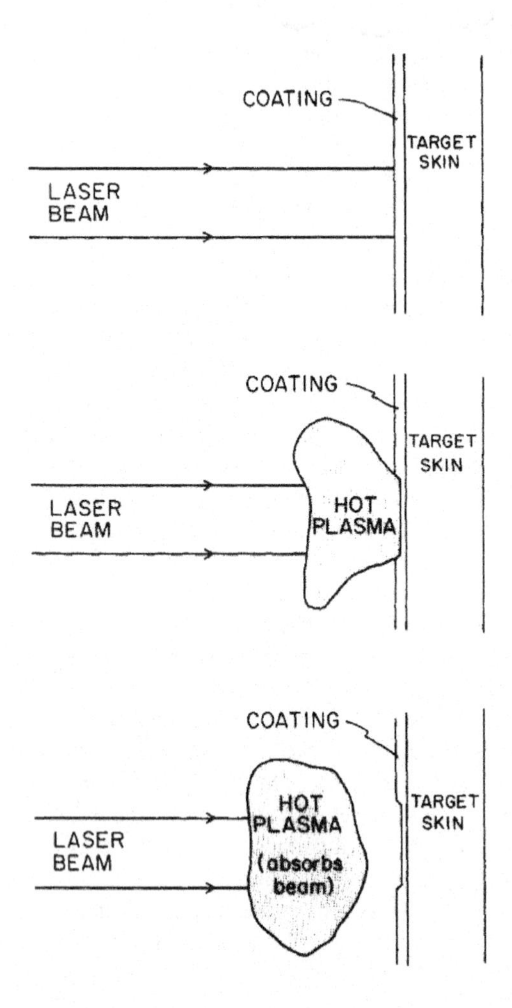

One conceptually simple laser countermeasure is to coat potential targets with a material that would evaporate when illuminated by a laser beam. The hot plasma produced would absorb the laser light, protecting the target by the sequence of events shown above. (Drawing by Arthur Giordani.)

Such plasma effects could serve as the basis for countermeasures against laser weapons by providing a hot plasma shield for the target. Military engineers have been looking at special coatings that would evaporate to produce just such a plasma shield when illuminated by a laser weapon. The details are complicated because formation of the plasma depends on an interaction with the surrounding atmosphere, which is strongly influenced by air pressure and speed of the target.

Plasmas are not always countermeasures, however. Under certain conditions plasmas can enhance the transfer of energy from the laser beam to the target, or cause other harm to the target. Ultraviolet lasers can continue heating a target even after a plasma has formed. High-temperature plasmas can also produce pressures up to 10 atmospheres, which can punch through a surface before the laser beam can melt through,[7] effectively reducing the amount of laser power needed to cause damage.

Another way to block a laser beam is to put something else into the atmosphere. A battlefield is normally a smoky place, and it is likely to become even smokier. The military has turned to infrared equipment because it can let soldiers see further through the dirty air of the battlefield than they can using visible light. However, military researchers trying to counter the advantages of infrared vision have updated the concept of the visible-light smoke screen to include ways of blocking infrared light as well. One example is "fog oil," produced by the inefficient burning of diesel fuel.[8]

The weather could also get in the way. The Pentagon is concerned about how fogs, haze, rain, snow, and cloudy weather, prevalent in Europe, would affect the operation of weapon systems that rely on optical or infrared sensors. The tiny droplets of water in fog obscure vision because they scatter light strongly, with the severity of scattering increasing as the wavelength decreases. Snow presents more complex problems; in the words of one military researcher, "it is not a nice simple phenomenon." Some observers have questioned the Pentagon's heavy reliance on optical and infrared systems, warning that weapons useful only in good weather are of limited use. This is particularly true of defensive weapons, because the enemy can decide under what weather conditions to attack.

The intense beams from high-energy lasers might be able to bum their way through smoke, fog, and other types of unfavorable weather that could block the vision of soldiers and sensors. The Pentagon has tested out some possibilities, including the use of scanning laser beams to dissipate fogs at airfields. There have also been tests of other types of hole burning. What is not clear, at least in the unclassified realm, is how effective the technique would be in improving the performance of weapon systems. The heating required to bore a hole through a fog bank could add to thermal blooming of the laser beam, or cause turbulence that might be hard to correct with adaptive optics. In the case of a rapidly moving target, it might be necessary to bore a new hole for each pulse in a series or to clear out a continuous swath of air between the laser and the target. Problems might also arise if the hole-boring process generated material that absorbed laser light, perhaps residues from smoke, or material ionized by the beam. And all this elaborate work in clearing a path for the high-energy beam might go for naught if the fire-control system couldn't "see" through the air to identify the target.

One intriguing potential "natural countermeasure" is lightning. Electrical discharges tend to travel through the atmosphere along paths with lowest electrical resistance, that is, those with the most ions (free electrons and atoms that have lost electrons, which conduct electricity because of their electric charge). If a high-power laser pulse generated enough ions, lightning could conceivably follow that ion path back to its source, the laser weapon. [9] The effect would certainly be striking (in more than one sense of the word), but there have been no reports of it happening in tests of laser weapons. Nonetheless, laser beams have been used to establish paths for electrical discharges and electron beams to follow in laboratory experiments.

If the range of countermeasures against laser attack was to include a counterattack against the laser, it would be helpful to know that a laser was being used and where it was located. Devices already exist that can tell whether a target is being illuminated by a laser beam, and can spot the direction from which the laser beam is coming. [10] Those capabilities would be of limited use if the laser could do lethal damage with a single shot lasting only a millionth of a second, but could be helpful if the damage took a matter of seconds. Then something else, perhaps another laser weapon, could launch a counterattack against the first laser weapon once its position was known.

The types of countermeasures that could protect sensors against laser attack are somewhat different from those needed to protect other potential targets. That is because sensors have to be protected against lower levels of laser light, which could blind or disable them without causing thermal or mechanical damage to other targets. There is a further complication in trying to protect optical and infrared sensors because they cannot do their jobs if they are sealed away where light cannot reach them. What is needed is a countermeasure that deflects light from an attacking laser but lets through the light the sensor needs to operate. Special types of optical filters do just that, reflecting virtually all of the light in one narrow range of wavelengths and transmitting most of the light elsewhere in the spectrum. This makes it possible to protect an infrared sensor from attack by the 10-μm wavelength of carbon dioxide lasers, while letting it "see" at other wavelengths so it can do its job. Such "rejection" filters are by no means invulnerable, but they are a simple way of providing some protection to the sensor itself. The problem is that you have to know the wavelength of the attacking laser in advance–when the sensor is being assembled–because the rejection wavelengths of a particular filter are fixed when it is made.

Another possible way to protect sensors is by covering them with materials that darken in response to intense light. Photochromic glasses used in spectacles that darken when exposed to full sunlight are familiar examples. The military has also developed "flash protection devices" that darken in a few millionths to a few thousandths of a second. The goal of such devices is to protect the eyes of bomber pilots and other soldiers so they are not blinded by the flashes of distant nuclear explosions. [11] This approach could be valuable in some cases but would be of only limited usefulness. Bomber pilots can be protected only against the flash blindness caused by light produced when the bomb explodes, an effect similar to the eye's response to a bright photo-flashbulb, *not* from nuclear radiation from the blast. The darkening of the glass (or other material) would block the sensor's vision until the laser beam went away. The light-absorbing materials would also be vulnerable to damage from a powerful laser because they protect the sensor by soaking up the light in the beam, not by reflecting it.

Particle-Beam Countermeasures

Countermeasures against particle-beam weapons have not been studied extensively, and are not as easy to envision as those against lasers. The major reason is that the particles in the beam would each carry a massive wallop, millions of times more energy than in a visible-wavelength laser photon. This energy is sufficient to let them penetrate deep into metals and most other potential target materials. If these energetic particles can make their way through the atmosphere or over the required distances in space, they could severely damage the target nearly instantaneously, and special coatings on the target surface could do little to stop them. In addition, clouds and rain are not expected to be any more serious a barrier to a charged-particle beam than the atmosphere in general, although smoke and foul weather could make the fire-control system ineffective.

There are some potential ways to block or divert a particle beam before it reaches its target. A thin layer of air could disrupt the travel of a neutral particle beam in space. One way that has been suggested to put such a layer of air high into space is to explode a small nuclear warhead near the top of the atmosphere, thus forcing some air up into the region between a space-based particle-beam weapon and its potential targets. Proposers of the concept say that after the neutral atoms had passed through the equivalent of only 1 μm (one millionth of a meter) of air at normal temperature and pressure, they would have lost their electrons, creating a beam of charged particles that could not propagate through space. [12] The explosion of a nuclear bomb in space could cause other problems, however–including some for the targets it was supposed to protect–so it is not clear if this approach is practical. It is worth noting, nonetheless, that there is normally a little air at altitudes above 100 km that could interact with a particle beam.

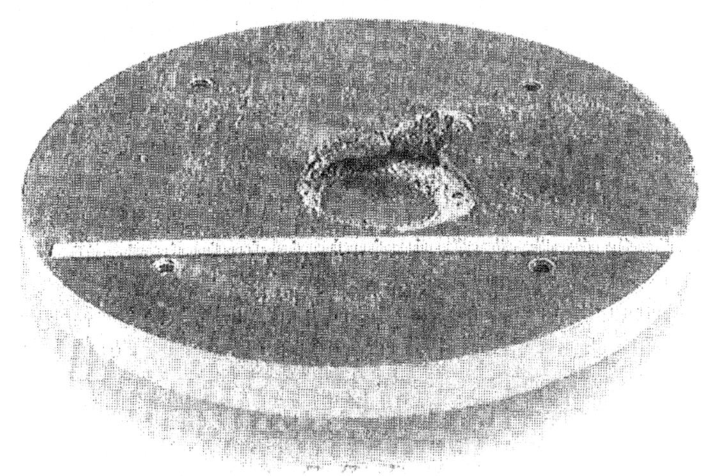

The damage caused to this 1-cm (0.4-in.) thick plate of aluminum shows why it is hard to develop countermeasures against particle beams. The bottom of this disk was hit by a beam of particles that penetrated about two-thirds of the way through, melting the metal and generating an intense shock wave, which ripped away pieces of metal from the back surface shown in the photo. (Courtesy of Air Force Weapons Laboratory, Kirtland Air Force Base, N.M.)

Beams of charged particles in the atmosphere could be deflected by strong magnetic fields. Nuclear explosions can generate such strong fields, although they would not be undertaken lightly. Another possibility is mounting magnetic devices inside potential targets.

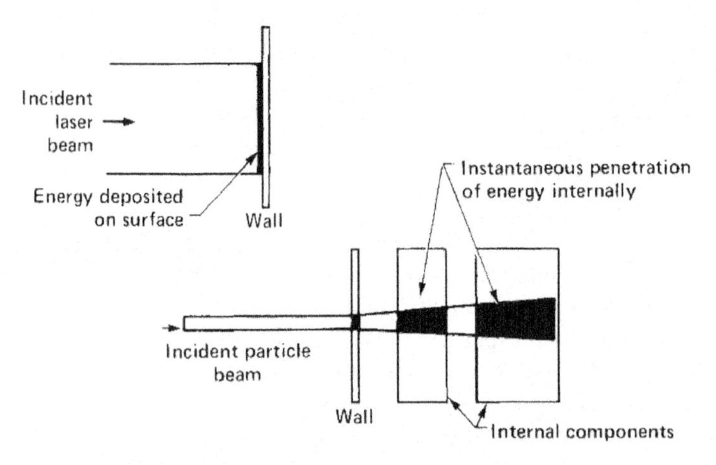

The reason it's so hard to defend against particle beams is that they penetrate much more deeply into targets than visible or infrared laser beams. Thus, the particle beam can pass right through the wall or skin of the target to damage internal components, while the laser beam must first bum a hole through the wall. (Courtesy of Lawrence Livermore National Laboratory.)

Most observers consider it futile to try to defend a target against a particle beam intense enough to zap a hole clear through a metal sheet. However, some shielding and "hardening" of military targets to resist attack by diffuse particle beams may be possible. The Pentagon's efforts to harden its hardware to resist "electromagnetic pulse" effects caused by nuclear explosions may pay some dividends in resistance to diffuse particle beams, although it is doubtful that the problem will be overcome totally.

X-Ray Laser Countermeasures

X-ray laser weapons would require their own distinct set of countermeasures. X-ray photons have roughly a 1000 times the energy of photons of visible light, but only about a thousandth of the energy of the energetic particles that would be used in particle-beam weapons. Unlike charged particles, X rays are not deflected by magnetic fields. Their penetrating power once they reach a target is intermediate between longer-wavelength lasers and particle beams. As mentioned in Chapter 6, ordinary materials essentially do not reflect X rays so they penetrate a modest to considerable distance, from a fraction of a millimeter to hundreds or thousands of times that distance depending on the wavelength and material. It should be somewhat easier to shield against *diffuse* X rays than against diffuse particle beams because X rays do not penetrate as deeply. Shielding against intense X-ray pulses is impractical because they would cause physical and shock damage to whatever they hit.

X rays do have a problem in penetrating air. X rays at the 14-Å wavelength of the supersecret Livermore laser would be able to make their way through only about 1 mm of air at sea-level density before half of them were absorbed. That absorption decreases somewhat as the wavelength decreases, but it remains a serious problem even at a wavelength of 1 Å, at the short-wavelength end of the X-ray region.[13]

The comparative rarity of countermeasures is one of the allures of X- ray laser weapons, although, as in the case of particle beams, that may reflect a general lack of attention to the problem as well as technical difficulties. However, the design of the weapon system described in Chapter 6 has an inherent weak point: you can't fire one X-ray laser at a time. To shoot a single X-ray laser shot, you would have to detonate the nuclear bomb intended to power a whole battle station's worth of X-ray lasers–in the process destroying the battle station. If X-ray laser battle stations are the only ones in space, an entire array of lasers would have to be fired just to destroy one target. This opens the possibility of a phased launch, one missile at a time, facing the laser controllers with the unpleasant choice of letting it go by, or of sacrificing an entire battle station that might be needed later to halt a larger volley of missiles. Getting around that problem requires a second missile defense system capable of picking off a few missiles at a time, something which advocates of X-ray lasers acknowledge would be a necessity.[14]

Microwave Countermeasures

The amount of work that has gone into countermeasures for microwave directed-energy weapons is comparable to that devoted to microwave weapons *per se*–in short, not much. It may be hard to defend against the cooking of targets with intense microwave beams, although gigantic antennas would be inviting targets for attack. At lower power levels there is a clear option: shielding. Microwaves are much easier to stop than X rays or particle beams. Military techniques for microwave shielding are fairly well advanced, a direct consequence of the military's heavy use of microwave equipment. Microwave shielding is even visible in consumer products–the sophisticated microwave ovens that put a microcomputer in the same case as a microwave oscillator producing a few hundred watts of power.

The Pentagon expects shielding to be used and realizes that ultimately the effects of microwave weapons will depend on how much microwave energy can leak through the shielding. But shielding isn't the only defense. Electronic signals can be converted into optical form and transmitted over fiber-optic cables that are immune to microwave interference. It may even be possible to replace some electronic elements with integrated optical devices now being developed in the laboratory.

Fire-Control Countermeasures

Fire-control requirements for beam weapons will be stringent, as noted in Chapter 5. The combination of these exacting requirements with the extensive work on electronic warfare on both sides of the Iron Curtain may make fire-control subsystems the part of a beam weapon most targeted by countermeasures.

Electronic warfare had its origin in World War I but didn't make its serious debut until the Battle of Britain during World War II. Britain and Germany engaged in a rapidly escalating struggle to improve the accuracy of–or to render harmless–air raids. At first the focus was on navigation systems, but attention quickly turned to radar as it was put into use first by Britain, then by Germany. [15]

Today, electronic warfare is a vastly more sophisticated field. Guidance, navigation, and radar equipment continue to be primary targets, along with communication systems and other equipment used to spot targets. There is extensive work both on jamming radar and communication systems and in devising jam-proof systems. As would be expected, much of the work is highly classified, as well as being sufficiently esoteric to be well beyond the purview of this book. Suffice it to say that as long as beam weapons use electronic fire-control systems, those systems will be the targets of electronic countermeasures, and will be defended by electronic counter-countermeasures.

Conventional microwave radars are likely to be used in some beam weapons, but stringent accuracy requirements will demand shorter-wavelength radars, operating in the millimeter-wave, infrared, or perhaps even visible regions. The shorter the wavelength, the more vulnerable the radar system becomes to bad weather, although the shift to shorter wavelengths gives better resolution. [16] Bad weather and smoke screens may turn out to be more deadly to fire-control radars than to directed-energy beams because the latter would have enough power to blast their way through. Indeed, on the battlefield beam-weapon fire control might have to rely on microwave radars, which could penetrate the murky atmosphere much better than shorter-wavelength systems.

Radars and sensors themselves are vulnerable to attack by both beam and conventional weaponry. There is a growing concern that microwave radars may be evolving into sitting ducks for enemy radar-seeking missiles. [17] Laser and millimeter-wave radars would emit narrower beams that would be harder for an enemy missile to home in on–but such radars would be vulnerable in other ways. Optical and infrared sensors can be overloaded by too much of the light they are looking for. Moderate-power lasers should be sufficient to knock out such sensors, particularly if the sensor uses a telescope arrangement to concentrate light. For some purposes it might be sufficient just to blind the sensors without causing permanent damage, in the same way that a pair of bright headlights can blind a person's night vision. A higher-power laser could put enough power into the sensor to cause permanent damage to the sensor and/or to the electronics that transform the sensor's output into a form meaningful to the rest of the system. As was mentioned earlier in the section on laser countermeasures, there are some ways to defend sensors against laser attack (which in this context would serve as counter-countermeasures), although they may temporarily take the sensor out of service.

Sensors and the electronic circuits attached to them are also vulnerable to other types of attack. High-power microwave beams, large electromagnetic fields generated by interference sources, and electromagnetic pulse (EMP) effects caused by nuclear explosions all could wreak havoc on electronics. Ordinary electromagnetic interference (EMI) is a continuing problem with electronic equipment because our environment is full of EMI noise sources, from CB radios and automobile spark plugs to electric pencil sharpeners and the proliferating variety of microwave communication systems. Semiconductor electronics are very vulnerable because they operate at low voltages; one friend lost a screen full of data from his personal computer when he used the electric pencil sharpener that was sitting beside the computer. Most military systems have at least some rudimentary shielding against these effects, but that may not be enough to protect against interference intentionally aimed at a sensor or electronic system.

Electromagnetic pulse effects were a quiet issue discussed in hushed tones in the Pentagon until discovered nearly simultaneously in 1981 by Janet Raloff of *Science News* and William J. Broad of *Science* magazine. [18] The basic problem is that the nuclear explosion generates intense electromagnetic fields, which radiate outward from the explosion site. The pulsed field is powerful enough to cause large, and potentially damaging, transient currents in electronic circuits far from the explosion. The effect is particularly severe for explosions in space, which could blanket large areas of the earth with high EMP levels and produce damaging surges in spacecraft.

The potential for temporary disruption of electronic circuits, such as the erasure of computer memories, is well known, as is the possibility of permanent damage to circuits. The worst problem, from the Pentagon's standpoint, is that the effect is not well quantified. In theory it seems possible to protect against EMP effects, but in practice very little equipment is protected. Noting the "almost universal dependence on electronic computers," nuclear physicist Edward Teller has written: "In the event of heavy EMP radiation, I suspect it would be easier to enumerate the apparatus that would continue to function than the apparatus that would stop."[19]

A few observers, notably William Broad, have jumped on EMP effects as a "killer" argument against putting beam weapons in space. In his words, "a single nuclear blast in outer space would instantly set up an electric pulse of up to a million volts per meter in hundreds of satellites and battle stations, zapping their solid-state circuits and ending their ability to wage war."[20] However, observers in the beam weapons community are skeptical that the EMP problem would prove fatal, pointing out that beam weapon developers have not bothered to address it because the Pentagon is concentrating on demonstrating technological feasibility, not on designing systems. Beam weapon developers will have the advantage of starting from scratch in dealing with the EMP problem. Some of the military's major concerns are focused on existing equipment that lacks any protection against EMP effects, an important distinction because one of the facts of life in engineering is that it is far more expensive to add a feature to existing equipment than to incorporate it into a new design. There have also been some recent reports that suggest EMP effects may not be quite as horrible as originally feared.[21]

A more conventional countermeasure against beam weapons would be to fool the fire-control system into shooting at the wrong target or to overload the system with more targets than it could possibly hit. A primitive example of this idea was the scattering of pieces of aluminum foil from British airplanes during World War II to confuse German radars. A more sophisticated approach would be to launch an attack that includes a large number of dummy or decoy targets as well as a smaller number of actual warheads. If a space-based beam weapon couldn't tell the difference between decoys and nuclear warheads during an attack, it would have to try to zap all potential targets, a much harder job than just knocking out the smaller number of bomb-carrying warheads. If the decoys overloaded the weapon system, some of the nuclear warheads would be able to get through.

Weapon Vulnerability

Beam weapons are not going to be omnipotent, whether on the battlefield or in space. Like any weapon system, they could be destroyed by an enemy with the willpower and the resources for a powerful enough attack. Beam weapons on the battlefield would be vulnerable to massive attack as well as to countermeasure blitzes. Space-based weapons would be particularly vulnerable during their launch and construction phases, presenting a serious temptation for the other side to attack before it would be threatened by the weapon system. After a beam-weapon battle station is completed, it would be inherently vulnerable to a more powerful beam weapon that could fire from outside of its range. There also may be other ways to disable space-based battle stations. Suggestions include putting ball bearings or a metallic mesh into the same orbit but going in the opposite direction, so they would collide with the battle station and destroy it,[22] or "mining" space with explosives that could be detonated on demand to disable the battle satellite.[23]

Counter-Countermeasures

An escalation from countermeasures to counter-countermeasures is inevitable if beam weapons are put into use, but much less has been said about the possibilities. A principal reason is that so far the most enthusiastic proponents of beam weapon (particularly laser) countermeasures have been the opponents of developing such systems, who have little reason to try to devise effective counters to their proposed countermeasures. The Pentagon has sponsored some work on counter-countermeasures, but predictably most of the details are classified. Thus, it is possible only to outline a few sketchy conceptual examples.

One possibility is the use of "tailored" laser pulses to penetrate efforts to block the beam, such as coatings that evaporate or that are very reflective. A series of short, intense pulses might be used to burn the coating off, or to dig more deeply into the material, with enough time left between pulses for the light-absorbing plasma to disperse. Or ultrashort pulses might be used to create some weak spots on the surface, which could be attacked more effectively by longer pulses or a continuous beam.

Another possibility that has been studied at low power levels is a "wavelength agile" laser; that is, one that can shift its wavelength. Pentagon researchers are considering such lasers for low-power target designators to make it harder to detect laser illumination because most existing detection methods require knowledge of the laser wavelength. The same tactic could be used at much higher powers to overcome efforts to filter out or reflect a particular wavelength. The ability to adjust the output wavelength might also help in "softening" the target, or in matching an absorption wavelength of the surface material. The best bet for the job seems to be the free-electron laser, the only type with good prospects for both adjustable wavelength and high output power.

It is clear that the game of countermeasures will be played with beam weaponry, but it is far from certain what the outcome will be, other than the expenditure of large sums to find out. The theoretical availability of simple countermeasures against laser weapons does little to convince military officials who can point to existing enemy weapons that have no protection whatever save for their aluminum skins. The practicality and effectiveness of countermeasures are clearly important issues, but many critical questions remain to be answered before the jury can return a verdict.

References

1. That is the major conclusion of a very pessimistic study of laser weapons: Michael Callaham and Kosta Tsipis, *High-Energy Laser Weapons: A Technical Assessment* (Program in Science and Technology for International Affairs, Massachusetts Institute of Technology, Cambridge, Massachusetts, November 1980).
2. Richard E. Fitts, ed, *The Strategy of Electromagnetic Conflict* (Peninsula Press, Los Altos, California, 1980), p. 39.
3. Figures taken from a chart in William L. Wolfe and George J. Zissis, eds, *The Infrared Handbook* (Office of Naval Research, Washington, D.C., 1978), pp. 7-81.
4. A. I. Braunstein and M. Braunstein, *Journal of Vacuum Science & Technology* 8, 412 (1971), cited in W. W. Duley, *CO_2 Lasers Effects & Applications* (Academic Press, New York, 1976), p. 107.
5. As might be expected, these proposals were made by an opponent of laser weaponry. Richard L. Garwin, "Are we on the verge of an arms race in space," *Bulletin of the Atomic Scientists* 37 (5), 48-53 (March 1981).
6. John F. Ready, "Material Processing-an overview," *Proceedings of the IEEE* 70 (6), 533-544 (June 1982).
7. Jack Daugherty, talk at Conference on Laser Systems & Technology, Boston, Mass., July 27-28, 1981, sponsored by American Institute of Aeronautics and Astronautics.
8. For a somewhat dated insight into military thinking on the subject, see: J. A. Boyd, D. B. Harris, D. D. King, and H. W. Welch, Jr., eds, *Electronic Countermeasures* (Peninsula Press, Los Altos, California, 1978) pp. 22-14-22-18. That book was originally classified SECRET when published by the government in 1961.
9. Laser weapons are not mentioned in the original proposal, Leonard M. Ball, "The laser

lightning rod system: thunderstorm domestication," *Applied Optics* 13 (10), 2292-2296 (October 1974). Ball suggested using lasers producing pulses only one billionth of a second long, considerably shorter than would be produced by laser weapons. In a recent letter to me, Ball stated that "full-scale demonstration of a laser lightning rod will be considerably more difficult than I had thought in 1974" because his earlier calculations had used some incorrect data. Although the concept has not been demonstrated, he believes it could be successful. Without experimental results, it is impossible to assess the likelihood of a laser weapon getting zapped by a lightning bolt, but if it happened it certainly would be impressive.

10. Paul T. Ballard, "Detecting laser illumination for military countermeasures," *Laser Focus* 17 (4), 72-80 (April 1981).
11. Flash protection devices are mentioned in passing in "Filter center," *Aviation Week & Space Technology,* June 14, 1982, p. 123; and in David Sliney and Myron Wolbarsht, *Safety With Lasers and Other Optical Sources* (Plenum Press, New York, 1980), pp. 138-141.
12. Concept proposed in G. Bekefi, B. T. Field, J. Parmentola, and K. Tsipis, *Particle Beam Weapons* (Program in Science and Technology for International Security, Massachusetts Institute of Technology, Cambridge, Massachusetts, December 1978), p. 53.
13. These are my calculations, based largely on data from Herbert L. Anderson, ed., *Physics Vade Mecum* (American Institute of Physics, New York, 1981), p. 73; for more details see Chapter 6, note 12.
14. Clarence A. Robinson, Jr., "Advance made on high-energy laser," *Aviation Week & Space Technology,* February 23, 1981, pp. 25-27.
15. For more historical details see: D. B. Harris, H. 0. Lorenzen, and S. Stiber, "History of electronic countermeasures," in J. A. Boyd , D. B. Harris, D. D. King, and H. W. Welch, Jr., eds, *Electronic Countermeasures* (Peninsula Press, Los Altos, California, 1978), Chapter 2.
16. Albert V. Jelalian, "Laser and microwave radar," *Laser Focus* 17 (4), 88-94 (April 1981).
17. Thomas A. Amlie, "Radar: shield or target?" *IEEE Spectrum* 19 (4), 61-65 (April 1982).
18. Both magazines published series of articles. Those by Janet Raloff in *Science News* were: "EMP: a sleeping electronic dragon," May 9, 1981, pp. 300-302; "EMP: defensive strategies," May 16, 1981, pp. 314-315; "EMP and electronic ignitions," May 30, 1981, p. 344; and "Would EMPs induce nuclear meltdowns?" June 6, 1981, p. 359. Those by William J. Broad in *Science,* all with the common title "Nuclear Pulse" were published in May 29, 1981, pp. 1009-1012; June 5, 1981, pp. 1116-1120; and June 12, 1981, pp. 1248-1251.
19. Edward Teller, "Electromagnetic pulses from nuclear explosions," *IEEE Spectrum* **19** (10), 65 (October 1982).
20. William J. Broad, "GAO ignores flaw in concept of space war," *Science* **216,** 499 (April 30, 1982).
21. "Radiation may be less dangerous to circuits than feared: DNA" *Electronic Engineering Times,* March 29, 1982 (note that DNA is an acronym for Defense Nuclear Agency, not the genetic material in living cells).
22. Michael Callaham and Kosta Tsipis, *High-Energy Laser Weapons: A Technical Assessment* (Program in Science and Technology for International Security, Massachusetts Institute of Technology, Cambridge, Massachusetts, November 1980), p. 76.
23. Richard Garwin, "Controversy flares on Soviet death ray," *Council for a Livable World Reports,* May 31, 1977, pp. 4-6.

10.
The Modern Military: High-Technology Warfare

About 15 years ago the late James Blish wrote a science fiction story in which there was a literal battle of Armageddon. Humankind had to fight a legion of demons from the traditional Christian hell. Mankind unleashed its machines of war, armies of robots, and automated weapon systems. The battle was long and hard, but ultimately Satan and his legions were banished from the earth. Then, as a reward for defeating the forces of evil, the mechanical armies that had fought the war were carried up to heaven, while the generals who had controlled the war from the sidelines were left behind in utter astonishment.

Blish's vision of mechanized warfare was less realistic than H. G. Wells's portrayal in *The War of the Worlds* written some 70 years earlier. Yet the compelling symbolism of Blish's story has stayed with me for a decade and a half, a potent reminder that warfare is far more than just a matter of hardware and tactics.

From a hardware standpoint science-fictional visions of robot warriors have missed the mark. No one seems to be talking about putting two-legged mechanical men such as C3-PO of *Star Wars* or Robby the Robot of *Forbidden Planet* onto the battlefield to replace human foot soldiers. Yet in a much more general sense, robots are very much on the minds of Pentagon planners. Military engineers talk about automatic fire-control systems, capable of spotting enemy forces and automatically shooting at them. They talk about "fire and forget" weapons, which need a soldier only to find a target and push a button; the weapon system does the rest, tracking down the target and destroying it. They devise elaborate schemes for "identification of friend or foe" that could automatically discriminate between enemy targets and friendly forces. They consider plans for putting "intelligent" computers on the battle- field to control weapons. They envision scenarios of war in space in which automated battle satellites take on fleets of self-guided intercontinental ballistic missiles. The coming generation of military hardware won't include any mechanical men, but there are plans for systems that could be labeled "robots" just as the term is used for bolt-tightening automatons on production lines.

What is happening is that warfare is being transformed from a battle of men into a battle of hardware. American strategists are seeking technological superiority to outweigh what they see as Soviet numerical superiority. Military officials are pushing for newer and ever more sophisticated technology, stimulating a massive arms race that is visible in most unsettling perspective in nuclear armament but stretching throughout the spectrum of military equipment.[1] The changes in military thinking are widespread. Military might is measured by looking at arsenals of nuclear bombs. Soldiers are equipped with an array of high-technology tools of destruction, not just rifles. Aircraft and a large family of sophisticated missiles are expected to play increasingly vital roles on the battlefield. Electronic, optical, and infrared systems guide missiles and other weapons to their targets. Massive investments are being made in electronic systems that will let commanders run a battle from the rear lines. The Reagan Administration is looking for new technologies that could provide a defense against nuclear attack.

Beam weapons fit naturally into this evolution of warfare. Not only are they sophisticated technologically, but their prime mission is destroying or disabling enemy hardware–not attacking soldiers. To paint a backdrop for the roles that beam weapons might play, this chapter briefly outlines current visions of high-technology warfare.[2]

Strategic and Tactical Weapons

Military science traditionally divides weaponry into two categories, strategic and tactical. New technology is permeating both areas, and directed- energy weapons are being considered for roles in both domains.

Strategic weapons are, loosely speaking, those intended to destroy the enemy's capability to make war. In practice, their function is largely to deter attack by threatening to cause widespread devastation in case the other side starts something. Traditionally, they have been aimed at strategic targets, defined as "any installation, network, group or buildings, or the like, considered vital to a country's war-making capability and singled out for attack."[3] However, the term "strategic" has come to be applied broadly to equipment designed to monitor or attack the other side's strategic weapons, a category that includes spy satellites, for example.

Tactical weapons are those intended to attack targets on a battlefield. They are generally used in a short-term effort to win a battle on the ground, at sea, or in the air. A tank or a rifle is a tactical weapon, while an intercontinental bomber or ballistic missile is considered a strategic weapon. There are some tactical missiles that are equipped with small nuclear warheads, so strictly speaking it is not correct to consider all nuclear weapons as strategic.

Beam weapons could be used for either strategic or tactical purposes. In practice, a beam weapon would be designed for one function or the other, not both. Long-range laser or particle-beam weapons proposed by President Reagan for defense against strategic bombers and missiles would fall under the "strategic" heading because their targets would be strategic weapons. Antisatellite weapons are also considered strategic because most satellites perform strategic functions, such as observing the other side's strategic forces or providing long-distance communications. Shorter-range beam weapons for use on the battlefield, in the air, or at sea would fall under the tactical heading, with the exception of ground-based systems for defense of fields of missile silos against nuclear missile attack.

Strategic Military Hardware

The key element in modern strategic arsenals is the nuclear bomb, intended for long-distance delivery to enemy territory by a bomber or missile. Defense strategists traditionally speak of a "strategic triad" with three "legs": land-based intercontinental ballistic missiles, submarine-launched ballistic missiles, and long-distance manned bombers.[4] Each of the three legs of the United States triad can deliver nuclear weapons to the Soviet Union, and each has its advantages and disadvantages. Land-based ballistic missiles are very accurate and can travel halfway around the globe in about half an hour, but the other side's spy satellites can ascertain their fixed locations, making them theoretically vulnerable to destruction, an issue that has been the focal point in the debate over plans for the MX missile. Submarine-launched ballistic missiles are less accurate, largely because of the inherent uncertainty in measuring a sub's position, but they are also very hard to detect and kill when hidden in a submerged submarine. Bombers, the oldest leg of the triad, are much slower to reach their targets than missiles and could be vulnerable to antiaircraft defense. But bombers can be launched on warning of attack to enhance their survival (without committing to an attack as is the case with missiles), have the flexibility inherent in a manned system, and can be adapted to carry traditional bombs or low-flying air-launched cruise missiles. A few observers have suggested calling cruise missiles a strategic defense "leg" because their ground-hugging trajectory is very different from the higher, arc-like path of a ballistic missile and renaming the triad a tetrad, but this idea has not caught on widely. The Soviet Union has a similar combination of forces, although it puts more emphasis on land-based ballistic missiles.

Satellites play a different, but nonetheless vital, role in strategic defense. Both the United States and the Soviet Union make widespread use of spy (or, in more polite language, surveillance) satellites to observe the activities of the other side and to monitor compliance with arms-control treaties. Cameras on low-altitude satellites, for example, have resolution as good as 15-30 cm (6-12 in.),[5] good enough to spot major construction projects on the ground. There are also electronic spy satellites that monitor communications signals transmitted by microwave links. Other satellites continually watch for any sign of nuclear attack, such as the hot exhaust gases that would betray the launch of a fleet of ballistic missiles. Satellites can also keep track of the other side's surface naval vessels so they can be targeted in the case of war. Both sides also make wide use of satellites to communicate with military facilities spread around the globe. Other military satellites perform functions such as aiding navigation, environmental monitoring, mapping, and making geodetic measurements to obtain information needed for missile guidance.[6] A study of a tabulation of satellites launched each year[7] shows that a large share are operated by military agencies. This heavy concentration of military satellites is of growing concern to some observers outside the United States and Soviet Union.[8]

President Reagan's proposal to develop new technology to defend against nuclear attack is only the latest in a series of schemes that have been suggested for attacking strategic missiles, bombers, and satellites. Most of these concepts have been shot down as impractical and/or costing far more than they are worth despite the high stakes involved. Before the Reagan proposal was made public in the spring of 1983, the United States and Soviet Union had been concentrating their public efforts on negotiating treaties to control strategic arms. The Soviet Union responded sharply to the Reagan proposal,[9] but both sides nonetheless continued negotiating over arms control. Those negotiations have continued over a number of years, with slow progress and many interruptions, but they have produced some results. The ratified SALT-I Treaty and a later addendum prohibit the United States and Soviet Union from deploying more than a single antiballistic missile system each, and ban the *use* of antisatellite weapons against the other side's spy satellites.[10] However, those limitations have not stopped research and development, or proposals for new approaches.

Beam weapons are only one of the candidates for such missions; actual spending for ballistic missile defense in fiscal 1983 fell short of the billion- dollar figure cited by President Reagan, and the bulk of that money went for technologies other than beam weapons.[11] Among beam weapons, the nearest- term prospects are for antisatellite lasers based on the ground, in aircraft, or in space, as described in Chapter 12. The development of higher-power lasers and of other directed-energy weapons as proposed by President Reagan for nuclear defense would take longer; the idea is described in more detail in Chapter 11.

Types of Nuclear War

Nuclear weapons play a pivotal role in modem military strategy. As strategic weapons, their most important role has been as signs of power to establish a balance of power. This balance of power might more appropriately be called a balance of nuclear terror. It is based on a strategy called Mutual Assured Destruction, or MAD for short. The object of the MAD game is to hold enough nuclear weapon chips to be able to wipe out the other side, even if that side attacks first. Thus, the Soviet Union would not dare to attack the United States first, for fear that even after a massive nuclear strike enough American weapons would remain to devastate the Soviet Union. (Soviet military planners rationalize their part in this strategy in the reverse–as a way to fend off possible American attack.)

President Reagan has said he wants to move beyond this MAD strategy to one based on active *defense* against nuclear attack. Other observers point out that although the MAD strategy may sound bizarre, it seems to have worked for a number of years, and it has staunch defenders inside and outside of the Pentagon. It lets both sides avoid the need for a "launch on warning" strategy that could result in a false alarm triggering a massive and irrevocable nuclear missile attack. And it explains why both nations have built up massive nuclear arsenals able to kill off the other side many times over.

There are other possible scenarios for nuclear warfare. One is a purely "tactical" war that would be limited in scope to a particular battle zone. This idea upsets many Europeans, who have a very legitimate concern that because of the proximity of NATO and Warsaw Pact forces, their homelands could become the battle zone in question. Other analysts also doubt that there could be any such thing as a limited nuclear war. Instead, they see the firing of the first nuclear shot as the trigger for increasingly severe reactions from both sides, culminating ultimately in a full-scale nuclear holocaust, with both sides throwing their entire arsenals at each other in a mutual orgy of destruction.

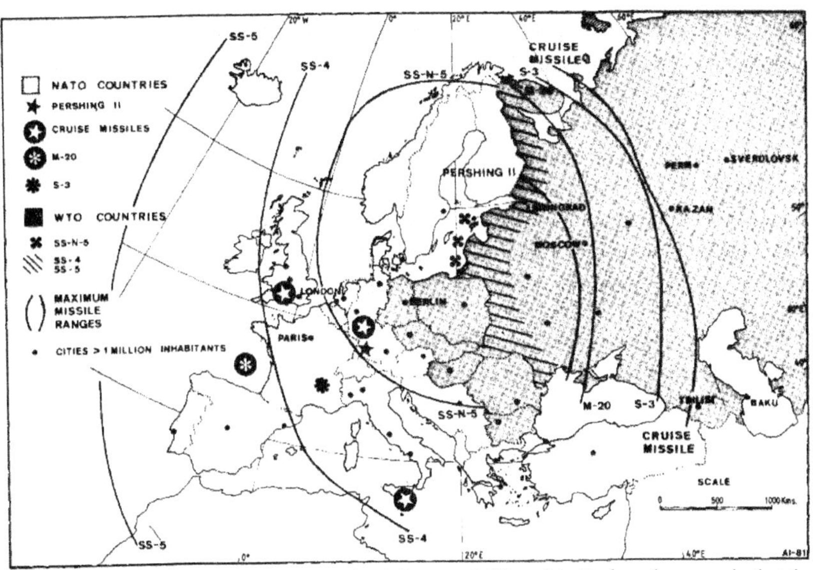

The reason why Europeans are upset about the possibility of tactical nuclear war is that they are surrounded by short-range or "theater" nuclear weapons. This map shows Warsaw Pact (SS series) and NATO (other) weapons deployed in Europe as of about 1981. (Courtesy of Stockholm International Peace Research Institute. From Ref. 1.)

Some military planners in the Reagan Administration have been talking quietly about another unsettling possibility–a protracted nuclear war. MAD-based strategy assumes that each side would use up all their nuclear weapons in the initial attacks, so that arsenals need only to survive the "first strike." In a protracted nuclear war those arsenals would have to last longer because the shooting wouldn't stop. Instead of launching their entire arsenals at once, one or both sides might launch only some of their nuclear weapons, perhaps in the hope that such a move would cause indecision or only limited retaliation. Thus, the Reagan Administration is putting "greater emphasis on obtaining enduring systems . . . to ensure that we are capable of retaliating with overwhelming strength under all circumstances . . . . we want to deny Soviet planners . . . every possible avenue of thinking whereby they could expect to obtain an advantage by initiating nuclear war," in the words of Fred C. Ikle, undersecretary of defense for policy.[12] Like the concept of a limited nuclear war, the idea of a protracted nuclear exchange is unsettling to many observers.

electronic equipment, with the usual stakes being information. It can involve attempts to jam or confuse radar and communication systems. It could also involve attempts to disable guidance or fire-control equipment so a weapon would miss its target, fail to fire, or detonate prematurely. Much of electronic warfare is devoted to electronic countermeasures (ECM) and electronic counter-countermeasures (ECCM), which are based on the principles of countermeasures described in Chapter 9. The proliferation of military electronics has made electronic warfare a vital area, which in honor of its importance is highly classified.

- *Fire-and-forget weapons* would automatically home in on a target after being fired, without any further help from a soldier. In a sense a bullet does that job, but this new generation of fire-and-forget weapons will contain equipment that lets them follow a moving target and automatically correct for going off course. Such automatic tracking is not yet practical, but the Pentagon hopes it could avoid the need for soldiers to guide weapons to their targets because the weapons can go off course if the soldiers duck.
- *Precision-guided munitions* are actively guided to their targets. An example is the "smart bomb," first used in the Vietnam War, which homes in on a laser spot on an enemy target marked by a friendly soldier with a low-power laser designator. This approach has its limitations because someone might start shooting at the soldier guiding the weapon, a problem which has prompted interest in fire-and-forget weapons. However, precision-guided munitions generally are much more accurate than unguided weapons, which has earned them a reputation for deadliness in military circles. Argentine commanders on the Falkland Islands reportedly decided to surrender after British forces let them overhear plans to drop laser-guided bombs on their headquarters. [19]
- *Avionic systems* control the operation of aircraft and the weapons they carry. In the past many have been mechanical or hydraulic, but electronic systems are becoming prevalent. Electronics are generally lighter and "smarter," but they are vulnerable to electronic warfare.
- *Guidance systems* tell an aircraft or missile where it is, where it's supposed to go, and how to get there. Impressive precision is possible over long distances, but these systems, too, can be the target of electronic warfare.
- *Ballistic missiles* follow what is called a ballistic trajectory, in which they soar high above the ground after launch, then plunge back toward the earth, pulled by gravity. Intercontinental ballistic missiles rise above the atmosphere. In this way they can travel long distances rapidly–between the United States and Soviet Union in about half an hour. Their trajectory above the atmosphere takes them out of the reach of conventional weapons that might be used against them.
- *Cruise missiles* fly low through the atmosphere, fairly close to the

earth's surface, making them hard to defend against and hard to spot on radar. Such low flight paths rely on a knowledge of obstacles (e.g., the terrain) between their launch point and their target. They can be launched from air, sea, or the ground and can carry nuclear or conventional warheads. Because it is simple to fly close to the flat surface of the ocean, they are considered particularly threatening to warships.

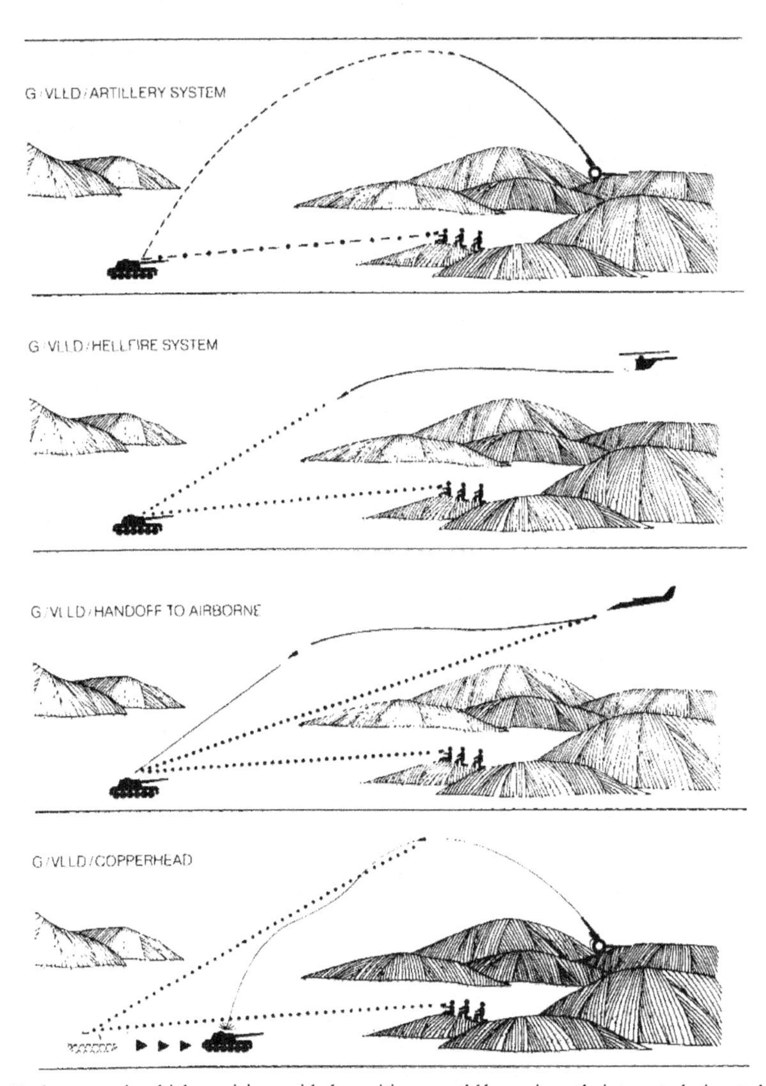

Various ways in which precision-guided munitions would home in on their targets designated by low-power laser beams (dotted lines). GLLD is an acronym for the Ground Laser Locator Designator, developed by Hughes Aircraft for the Army; Hellfire is a laser-homing missile launched from helicopters. (Courtesy of Hughes Aircraft, Electro-Optical & Data Systems Group.)

The Weapons Technology Issue

Where is all this technology taking us? High-level planners in the Pentagon say they hope it will help to deter warfare. Robert S. Cooper, director of the Defense Advanced Research Projects Agency, has written:

> The next generation of advanced surveillance systems and precision-guided stand-off weapons may provide a conventional military power so formidable as to rival in the tactical arena the deterrent effect nuclear weapons have had on strategic war.[20]

He added that the potential for attaining extremely high accuracy with some new weapons has implications "so profound that the tactic of massing forces, employed by military commanders for centuries, may no longer be feasible or effective on the battlefield."

By year 2000, Cooper considers it likely that the available technology will deter tactical warfare. This "will be based on advanced standoff weapons of superior intelligence vectored by airborne and spaceborne automatic surveillance sensors with near-human qualities of cognition."[21] Translated from the Pentagonese, that means that weapon and surveillance systems would have enough electronic brainpower to pick out targets and kill them with nearly 100% efficiency. That inevitability of destruction presumably would convince armies that they'd rather stay home.

Needless to say, not everyone shares this rosy view of new military technology. Critics of the military cite a variety of problems. They doubt that there is any such thing as an "ultimate" weapon that can prevent war. They see large military expenditures as a dangerous and nonproductive drain on the economy, which can divert money from more urgent human needs. There is particular concern about Third World countries, where heavy military expenditures are draining severely limited resources. There are also concerns that this technology can be used as a tool of oppression in less developed countries, either by their rulers or by the rulers of other countries equipped with the technology. These concerns are important, but this is not the place to go into them in depth.

Other military critics have taken a different view, saying that the problem with sophisticated technology is that it may not be able to do the job.[22] In their eyes the more complex the weapon system, the more likely it is to fail. There have been some notable problems lately, as elaborately designed weapon systems have failed to live up to promised specifications. One such case was the Copperhead laser-guided artillery shell, which the Army halted production of in late 1982 at the request of Congress. In six years of development and two years of production, the project had consumed some $630 million. Original plans called for production of 44,000 shells; with development costs figured in, the price tag would have run well over $20,000 each. That's a lot of money for a tank-killing artillery shell, and to justify it, Army specifications called for Copperhead shells to hit their target 80% of the time at a 16-km (10-mile) range–so one or two such shells could do a job that otherwise would take hundreds of conventional shells. In mid-1982 tests only 67% of the shells hit their targets, results poor enough that Congress felt they weren't worth their price.[23] In the spring of the following year, Secretary of Defense Weinberger and prime contractor Martin-Marietta managed to revive the program, citing better results from later tests.[24]

Problems with the Army's new M1 tank have attracted wide attention in the general press. A report from the General Accounting Office complained that the tank kept breaking down, and some observers labeled it a "lemon." The tank is so full of sophisticated hardware that the *Washington Post* quipped that the M1 "does everything except run reliably." However, military officials and important Congressmen continue to defend the program, and the tank remains in production.[25]

The inside of a modem tank is filled with sophisticated hardware, as indicated by this photo of a crewman contemplating the fire-control system in an M60A3 tank. This system uses a laser to help aim the tank's cannon at targets. This is not the Army's most modem tank; that is the new MI. (Courtesy of Hughes Aircraft, Electro-Optical & Data Systems Group, maker of the fire-control system.)

Concern that the Pentagon is becoming over-reliant on oversophisticated hardware was raised by James Fallows in his book *National Defense.*[26] Fallows advocated buying many simple and inexpensive weapons rather than concentrating resources in a small quantity of sophisticated and expensive equipment. He asserted that because of their complexity, the more sophisticated systems were more likely to break down under the harsh conditions of the battlefield, leaving soldiers without vital equipment. Simpler weapons, however, would be less likely to fail and easier to repair if they did. What's more, the Pentagon could afford many more such weapons than they could sophisticated systems. Fallows is by no means the only person to note that the price tags of weapons have soared into the stratosphere as military engineers have added more features, sometimes called "bells and whistles."

Although there seems to be some sympathy for Fallows' position within the Department of Defense, there has been little sign of any significant shift in direction. The type of high-technology missile defense advocated by President Reagan, for example, would be extremely complex.

Weapon system designers have also been criticized for becoming so enamored of technological gimmickry that they forget the importance of the relationship between soldier and weapon. Fallows cites some weapons that soldiers disliked enough that they tended to "forget" them. That's something of a timeless phenomenon. Military commanders over the ages have noted that soldiers often tend to lose equipment they don't like. There also have been concerns about equipment such as certain laser designation and fire-control systems, which require an operator to stand at a spot where the laser beam could travel a straight line to designate a target such as a tank, a spot at which the soldier could find *himself* exposed to hostile fire. However, this equipment has yet to be tested in a real war.

There is also a more subtle problem associated with the Pentagon's strategy of relying on technological superiority in some areas to offset what is seen as Soviet superiority in military manpower. Keeping a technological edge requires that the new technology be kept out of the hands of the Soviet Union. This gets difficult because much technology used in sophisticated weapon systems also exists in the civilian world, where it is not subject to classification rules. For example, sophisticated electronic circuits that can be bought commercially in the United States, Japan, and much of Western Europe may also be useful in missile guidance and fire control.

Government officials believe that the Soviet Union is making a major effort to acquire American technology. To try to keep critical military technology from leaking to the Soviets, the United States and many of its allies impose controls on what hardware can be exported. This is a controversial and not always effective move because much of the technology is also available from other countries that do not have such strict controls. Some businesses complain that the main effect of the controls has been to shift Soviet business to countries such as Japan and France, hurting American business but not preventing the Soviets from acquiring new technology.

The Department of Defense has also made some periodic crackdowns on the presentation of unclassified papers at technical symposiums. [27] These restrictions are based on government claims that export control rules can be applied to technical information as well as hardware–if the information is disclosed to foreign nationals attending an unclassified symposium in the United States. So far these threats have been effective in persuading many people to limit distribution of technical papers, presumably reflecting a general unwillingness of developers to upset the military agencies that pay for most of their work. Legal sanctions have been threatened, but at this writing none have been imposed, and their legal validity remains untested in the courts.

The Arms Race and Arms Control

Precise figures on world military spending are impossible to come by, but it has been estimated that the world's governments spent some $600 to $650 billion on military programs in 1981.[28] Many economists claim that this heavy spending is draining the economies of both the United States and Soviet Union, and some point to Japan's low military budget as a major reason for that country's rapid economic growth. They say that economic concerns alone are reason to cut back the arms race.

There are many other reasons as well, but halting the arms race is much easier said than done. There are powerful forces in both the United States and Soviet Union that point at the other side and say they can't be trusted. It has proved very difficult to get the two sides to sit down at a bargaining table and negotiate in good faith toward meaningful arms control.

Some tentative steps toward arms control have been taken.[29] Many of them are outside the scope of this book, but a few (as described in Chapter 16) are relevant to beam weaponry. That was not their original intent; indeed, much of the meaningful progress was made before beam weapon concepts had been developed. Still, the wording of arms-control agreements such as the limitations on antiballistic missile systems does appear relevant to beam weaponry. Agreeing on a precise interpretation of this wording may be difficult.

The Role of Beam Weapons

The technological feasibility of beam weaponry can be considered by itself. However, it is impossible to consider the military usefulness of such systems in isolation. The development of beam weapons clearly will be a major step in the arms race, particularly as it relates to defense against nuclear attack. Beam weapons may bring important new capabilities to military strategists. Their technology differs from most existing military equipment and to some seem best suited for strategic defense or battlefield use. Yet the adaptation of this new technology for military purposes is only a logical step along the road to ever-more-sophisticated weaponry for both offense and defense. The following chapters look at the military roles this new technology could find and the impacts it could have.

References

1. See, for example: Stockholm International Peace Research Institute, *The Arms Race and Arms Control* (Oelgeschlager, Gunn & Hain, Cambridge, Massachusetts, 1982).
2. The development of military technology is a serious issue in its own right that cannot be treated in any depth here. For a starting point, see some of the other references in this chapter.
3. Daniel N. Lapedes, ed, *McGraw-Hill Dictionary of Scientific and Technical Terms* (McGraw-Hill, New York, 1974).
4. The Organization of the Joint Chiefs of Staff, *U.S. Military Posture for Fiscal Year 1983*, pp. 19 and 71. This document was published at the end of the first volume of the transcripts of Hearings before the Senate Armed Services Committee, *Department of Defense Authorization for Appropriations for Fiscal Year 1983, part 1; Posture Statement and Economies and Efficiencies in the Defense Budget* (U.S. Government Printing Office, Washington, D.C., 1982).
5. Bhupendra Jasani, ed, *Outer Space-A New Dimension of the Arms Race* (Oelgeschlager,

Gunn & Hain, Cambridge, Massachusetts, 1982), pp. 46-47.
6. Joint Chiefs of Staff, *op. cit.,* p. 83.
7. A tabulation of satellites launched appears in J. N. Matthews, ed, *TRW Space Log* (TRW Defense and Space Systems Group, Redondo Beach, California, 1981), which is updated annually.
8. Bhupendra Jasani, "How satellites promote the arms race," *New Scientist* **96**, 346-348 (November 11, 1982).
9. Dusko Doder, "Andropov calls ABM proposal 'insane'," *Boston Globe* March 27, 1983, pp. 1.
10. For brief summaries of arms-control agreements in effect, see Stockholm International Peace Research Institute, *The Arms Race and Arms Control* (Oelgeschlager, Gunn & Hain, Cambridge, Massachusetts, 1982), pp. 221-233; for the full texts and details of implementation see J. Goldblat, *Agreements for Arms Control: A Critical Survey* (Taylor & Francis, London, 1982).
11. Christopher Joyce, "Reagan's ray guns are decades away," *New Scientist* 97, 871 (March 31, 1983).
12. *Strategic Force Modernization Programs,* Hearings before the Subcommittee on Strategic and Theater Forces of the Senate Armed Services Committee, 97th Congress, first Session, p. 73; the testimony was given October 27, 1981.
13. Caspar W. Weinberger, *Secretary of Defense Annual Report to Congress Fiscal Year 1983* (U.S. Government Printing Office, Washington, D.C., February 8, 1982), p. 1-28.
14. *Ibid.,* p. B-5; final figures approved by Congress differed slightly from these numbers.
15. "DoD official: technology Jag now double what it once was," *Electronic Engineering Times,* November 22, 1982, p. 31.
16. An interesting starting point for in learning more about battlefield technology is Paul Dickson, *The Electronic Battlefield* (Indiana University Press, Bloomington, 1976). The field is also covered regularly in trade magazines such as *Aviation Week & Space Technology* and *Defense System Review.* There is a wealth of information in government documents such as the Annual Report of the Secretary of Defense and transcripts of hearings before the House and Senate Armed Services Committees, but it takes some digging to extract it. Be wary, however, of accepting statements of opinion as facts; this is a field full of controversy, and opinions vary widely.
17. For an introduction to this complex field, see David R. McMillan, "Introducing C^3I," *Astronautics & Aeronautics* 20 (7/8), 48-54 (July/August 1982).
18. Caspar W. Weinberger, *op. cit.,* p. III-77.
19. Mark Hewish, "Laser bombs forced the surrender at Port Stanley," *New Scientist* 96, 483 (November 25, 1982).
20. Robert S. Cooper, "The coming revolution in conventional weapons," *Astronautics & Aeronautics* 20 (10), 73-75 & 84 (October 1982).
21. *Ibid.*
22. See, for example: James Fallows, *National Defense* (Random House, New York, 1981); similar arguments appear in Mary Kaldor, *The Baroque Arsenal* (Hill & Wang, New York, 1981).
23. "Army stops Copperhead laser-guided shell production," *Lasers & Applications* **1**(3), 20-22 (November 1982).
24. "Late news," *Lasers & Applications* 2 (5), 4 (May 1983).
25. Hearings before the Senate Armed Services Committee, 97th Congress, 2nd session, on DoD Authorization for Appropriations for Fiscal Year 1983, Part 4 Tactical Warfare, pp. 2368-2433.
26. James Fallows, *National Defense* (Random House, New York, 1981); similar arguments appear in another book which was published at about the same time but got much less attention, Mary Kaldor, *The Baroque Arsenal* (Hill & Wang, New York, 1981).
27. The most visible sign of the crackdown was the withdrawal of about 100 unclassified papers from an August 1982 symposium sponsored by the Society of Photo-Optical Instrumentation Engineers, described in " 'Remote censoring:' DOD blocks symposium

papers," *Science News* **122,** 148-149 (September 4, 1982). There were also a number of less-publicized incidents, some involving individual scientists; see, for example: C. Breck Hitz, "Editorial," *Lasers & Applications* **1** (3), 16 (November 1982), and "Restrictions on technical papers raise concerns," *Aviation Week & Space Technology,* January 17, 1983, pp. 22-23. Most of the concern focuses on unclassified work performed by defense contractors, who usually are willing to forego public presentations of their work if they fear such discussion might hurt their chances of getting future contracts.

28. Stockholm International Peace Research Institute, *The Arms Race and Arms Control* (Oelgeschlager, Gunn & Hain, Cambridge, Massachusetts, 1982), p. 20.
29. For an overview of progress in arms control, see *The Arms Race and Arms Control* (Ref. 28).

11. Defense Against Nuclear Attack

Today's balance of strategic power is an uneasy balance of nuclear terror. Both the United States and the Soviet Union have nuclear arsenals more than adequate to wipe the other side out. What seems to have kept the peace is a theory called Mutual Assured Destruction, or MAD for short. Its central thesis is that neither side dares attack the other because even after such an attack, the victim would have enough nuclear firepower to blow the attacker off the map. However unsettling that strategy may sound, in the absence of an effective way to defend against nuclear attack, there really hasn't been much of an alternative. New technology may change that.

President Reagan's call to develop systems capable of defending against strategic ballistic missiles is an effort to move away from the MAD strategy. There is an undeniable philosophical appeal to a military strategy that relies on defensive power to convince an opponent his weapons will be useless rather than on offensive power that only threatens mass destruction. In Reagan's words, "my advisors . . . have underscored the necessity to break out of a future that relies solely on offensive retaliation for our security. I have become more and more deeply convinced that the human spirit must be capable of rising above dealing with other nations and human beings by threatening their existence."[1] He concluded his March 23, 1983, speech by saying that he was ordering "a comprehensive and intensive effort to define a long-term research and development program to begin to achieve our ultimate goal of eliminating the threat posed by strategic nuclear missiles. This could pave the way for arms control measures to eliminate the weapons themselves."[2]

Even Reagan acknowledged that his goal might not be reached by the end of the century. His critics charged that it was much farther away, and perhaps even impossible.[3] They countered his promises of defensive security by saying that offensive weapons were much cheaper to build. And they warned that any attempts to move away from the present MAD posture might dangerously destabilize the balance of power. One went so far as to say that the Reagan proposal was part of an effort by "Nuclear Warfighters" in the Administration to build capabilities for an American nuclear first strike against the Soviet Union.[4]

There are undeniably major technological obstacles to building systems for nuclear defense. Years of effort and some $10 billion of the taxpayers' dollars[5] have failed to produce an effective defense against the current generation of offensive nuclear weapons carried by missiles and manned bombers. In the face of these problems MAD has been semiofficially institutionalized, with initial efforts to limit strategic arms focusing on limiting the deployment of antimissile systems. For the United States, at least, the decision to limit antimissile systems stemmed from a conviction that such hardware would be impractically expensive to put into operation, and perhaps ineffective as well. Existing treaties limit each side to one operational antimissile system. The Soviet Union maintains a system it built around Moscow, but the United States abandoned in 1976 the Safeguard installation it built around missile silos near Grand Forks, North Dakota. Under the treaties *research* on (as opposed to actual deployment of) ballistic missile defense is permitted, and both the United States and Soviet Union have vigorous research programs. The U.S. effort existed long before Reagan's speech. The program is fragmented, and thus hard to count precisely, but the total budgeted for fiscal 1983 seems closer to $700 million than the $1 billion figure estimated by the Reagan administration.[6]

Directed-energy weapons are an important part of the ballistic missile defense research program, although they account for only one-quarter to one- third of the current spending. Their allure is their potential to strike swiftly over long distances against such targets as intercontinental ballistic missiles in their "boost" phase when they are being accelerated up out of the atmosphere. Other potential targets include high-flying long-distance bombers and submarine-launched nuclear missiles. Estimated timetables for actually building effective missile defense systems using beam weapons vary widely. Advocates in the Reagan administration are looking toward the year 2000, although some others are much more optimistic and would like to see a crash program to build beam weapons within a decade. There are also many pessimists, particularly outside the Pentagon, who tend to dismiss the likelihood of beam weapons operating in the next few decades, or even consider the whole idea impossible.

Missile Defense Concepts

The technological difficulties of ballistic missile defense are inherent in the nature of the task. The targets, particularly nuclear warheads on their way to their targets, are small and fast, making them hard to hit with projectile weapons, such as other missiles. In theory, missiles could be destroyed anywhere along their path: as they are boosted into space, as they travel through space toward their targets, or as they speed through the atmosphere just before reaching their targets The picture is complicated by the use of one missile to launch multiple warheads (MIRVs, or Multiple Independently-retargetable Re-entry Vehicles), which separate after the launch to create multiple targets for a defense system. practice, there are serious problems trying to hit a missile or warhead anywhere along its path.

The United States began studying ballistic missile defense ideas around 1960. Because space technology was still embryonic, efforts focused on destroying the missiles near their targets, either shortly before they re-entered the atmosphere, or as they plummeted through the air. What started as the Nike-X program eventually evolved into the Safeguard system, which was briefly put into service around U.S. missile silos in North Dakota in the mid 1970s. The system was far from the nuclear umbrella proposed by the Reagan administration.

Safeguard consisted of large radars to precisely track the incoming warheads and two types of defense missiles. A large Spartan missile carried a nuclear warhead and was designed to fly several hundred miles out into space to destroy attacking missiles. The smaller, shorter-range Sprint missile was intended for defense at altitudes below about 16 km (50,000 ft on 10 miles). This was the first example of what has come to be called a "layered" defense in which attacking missiles that survive the first layer of defensive missiles would be targeted by a second layer.

In 1971 the Pentagon began working on another missile defense system, specifically intended to protect Minuteman missile silos, which had been "hardened" to survive a nearby nuclear explosion. The program relied on low-altitude defense with an upgraded version of the Sprint missile, and by the time it was completed in September 1980, it would teach the Pentagon a lot that wasn't known when Safeguard was deployed in 1975.[7]

Meanwhile, the Strategic Arms Limitation Talks were bearing fruit. In 1972 the Anti-Ballistic Missile Treaty limited both the United States and Soviet Union to building two missile defense systems, one around a missile launch site and the other around the national capitol. A 1974 addendum reduced this to one site per country.[8] The United States chose to protect its missile launching silos in North Dakota, while the Soviet Union built a system around Moscow. In 1976 American officials concluded that the Safeguard antiballistic missile system was not worthwhile and abandoned it.

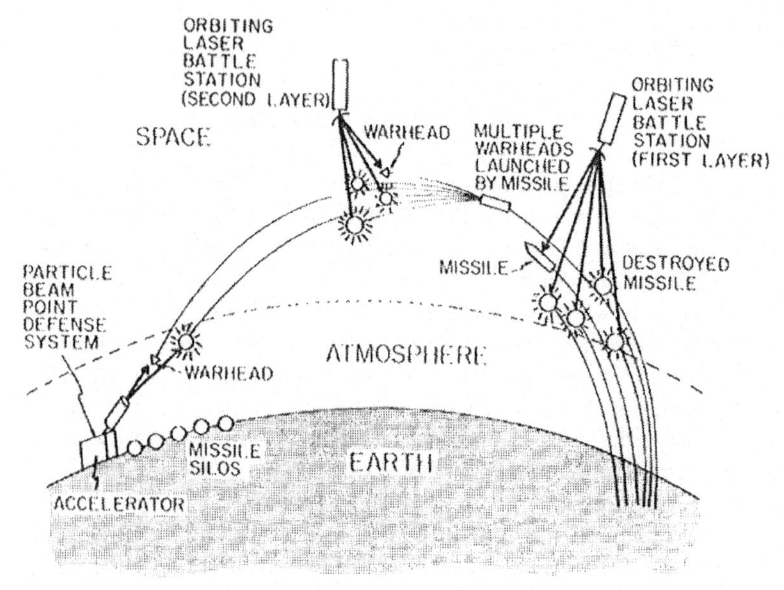

The basic idea of a "layered" missile defense system is to make attacking missiles run a gauntlet of defensive systems. As shown in this sketch missiles would be attacked as they rose out of the atmosphere by a laser battle station. Missiles that survived this attack would then launch their multiple re-entry vehicles, which could be attacked during flight by a second layer of battle stations. Any warheads that survived might have to face ground-based point defenses around strategic targets such as fields of missile silos. The central theory is that even if no single layer could stop all the warheads, together they should be able to stop most of the warheads. This diagram assumes lasers and particle beams would be used, but in theory other types of defense could be used in space or on the ground. (Drawing by Arthur Giordani.)

That decision was not without reason because the antimissile missile approach has inherent problems. Very large, sophisticated, and vulnerable radars are needed to track the incoming missiles with the required precision. If radars are knocked out, the antimissile missiles are useless. The defensive missiles have to react quickly, but accelerating them to the required velocity takes time, as exemplified by the long time it seems to take a spacecraft to lift the first meter or two off the ground. The incoming re-entry vehicles would be numerous and hard to hit. And explosions of nuclear warheads on the antimissile missiles would generate intense electromagnetic pulse effects across large areas, perhaps enough to have disabled Safeguard itself and other weapon systems.

Safeguard was also limited to defending missile silos, which are much easier to protect than cities. Silos are "hardened" so they can survive the effects of a nuclear blast in the vicinity (although not a direct hit). Cities, in contrast, are fragile targets. Where a Minuteman-type missile silo can with- stand nuclear blast shock waves with forces of 1000 to 2000 pounds per square inch, cities and their inhabitants can survive pressures of no more than 1 to 2 pounds per square inch (0.7 to 1.4 Newtons/cm^2). That is important because antimissile missiles might detonate the warheads on attacking re-entry vehicles, or have nuclear warheads themselves. For a city to survive nuclear explosions, the bombs would have to be detonated at least 14-23 km (9-14 miles) away, while comparable-size explosions could occur within 370-460 m (1200-1500 ft) of a hardened silo without causing serious damage to the missile inside.[9] This, of course, neglects long-term effects of radioactive fallout or short-term effects on anyone unfortunate enough to be 14 miles in the wrong direction from the city being defended.

The Pentagon has been working on a new ground-based missile defense system called Low Altitude Defense (LoAD). The goal has been to have technology ready to defend silos containing MX missiles, if military planners decide such defense would be needed to make sure the missiles could survive an attack, and if Congress and the Pentagon finally agree on a place to put MX. Official statements say, "the current LoAD preprototype development program is being accelerated within the constraints of the ABM treaty,"[10] but do not explain how.

Plans call for LoAD to use modest-power radars. The defensive missiles would probably be equipped with nuclear warheads because they are more likely to "kill" attacking re-entry vehicles. Pentagon officials have estimated there is about a 50-50 probability of developing suitable nonnuclear warheads in the near future[11]–a desirable achievement in order to avoid the EMP and radioactive fallout from nuclear warheads. Military planners have tried to avoid estimating the cost of a system to defend MX, with rough guesses running in the $10 billion to $25 billion range.[12] All this money would only indirectly defend civilian populations, by threatening to foil a Soviet attack intended to destroy American capabilities to devastate the Soviet Union. The logic may sound tortuous, but that is the type of thinking involved in defense strategy.

Space-Based Defense Systems

Military analysts inside and outside of the Pentagon have also been considering basing ballistic missile defense systems partly or entirely in space. Although the technology is not easy, the concept does have real attractions. Space-based systems could defend cities as well as hardened military sites because they would destroy nuclear warheads in space, far from the vulnerable cities. They could also serve as more layers in a "layered" defense system.

Space-based systems do not have to rely on beam weapons. An ambitious proposal based on projectile weapons has been made by a Washington-based group called High Frontier, which was supported by a grant from the conservative Heritage Foundation. The High Frontier proposal[13] for a "global ballistic missile defense" system includes an array of orbiting "trucks," each carrying 40 to 45 attack vehicles. In the example cited in the study, 432 such "trucks" would be in orbits 300 nautical miles above the earth's surface, inclined 65° to the equator to let them cover most of the Soviet Union's land mass. Each of the attack vehicles would include a rocket to accelerate it to a speed of about 1 km (3000 ft) per second relative to the "truck." The "kill vehicles" would destroy their targets by the force of their impact without need for a nuclear explosion.

The ideal targets for this system would be missiles in their boost phase that have yet to launch warheads and decoys that would multiply the number of targets. By putting many of the defensive satellites in orbit, the High Frontier proposal would assure that some of them would always be above the Soviet Union, ready to counter a missile attack. This boost-phase attack would thin out the ranks of missiles.

Warheads and decoys launched by missiles that survived the first layer of defense could be attacked during their flight through space by other kill vehicles that were at points in their orbits between the United States and Soviet Union. Hitting such targets is thought to be more difficult than hitting boost-phase missiles, but it could further thin out the ranks of attackers.

There would also be a ground-based system to defend selected prime targets such as fields of missile silos. The original proposal called for an attack on such a site to trigger the launch of a swarm of some 10,000 small rockets. These rockets would rapidly reach velocities of around 1.5 km (5000 ft) per second and destroy incoming warheads by the force of their impact.[14] In an early-1983 telephone interview with a High Frontier spokesman, I was told that this part of the proposal had been changed. Instead of a swarm of rockets, High Frontier now recommends using a Gatling gun system that could rapidly fire bullets to create a defensive "wall of lead" around the ground targets. Similar guns are now said to be used to defend ships against air attack.[15]

Authors of the High Frontier study envision their approach as an "end run" that would let the United States surpass Soviet capabilities and say it should be possible to build the system using "off-the-shelf components." They predict that implementation of their proposal would in two or three years produce a ground-based point defense system "adequate to protect our ICBMs [intercontinental ballistic missiles] in silos and avoid the high cost deployment modes for MX [an ICBM]. An initial spaceborne global ballistic missile defense can be acquired in five or six years given adequate priority. "[16] Overall cost is put at $24 billion over the next five or six years, that is, through the mid-1980s, and at some $40 billion through 1990 assuming the program was started in 1982 when proposed. The lower figure is comparable to the upper end of the cost range Pentagon officials have estimated for a ballistic missile defense system to protect MX alone.

Beam weaponry was not neglected in the High Frontier report, but the authors did not see it as meeting their goals for a quick fix for American defense problems. "While significant beam weapons capabilities have been demonstrated in the laboratory, their deployment in global defensive systems is too far in the future to meet the urgencies of the High Frontier study," the authors wrote. They did cite the long-term promise of the technology and urged that "a well-planned and funded U.S. R&D program, of at least coping proportions, be conducted."[17]

Issued with high hopes by a group composed largely of retired generals and space-development advocates, High Frontier did receive a hearing in the Pentagon. [18] Study director Daniel O. Graham, a retired general, was quoted as saying that Boeing had evaluated the proposal and concluded that it was feasible, although doubting the cost estimates. The press largely ignored it, perhaps because its goals were so ambitious as to strain the credulity of many observers. The British magazine *New Scientist* published a brief account, and a longer one appeared in *Aviation Week & Space Technology,* although re-strained by that magazine's standards. The idea was ignored altogether by *Science,* with nary a mention of High Frontier in the several issues following the March 4, 1982, press conference announcing completion of the study.

There are ample reasons for skepticism about the High Frontier proposal. Given the decade-plus lead times generally needed to translate concepts into operational military hardware, the time scales of a few years that were pro- posed in the study seem like wishful thinking. Defense Advanced Research Projects Agency director Robert Cooper told the Senate Armed Services Committee that despite sympathy with High Frontier's goals,

> we do not share [the authors'] optimism in being able to develop and field such a capability within their time frame and cost projections. We have . . . experienced some difficulties in ratifying the existence of "off-the-shelf components or technologies" to provide [the required capabilities]. Our understandings of systems implications and costs would lead us to project expenditures on the order of $200 to $300 billion in acquisition costs alone for the proposed system.[19]

Although the proposal is ambitious, some of the details seem to have been around for a while. The ballistic missile defense concept is similar to a proposal called BAMBI originally made in the early 1960s, according to Eberhard Rechtin, president of the Aerospace Corp. and a former high-level

Pentagon official. He explained: "The BAMBI scheme turned out to be impractical technically in the early 1960s. It is somewhat better now with some of the newer technologies, but it is still quite a way from being a practical thing, requiring technologies that are still unavailable." He also believes that "the system itself has some basic flaws in it: how the enemy would counter it, what happens if some of the satellites are not working or if you haven't replenished them at the right rate."[20]

High-Energy Laser Weapons

The idea of using high-energy laser beams for defense against ballistic missile attack apparently has been kicking around for a number of years in military circles. It received little attention in the outside world until Senator Malcolm Wallop (R-Wyo) made a public proposal in late 1979. Even then, some people in the scientific press tended to downplay the idea, apparently because it was proposed by a junior member of the minority party in the Senate. Nonetheless, Wallop's proposal marks a landmark of sorts: the opening of serious debate on concepts for space-based laser missile defense.

The proposal originated with a group called "the gang of four," engineers from aerospace companies deeply involved in laser weapon development: Maxwell W. Hunter II, assistant to the vice president for research and development of Lockheed Missiles & Space; Joseph Miller of TRW Inc., which has built large chemical lasers; Norbert Schnog of Perkin-Elmer, which makes large optics; and Gerald Ouelette of Charles Stark Draper Laboratory, which develops pointing, tracking, and fire control.

The goal was an "initial operating capability" that could be achieved in the "relatively near term." The proposal for a "first-generation" laser defense system is worth repeating, both because it was the first public airing of the concept and because the overall idea has remained unchanged (although there have been modifications to some of the details). In Wallop's words:

> [The proposal] assumes that, by the time we are ready to deploy, we will have lasers capable of five megawatts [million watts] power and mirrors with a four meter diameter. With such components we could protect against all Soviet heavy missiles, about 300 other ICBMs [intercontinental ballistic missiles], nearly all SLBMs [submarine launched ballistic missiles], and all long-range bombers and cruise-missile carriers. This would be in the event the Soviets launched all of their strategic systems, worldwide, within 15 minutes. If they took longer to launch, the system's results would be better. If they could manage this in less time, the system's results would be degraded somewhat.
>
> It's possible that by the time of first deployment we may have achieved greater power or bigger mirrors, in which case the results would be better. We surely will build bigger lasers and mirrors in the future. In short, it seems as if this purely defensive technology is moving faster than the technology of offensive weapons of mass destruction.
>
> The system would be composed of 18 "battle stations" with six in each of three polar orbits. They would cover the Earth. Their range would be somewhat Jess than 3000 miles. Each station could fire about a thousand shots. The stations would be composed of a system for detecting and acquiring targets, a large laser, a large mirror, a pointing-tracking device, and command and control

gear. Each station would be armored enough to withstand a nearby nuclear explosion. It could attack any satellite which threatened to come too close. It could be defeated only by a more powerful laser, which could "shoot" from farther away. There is no ultimate weapon, but this one holds great promise for a decade or two of real protection for Americans.[21]

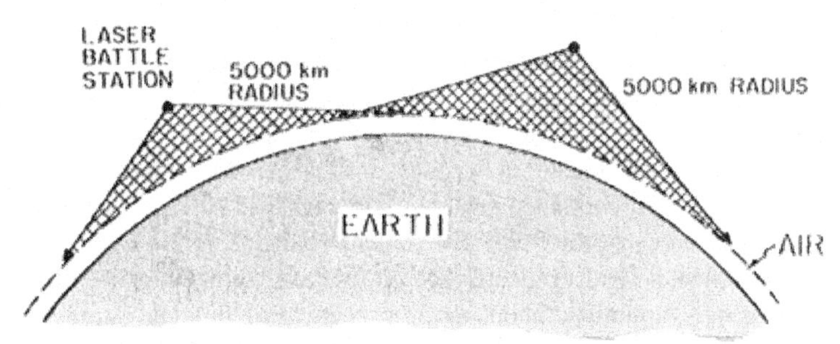

How laser battle stations would blanket the earth. If each laser had a range of 5000 km (3000 miles), as proposed by Senator Wallop of Wyoming, they would be able to zap any missiles launched from the shaded area. Some overlap of coverage would be likely at the fringes of the fields of the battle stations. The idea is the same even if the distances and the spacing of the satellites are altered. (Drawing by Arthur Giordani.)

The satellites proposed in the original plan would orbit at an altitude of 1300 km (800 miles). If they could achieve a 5000-km (3000-mile) range, each one would be able to cover about 10% of the Earth's surface, or roughly 20 million square miles (50 million square kilometers) according to Wallop.[22] The lasers would be high-power chemical hydrogen fluoride devices, 6-8 m long and weighing about 17,000 kg (19 tons), designed for Space Shuttle launch. Each would carry enough fuel for about 1000 shots, at roughly 25-50 kg (50-100 lb) of fuel per shot. Wallop's estimate was that the total system would carry a $10-billion price tag, with $3.6 billion for research and development.

Defense Department analysts reportedly concluded that the general plan seemed feasible but differed on some specifics. For example, they felt that 25 battle stations would be needed to intercept 1000 Soviet missiles, assuming that it took 10 to 20 sec to destroy each booster.[23] Because the current Soviet inventory includes 1400 land-based intercontinental missile launchers, 950 submarine-based launchers, and 156 long-range bombers,[24] this assessment reflects a belief that the system would have much more limited destructive capacity than Wallop's advisors indicated in saying it could take care of all long-range Soviet missiles.

To put teeth into his proposal, Wallop introduced in the Senate an amendment to the fiscal 1981 defense budget calling for crash development of space-based laser weapons. The amendment, calling for spending $10 billion over a period of three to five years, was defeated by a 50 to 39 vote on the Senate floor.

Wallop is the Senate's most enthusiastic supporter of space laser weapons. A more moderate position is held by Senator Howell Heflin (D-Ala), who held a series of hearings on laser research and applications in late 1979 and early 1980. Wallop described his laser battle station concept to Heflin's subcommittee. Heflin's final report notes Wallop's concept but includes a footnote that drily observes: "It is not possible to develop and deploy a space based laser ABM missile [sic] system within five years."[25] Nonetheless, Heflin did urge development of a "a DoD strategy for high- energy laser technology development and the deployment of high-energy laser weapon systems." Three possible uses of laser weapons were identified:

- Near-term applications to disable sensors and satellites
- Intermediate-term use for low-altitude air and ship defense
- Long-term defense against intermediate-range and intercontinental ballistic missiles[26]

Although Wallop's Senate colleagues don't share his enthusiasm for a crash program, he has continued pushing. He managed to add $50 million for space laser research to the fiscal 1982 defense budget, only a small fraction of the $250 million he had originally sought, but still something. He also has grown increasingly optimistic about the prospects for building large lasers and optics. He originally talked about a 5-million-watt laser with a 4-m (13-ft) mirror. During debate over the fiscal 1982 budget, he talked about a 10-million-watt laser with a 10-m (33-ft) mirror and beam-pointing accuracy of 0.2 millionths of a radian as technology that was within reach.[27]

Laser proponents concede that it would be possible to "harden" ballistic missiles so they could better resist laser attack but say this would require a new generation of missiles. They also believe that it would only take higher-power lasers to overcome such hardening. A Department of Defense study is said to have concluded that 100 orbiting laser battle stations, each equipped with a 25-million-watt laser and a 15-m (50-ft) mirror, could destroy 1000 intercontinental ballistic missiles, even if they were launched simultaneously and hardened to resist laser irradiation of 10,000-20,000 J (joules) per square centimeter.[28] Existing missiles are said to be able to withstand energy densities of around 1000 J/cm^2.[29] Laser advocates tend to be vague in their estimates of how many attacking missiles could be destroyed. Although President Reagan apparently would like to stop all of them, military analyses typically assume that a few would slip through, requiring other defensive "layers" to stop them. Even if the defensive system is not perfect, the theory goes, it would stop enough enemy missiles to leave the United States reasonably intact and with more than enough nuclear weapons to devastate the attacking country–increasing the uncertainty of the results of a first strike, and hence serving to deter a war.[30]

Skepticism over Laser Battle Stations

Many observers inside and outside the Department of Defense have been flatly skeptical of the feasibility of Wallop' s proposed timetable. In his speech calling for development of technology for missile defense, President Reagan said that while "current technology has attained a level of sophistication where it is reasonable for us to begin this effort, [it] will take years, probably decades, of effort on many fronts" to perfect missile defense systems.[31] The definition of a long-term research and development program that President Reagan directed in March 1983 was not complete when this book was written, but the ground rules stressed the need for looking at the long term. In the words of Major General Donald Lamberson, then head of the Pentagon's directed- energy program, "We are not interested in premature commitments to near- term weapon system concepts. Their pursuit will detour us from developing those technologies that may eventually be truly effective. "[32] That represents little change from the official Pentagon position before Reagan's speech and is in line with Heflin's recommendations: that space-based laser defense against missile attack is a long way off, but is nonetheless important enough to deserve support.

Department of Defense View of the Potential Use of Space Laser Battle Stations

Theater (i.e., local) conflict

• Permits projection of U.S. power around the world, potentially instantaneously

Nuclear crisis

• Creates doubt of achieving a successful surprise attack

• Deters attack by limiting prospects for success of a "counterforce" strike to disable U.S. weapon systems

Strategic nuclear exchange

• Deters countervalue attack by increasing uncertainty about success of an attack on U.S. strategic forces, with or without surprise

Protracted nuclear war

• Markedly reduces value of remaining enemy strategic forces while increasing the value of U.S. strategic forces

(Adapted from a list of "Potential contributions of laser weapons in space," presented by H. Alan Pike, Deputy Director of the Directed-Energy Office of the Defense Advanced Research Projects Agency, in a talk at the Laser Systems & Technology Conference, July 9-10, 1981, Washington, D.C., sponsored by the American Institute of Aeronautics and Astronautics.)

There are probably some skeptics about the whole idea of beam weapon defense against ballistic missiles within the Department of Defense, but the most visible dispute in military circles is over the type of beam weapons and the timetables for developing them, topics which will be discussed later in this chapter. Outside critics have taken aim at the whole Reagan proposal for developing ballistic missile defense, saying it requires technology that is not ready now, and may not ever be feasible.

Plans for developing lasers for ballistic missile defense have come under sharp criticism from military analysts outside the Pentagon, notably Richard L. Garwin, an IBM Fellow who works at the IBM Thomas J. Watson Research Center in Yorktown Heights, New York, and Kosta Tsipis, a physicist on the staff of the Massachusetts Institute of Technology. Garwin has written that the laser capabilities required for orbiting missile defense battle stations "are far from available. They require an increase in laser brightness of at least a factor of 10^6 [one million] over that which has been demonstrated in a ground-based system."[33] He also cites the possibility of what he considers simple and inexpensive countermeasures such as rotating the target missile or surrounding it with a reflective plastic balloon. Much the same arguments have been made by

Tsipis and Michael Callaham, now an assistant professor of electrical engineering at Carnegie-Mellon University, who have published a report that goes into more detail than Garwin's.[34]

Critics have also cited several other possible problems:

- The difficulty in orbiting a large laser battle station and the fuel it would require
- The vulnerability of the laser weapon system to attack during construction or to "mining" with small satellites in close orbit containing bombs that could be detonated to destroy the laser
- The severe difficulty in focusing the beam onto a tiny spot on a target thousands of kilometers away
- Focusing difficulties arising from the long wavelength of chemical lasers and the large optics that would be needed

Exactly how serious these problems will be is a matter of debate. Many prominent researchers who work with low- or moderate-power lasers are skeptical about the prospects for laser weaponry. [35] Developers of high-power lasers generally acknowledge that there are major obstacles to be overcome. At times some will concede that they are not certain they will be able to solve all the problems, yet they still feel the effort is worthwhile. Admittedly, such a feeling may be a prerequisite to staying in the business, since few scientists like to work on projects they don't feel are worthwhile.

One striking thing is how few people active in laser weapon development have anything good to say about Tsipis's study. Barry J. Smernoff, an arms-control analyst now at National Defense University, wrote:

> In the spirit of the anecdotal engineer who proved that bumblebees cannot fly, Tsipis selected the most difficult mission of damage-denial [ballistic missile defense] for laser weapons and then proceeded to unwind his numbers game to demonstrate that [space-based lasers] would make no sense for the [ballistic missile defense] (or any other) mission.[36]

Others in the field have told me that Tsipis's calculations are riddled with errors and overly pessimistic assumptions that make laser weapon development seem unrealistically difficult. They also have indicated that although it was possible to disprove at least some of Tsipis's results, it would require disclosing classified information and/or was not worth the effort. I have yet to meet anyone active in laser weapon development who would confess to taking Tsipis's calculations seriously, even some who have their own doubts about laser weapons.[37]

I have not attempted the detailed study and calculations needed to verify those claims in entirety. One reason is that my sources have much more expertise in the field. Another is that I–like Tsipis–lack access to classified information that would allow me to make assumptions accurate enough to get reasonable results. In addition, the calculations in the Tsipis-Callaham study are often hard to follow, and the assumptions on which they are based sometimes are explained so poorly that they seem arbitrary and unjustified. Finally, without spending a great deal of time, I found serious underestimates of the capability of mirror-making technology, which translate into unrealistically high demands for laser power. One obvious mistake is the statement that "solid mirrors over four meters [in diameter] are not within the current U.S. technological capacity"[38]–a surprising statement indeed since an American-made mirror 5 m in diameter, the 200-in. Hale telescope, has been in use since 1947 for astronomy at Mount Palomar, California, although it could not handle a high-power laser beam. The calculations of power required from a hydrogen fluoride laser weapon (and hence the number of Space Shuttle trips that would be needed to supply the fuel) assume that the laser's beam-directing mirror is only 1.4 m (4.6 ft) in diameter.[39] That's barely half the size of the 2.4-m (8-ft) space telescope mirror built for astronomy. Such a small mirror diameter would cause the laser power to be dispersed over a large area at the target, hence requiring higher laser powers for a "kill" than if a larger mirror were used. This, in turn, translates into hefty fuel requirements. The study would have been far more valuable if its assumptions were at least comparable to those made by Wallop's advisors.[40]

The sharply differing conclusions arising from different pencil-and-paper studies of laser weaponry serve to point out the limitations of that kind of research. It's worth remembering the lesson of the scholars mentioned in Chapter 2 who proved that Archimedes couldn't have used mirrors to ignite Roman ships, although experiments show that he could have. That's why the Pentagon is building some experimental hardware to test the feasibility of space-based laser weaponry.

The Space Laser Triad

The Pentagon's experiment in space laser weaponry may not be the crash program that Senator Wallop wants, but neither is it a modest effort being conducted in the corner of an obscure laboratory. From 1982 through 1987 the Department of Defense plans to spend some $1 billion on a program called the "Space Laser Triad," which is being run by the Defense Advanced Research Projects Agency. It is called a triad because it is aimed at separate demonstrations of the three key capabilities needed for space-based laser weapons: beam pointing and tracking, high-power chemical lasers, and large optics. Plans in effect before President Reagan's call for development of antimissile systems call for pointing and tracking to be demonstrated in space, with the laser and optics demonstrations on the ground. Only after those tests are completed does the Pentagon plan to decide whether to start actual "weaponization" of space-based laser systems.

The program in target acquisition, tracking, and pointing, called "Talon Gold" is budgeted for $35 million in fiscal 1983, is being conducted by Lockheed. To demonstrate pointing and tracking capabilities, a low-power laser with a scale-model system for target acquisition, beam pointing and tracking will be carried on one or two Space Shuttle flights. In these tests the system will be pointed at high-altitude air and space targets to assess performance with "realistic" types of targets and backgrounds. That is vital because, in general, pointing and tracking systems tend to have problems in sorting the targets from the background, for example, in picking a satellite out against the backdrop of the earth. The test is intended to demonstrate "a significant improvement over current capabilities for pointing and tracking," according to DARPA.[41] Space Shuttle tests reportedly are now set for June 1987, later than originally planned.[42]

Although Talon Gold is considered an important milestone, space-based laser weapons will require even better acquisition, tracking, and pointing capabilities. Thus, DARPA is sponsoring development of better techniques for target acquisition and tracking, target identification, selecting the point at which the beam will be aimed, and optically mounting precision pointing systems. Progress so far includes improvements in the tracking of faint targets and demonstration that a simple method for selecting "aimpoints" can make weapon systems much more effective. Also in the works, at this writing, are development of techniques to improve beam stabilization and limit beam wander off the target and improvements in fire-control systems.

A 5-Million-Watt Laser

The laser portion of the triad is Alpha, a 5-million-watt hydrogen fluoride chemical laser being built by TRW Inc. in California. The fiscal 1983 budget allocated $22 million to work on Alpha, with TRW Inc. the prime contractor. Goals include test and evaluation of a laser that operates at low internal gas pressures and efficiently turns input energy (in the form of laser fuels) into a laser beam. The system is designed to be "scalable" to much higher output powers by increasing its size.

The Alpha program is intended first to demonstrate the capabilities needed for long-range kills either to attack or to defend satellites. Then the program will turn to longer-term objectives including tests of concepts for defense against strategic bombers and nuclear-armed missiles. Detailed design began in fiscal 1982. Actual construction of the test laser was to start after a 1983 design review, but the laser probably won't operate until around 1985 or 1986 (no formal timetable has been made public). Angelo Codevilla, a member of the Senate Intelligence Committee staff who works with Senator Wallop, is optimistic that improvements in nozzle technology will let Alpha produce powers of 10-million watts, but other observers are less optimistic.

Alpha is not the Defense Department's only chemical laser program, although it is the biggest demonstration laser known to be in the works. As mentioned in Chapter 4, military researchers are also working on shorter-wavelength lasers, specifically chemical iodine lasers emitting at 1.3 μm in the infrared, excimer lasers emitting around 0.35 μm in the ultraviolet, and free-electron lasers with adjustable wavelengths. So far these tests seem encouraging, but output powers are still far below those already demonstrated by chemical lasers.

Large Optics Demonstration Experiment

The third part of DARPA's space-laser triad is the Large Optics Demonstration Experiment, LODE. Goals of this ground-based program "include demonstrating the ability to manufacture a large-aperture mirror that includes complex interactive control systems, and energy management in an overall beam control system that yields high optical quality and is consistent with the required pointing accuracy in a simulated operational environment."[43] Although the tests are intended to evaluate the feasibility of using such technology in space, the actual experiments will be performed in a space-simulation facility on the ground, using a low-power laser.

Plans called for design of the primary mirror and experimental facilities to be completed in fiscal 1983, tasks for which DARPA has budgeted $14 million. The ultimate goal is a 4-m segmented mirror which Itek and Kodak will make of low-expansion glass. It will use adaptive optical techniques to control characteristics of the output laser beam. The optics should hold the beam on the target with a "jitter" (wandering around the ideal point) of 0.2 millionths of a radian.[44] The LODE tests will take place at about the same time as the Alpha and Talon Gold demonstrations.

Controversy over the Triad

DARPA's space laser program got a thorough going-over as Congress debated the fiscal 1983 defense budget. Senator Wallop managed to insert an amendment in the original Senate version that directed the defense department to conduct all three of the Triad experiments in space. The addition was made as the Senate worked through an all-night session and may have been passed simply to get Wallop to vote for the bill. It did not provide the extra money that would be needed for tests in space, but Wallop may have hoped to add that later.

The House Armed Services Committee, in contrast, attacked the space laser budget. Its version of the budget chopped out the Alpha and LODE portions of the triad, as well as DARPA's program in chemical laser technology, and $41 million that had been allocated to the Air Force for research on space-based laser weapons. Instead, the Committee allocated $50 million for its pet enthusiasm, research on unidentified short-wavelength lasers, evidently including free-electron, X-ray, and excimer types. Talon Gold, was left untouched as being applicable to short-wavelength lasers, a rationale which critics say also extends to LODE.

The House proposal nearly doubled the $26.6 million the Pentagon had originally allocated for short-wavelength laser research in fiscal 1983. The reason, according to the committee's report, was that "emphasis is being focused on the wrong laser technology." Long-wavelength chemical lasers "not only present extreme technical challenges but ultimately offer lower military utility than lasers operating in [sic] shorter wavelengths."[45]

The House committee also added $2 million to the budget for the Advanced Test Accelerator being built for particle-beam weapon tests at the Lawrence Livermore National Laboratory, to modify it for use in free-electron laser experiments. Enthusiasm for free-electron lasers had been stimulated by encouraging demonstrations in late 1981 and early 1982. One of the strongest advocates was John D. Rather, a high-energy laser consultant who then headed a small firm, Pan-Scion Research. He told the Senate Armed Services Committee that, "it is a national travesty that free-electron lasers are not being pursued on a crash basis, because there is ample evidence that they hold a probable solution for the need for sustained high power at short wavelengths." He claimed that given a crash priority, a short-wavelength free-electron laser could be developed on about the same schedule as the chemical laser triad.[46]

The Pentagon and the Senate were not happy with the House version of the budget. DARPA director Robert Cooper told the Senate Defense Appropriations Subcommittee that the House action would end development of space-based laser weaponry for the time being, and would probably delay U.S. deployment of laser weapons in space until the mid- to late 1990s, some five to ten years later than what the Pentagon believed would be possible with chemical lasers.[47,48]

The eventual compromise was closer to the Senate version. Funding for the DARPA programs was restored. The money originally set aside for the Air Force's space laser research was instead split between a $20 million addition to DARPA's short-wavelength laser research program and a $20 million program to evaluate the vulnerability and lethality of laser weaponry, two of the issues raised by critics of the Pentagon's original plans. The former move brought DARPA's short-wavelength budget up to $47.6 million, nearly what the House had proposed. The $2-million addition to the particle-beam budget for free-electron laser tests was left in the final version. Wallop's amendment specifying that LODE and Alpha be tested in space was dropped.[49]

The differences in opinion that led to the conflict remain, and politics undoubtedly will continue to rear its head–if anything, more loudly now that President Reagan has brought the idea into the public eye. The basic conflict is between those who believe chemical lasers would suffice for a first-generation missile defense system in space and those who hold that chemical lasers are simply not up to the task. There is a general agreement that in the long term, short-wavelength lasers seem more attractive. However, chemical laser technology is generally considered to be several years ahead of shorter-wavelength lasers, and there are some who remain skeptical that some of the brightest short-wavelength hopes such as free-electron, excimer, and X-ray lasers will ever prove practical for use in space weapon systems. Some observers see the central issue as trying to decide whether to pursue chemical lasers as a technology "good enough" for the immediate future or wait until an "ideal" laser can be developed that meets long-term requirements. Recently, new interest has emerged in a different approach to laser ballistic missile defense in which the laser itself would stay on the ground, but the beam would be steered by large mirrors in space.[50] The laser would be mounted on a large mountain with steady air, the type of site now used for large telescopes, to minimize atmospheric transmission losses. It would transmit a powerful beam to a satellite in low orbit, which would then redirect it to other relay mirrors, or aim it directly at a target.

This approach offers a simple way around the major problem of getting a massive laser off the ground; a problem that seems particularly severe for the free-electron laser. However, it raises other difficulties. The orbiting mirror would not stay obligingly straight overhead unless it was in geosynchronous orbit and the laser was on the equator. Low earth orbit is a much better choice because the mirror would be much closer to the laser, making it possible to use a much smaller mirror and avoiding the trouble and expense of putting a large mirror in the high geosynchronous orbit. However, that would require a family of mirrors in low orbit, with one moving into position as another leaves the vicinity of the laser.

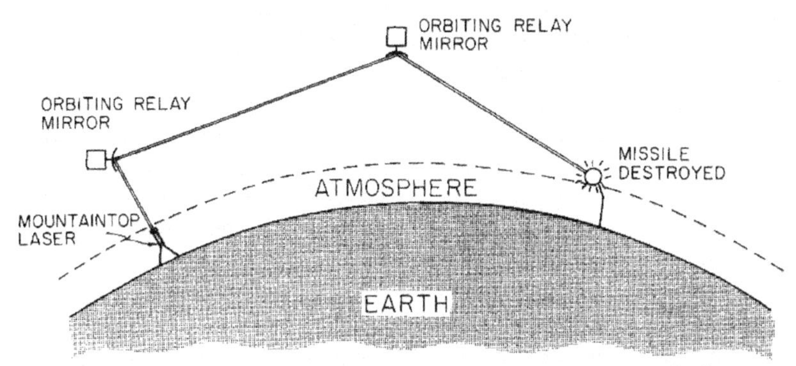

Leaving the laser on the ground would keep the heaviest component of a laser weapon system out of orbit, but it would mean that the beam would have to make its way through the atmosphere. This diagram shows how the beam from a mountaintop laser might be focused by two relay mirrors to destroy a missile. Sophisticated control systems would be needed to pass the laser beam from one mirror to another, and many mirrors would be needed because orbital motion would quickly take them out of reach of the ground-based laser. Despite these problems, this approach has recently gained considerable support because it promises to avoid the need for putting massive lasers in orbit. (Drawing by Arthur Giordani.)

Leaving the laser on the ground does give up one major advantage of having the laser in space: it makes the beam subject to atmospheric absorption, scattering, and thermal blooming. It is at least theoretically possible for a large enough mirror on the ground to compensate for some effects that cause the beam to spread out and wander while traveling through the air, but there is no way around absorption. The wavelengths of certain promising high- energy lasers, notably krypton fluoride at 0.25 μm and hydrogen fluoride at 2.5 to 3 μm, are completely soaked up by the air. Others, such as the xenon fluoride excimer at 0.35 μm and the chemical oxygen iodine laser at 1.3 μm, may lose around half of their power. The best choice for such a ground-based laser would probably be the free-electron laser because, at least in theory, its output wavelength can be adjusted to match the wavelengths at which air transmits light best. In addition, free-electron lasers also seem to have the most severe weight and bulk problems among major laser weapon contenders. Once the laser beam made its way into space, the weapon system would function in much the same way as an orbiting laser battle station, with the size of the focal spot on the target dependent on laser wavelength and transmitting mirror diameter in the same way. Hypothetically, it should be possible to relay the beam from a laser on the ground in the United States all the way around the world if there were enough relay mirrors in the right places. In practice, there would be some losses, probably a factor of 5 to 10, increasing with the number of mirrors involved.

Soviet Space Laser Programs

The Soviet Union is widely believed to have its own large program in space-based laser weaponry; but, not surprisingly, few details are available in the public domain. Dire warnings of imminent Soviet deployment of laser weapons in space almost inevitably focus on antisatellite lasers, a topic described in more detail in Chapter 12. Officially, American intelligence sources have little to say about the entire Soviet program other than an assessment that, "the Soviet high-energy laser program is three to five times the U.S. level of effort and is tailored to the development of specific laser weapon systems."[51] That estimate is usually accompanied by a statement that U.S. officials believe development of specific weapons to be premature; there have been no public attempts to divide the estimate among subprograms such as tactical weapons for the battlefield and antimissile lasers for space. Interestingly, that estimate has remained unchanged despite a sharp increase in the American effort, which has caused spending to more than double between fiscal 1981 and fiscal 1983.

Assessing the progress of Soviet military technology has always been difficult. For example, despite a dozen years of reports that Soviet engineers are developing a space booster similar in size and performance to the Saturn V used in the American Apollo program, the rocket evidently has yet to operate successfully.[52]

One possibly intentional leak of classified information in early 1982 did sound an unsettling note. Richard D. DeLauer, undersecretary of defense for research and engineering, was quoted as saying in classified testimony to Congress: "We expect a large, permanent, manned [Soviet] orbital space complex to be operational by about 1990 . . . capable of effectively attacking . . . ground, sea, and air targets from space" with high-energy lasers.[53] Presumably that capability might also be aimed at satellites and missiles, although its full meaning is unclear. The mention of ground and sea targets is surprising because most observers appear to believe that it is impractical–or simply not cost effective–to shoot at them from space.

It should be noted, however, that DeLauer's verbal train of thought sometimes does not seem to follow any obvious tracks, which makes it hazardous to quote him out of context. One example is an excerpt from his testimony on the laser weapon program before the House Armed Services Committee's Research and Development Subcommittee on March 11, 1982, a week after the leak mentioned above:

> If you look at the four times [the Soviet Union is spending more on laser research than the United States], and then look at the split between what we are doing in lasers in general and in the space-based stuff, and say that maybe we spent a quarter to half of ours for that, and they do the same thing, then perhaps they are going to be able to beat us by 4 or 5 years, maybe 7 to 10 years perhaps,

if they have been making this effort.

And this is the one I keep saying. They are very determined when they want to go after it. We have some assessment of what they might be doing, and it is not clear they are going to be successful, but certainly if resource allocation is any indication, why they are doing a little bit better than we are. But I think we have got the properly paced program, I really do.[54]

Particle-Beam Weaponry

Particle-beam weapons are being considered for two types of ballistic missile defense. In one approach a charged particle generator on the ground would send bolts through the air to destroy missiles approaching the targets protected by the particle-beam system. In the other a neutral-beam generator would be put into an orbiting battle station similar in concept to what Senator Wallop proposed for laser systems.

The basic idea of short-range defense with a charged particle generator on the ground is similar to one of the concepts proposed by High Frontier. Instead of shooting rockets at the incoming warheads, the particle-beam system would shoot charged particles. The intense beam would make its way through the air at about one kilometer per thousandth of a second, much slower than light, but far faster than a rocket can accelerate. It would then zap its target nearly instantaneously with a pulse carrying energy measured in millions of joules, advocates say. If the fire-control system can handle the job, the generator should be able to zap tens of targets per second. That makes charged particle-beam weapons "particularly well-suited to counter small, very fast, highly maneuverable threats."[55] That category includes incoming cruise missiles as well as re-entry vehicles launched by ballistic missiles.

A tightly focused and very intense particle beam can penetrate deeply into a target, depositing its energy in a long conical volume and making the possibility of shielding impossible. Damage is said to be immediate and severe, including detonation of chemical explosives, structural damage to the target, and damage or disruption to electronic equipment.

As indicated in Chapter 7, a space-based particle-beam weapon would have to generate a neutral beam. Like an antimissile laser satellite, it would be placed in low earth orbit, designed to shoot at missiles during their vulnerable boost phase. If the beam could be kept tightly focused over the required distances, it could extensively damage targets in ways similar to a charged particle beam in the atmosphere. Unimpeded by the atmosphere, the beam itself would travel at nearly the speed of light.

Either type of particle-beam weapon has some potential advantages over lasers. A major one is the potential for nearly instantaneous "kills" with a single pulse. Advocates of particle-beam weapons also say that the technology can generate pulses with higher power and more efficiency than lasers.

There clearly are some potential problems, however, including the likely large bulk of such a system, potential high costs, and uncertainty over beam- and fire-control technology, as described in Chapter 7. The harshest critics of the concept have been Richard Garwin and Kosta Tsipis, with probably the harshest assessment coming in a 1978 report from an MIT group that included Tsipis. That report concluded that it appeared possible to build a particle-beam accelerator with the characteristics needed for a weapon, but that the operational problems needed to turn such a massive generator into an effective weapon "appear insurmountable." The authors also concluded that "there is serious doubt whether the necessary fire-control and beam-control systems needed for a weapon are technically feasible."[56]

The Pentagon has tended to steer a cautious middle course between opponents of particle-beam weapons and those calling for crash development. The initial controversy in the late 1970s, described in Chapter 2, resulted in some sharp increases in what had been a nearly dormant program. In 1983 General Lamberson called particle beams "the least mature of the directed energy technology efforts."[57] During fiscal 1983, the Pentagon budget for particle-beam research increased slightly beyond the $50 million mark, and accounted for roughly 10% of the beam weapon budget as a whole. Roughly half that total was allocated for the Advanced Test Accelerator at the Lawrence Livermore National Laboratory, with another $4.5 million going to the White Horse program at the Los Alamos National Laboratory (programs described in Chapter 7). The remaining $20 million or so is slated for "technology base" research sponsored by the three armed services.

The Pentagon characterizes its particle-beam program as being "in the very early research and exploratory development phases with fundamental issues of feasibility to be resolved. There is an enormous gulf between the technology required for fulfillment of the conceptual payoffs and the state of the art."[58] Translated from the bureaucratese, that means weapon systems are a long way off, and may not even be possible.

The current program is aimed at finding out if it's possible to build a particle-beam weapon. Engineering will come later, if ever. The major issues are questions of physics:

- Can charged particle beams be made to travel predictable paths through air?
- Can neutral beams be produced with the extremely high quality needed to travel long distances in space?
- Can powerful enough particle-beam accelerators be built?

There are also some fundamental technology issues:

- Can fast and reliable switches be built for rapid firing?
- Can the beam be sensed and controlled well enough to kill targets?
- Can beam generators and power supplies be built compactly enough for practical use in space?
- What types of damage do particle beams do to real military targets?

Even if feasibility can be demonstrated, several years of development would be needed to build a prototype, which itself would be a preliminary to building actual weapon systems. Following the current schedule would put particle-beam weapons into the late 1990s at best.

There is little doubt that the Soviet Union is working on technology that could be used for particle-beam weapon research, but how far they are going beyond that is open to doubt. Articles in *Aviation Week & Space Technology* have gone as far as claiming that actual tests are being planned or conducted.[59] There does not seem to be a consensus within the Pentagon. One public statement, the brochure *Soviet Military Power,* says that, "the Soviets have been interested in particle beam weapons (PBW) concepts since the early 1950s. There is considerable work within the USSR in areas of technology relevant to such weapons." However, the fact sheet that the defense department issued on the particle-beam program takes the more cautious position that "no direct correlation between Soviet particle-beam work and weapons related work has been established."[60]

X-Ray Laser Weaponry

A third beam weapon alternative for space-based defense against nuclear missiles is the X-ray laser battle station described in Chapter 6. The basic idea is to arrange rods of material that could serve as X-ray lasers around a small nuclear bomb, then orbit the array when needed. Once the satellite was in orbit, the rods could be pointed at potential targets. Detonation of the bomb would stimulate the rods to produce intense pulses of X rays that would disable targets either by causing mechanical or radiation damage. That is, if it works– something which some observers doubt for reasons described in more detail in Chapter 6.

The X-ray laser program is in a peculiar political position. Because its energy comes from detonation of a nuclear bomb, most of the funding for its development has come from the portion of the Department of Energy responsible for nuclear bomb research. The Department of Defense has contributed a smaller share.

The most vocal public advocate of X-ray laser battle stations is the controversial physicist Edward Teller, who is credited with a major role in developing the American hydrogen bomb. As far back as October 1981, eight months after the program was first described in public, Teller sent a classified letter to Congress that evidently urged support for a crash program in X-ray laser development. That generated some interest in the House Armed Services Committee, but evidently not in the Pentagon. In the summer of 1982 Teller went directly to President Reagan to urge a crash program of $200 million a year for X-ray laser research. That tactic also did not work and is said to have offended Secretary of Defense Caspar Weinberger.[61] Although the Pentagon's directed-energy office is looking at short-wavelength lasers, the formidable technical obstacles described in Chapter 6 appear to have left officials somewhat skeptical about X-ray lasers.

Although Teller's name carries prestige, he has also earned a reputation in some circles for overenthusiasm. One outspoken advocate of space-based chemical laser weaponry scoffed at the X-ray laser as "Teller's toy."[62] In an account of the development of the hydrogen bomb, another participant in that effort, Nobel laureate Hans Bethe, wrote that "nine out of ten of Teller's ideas are useless. He needs men with more judgment, even if they be less gifted, to select the tenth idea which often is a stroke of genius."[63]

There has also been some speculation that the X-ray laser might not merely be an end in itself. One is that development of the X-ray laser is only part of an effort to produce a "third generation" of nuclear weapons, a term which Teller used to describe the X-ray laser before Department of Energy officials reportedly classified the term itself. Pursuit of such a development program would make it impossible to move to a complete nuclear test ban, and in fact might be used to pressure for a resumption of testing in the outer atmosphere or a start of space tests–both banned by existing treaties.[64] At least one observer would go farther than that. Writing in *The Progressive,* Michio Kaku, a professor of nuclear physics at the graduate center of the City University of New York, asserted that the whole effort to develop laser ballistic missile defense is part of a plan by "nuclear warfighting strategists" to build up the capabilities the United States needs to win a nuclear war by launching a pre-emptive first strike against the Soviet Union.[65]

Other observers suspect that there may be other ulterior motives that are much less sinister and have more to do with protecting bureaucratic turf than fighting a nuclear war. Some see Teller as trying to protect and enhance programs at the Lawrence Livermore National Laboratory, which he helped found in the 1950s. One particular related program, laser fusion (which Teller helped start well over a decade ago), has encountered more technical difficulties than expected and recently has had to battle to save its budget from serious cuts. Some observers have wondered if the X-ray laser program could be used to bolster support for laser fusion (because the same large ultra-short-pulse lasers used for fusion could be used in some X-ray laser experiments) or was merely a haven for researchers whose jobs would be threatened if the laser fusion budget were cut.

Different Approaches to Nuclear Defense

In some ways the various beam weapon approaches to ballistic missile defense would offer similar capabilities, although the potential role of microwaves is as this point unclear. However, there are also fundamental and important differences among the various approaches. Despite common requirements for capabilities in fire control, tracking, power supplies, and putting large structures in orbit, the different approaches require different technologies. None of them may work, but the fact that one of them won't work does not necessarily rule the others out. Problems with putting large lasers or particle-beam generators into space would not rule out ground-based or "pop-up" defense systems. Thus, while there are important ways in which the separate beam weapon programs can complement each other, they will not necessarily have to stand or fall together.

An effective system for missile defense, particularly one that would attack the missiles during their boost phase, would dramatically change the rules of the international strategy game. Even an attempt to build an ineffective system could, according to some observers, have a serious impact. These impacts are described in Chapter 14.

References

1. "President's speech on military spending and a new defense," *New York Times,* March 24, 1983, p. A20 (transcript of President Reagan's speech of March 23, 1983).
2. *Ibid.*
3. Fred Kaplan, "Reagan's strategic surprise," *Boston Globe,* March 27, 1983, pp. A-13, A-18.
4. Michio Kaku, "Wasting space, countdown to a first strike," *The Progressive* 47 (6), 19-22 (June 1983).
5. Kaplan, *op. cit.*
6. Figures in the fiscal 1983 budget approved by Congress were slightly over $500 million for projectile-based research on ballistic missile defense, according to a Pentagon spokesman who supplied information for a *New Scientist* report I helped prepare: Christopher Joyce, "Reagan's ray guns are decades away," *New Scientist* 97, 871 (March 31, 1983). Beam weapon spending is not explicitly broken down into portions for missile defense and other applications, but a reasonable estimate would put the part for missile defense and other space applications between $100 million and $200 million.
7. History of the ballistic missile defense program is taken from testimony of Maj. Gen. Grayson Tate, *Department of Defense Authorization for Appropriations for Fiscal Year 1983,* Hearings before the Senate Armed Services Committee, 97th Congress, 2nd session, Part 7, pp. 4441-4444.
8. *Fiscal Year 1983 Arms Control Impact Statements* (U.S. Government Printing Office, Washington, D.C., March 1982), p. 391.
9. Harold Rosenbaum, "A stabilizing BMD not technically ready," *Astronautics & Aeronautics* **19** (5), 36-40 (May 1981); his figures were in English units.
10. Joint Chiefs of Staff, *U.S. Military Posture for Fiscal Year 1983,* p. 77; see Ref. 14, Chapter 10.
11. Testimony of Richard DeLauer, in Hearings on Military Posture and Department of Defense Authorization for Appropriations for FY 1983 before House Committee on Armed Services, 97th Congress, 2nd session, R&D Title 2, part 5, p. 159.
12. *Ibid.,* p. 387.
13. Daniel O. Graham, *High Frontier: A New National Strategy* (High Frontier, Washington, D.C., 1982), pp. 119-128.
14. *Ibid.,* pp. 115-117.
15. John J. Coakley, High Frontier, private communication.
16. Graham, *op. cit.,* p. 4.
17. *Ibid.,* p. 135.
18. Christopher Joyce, "America debates extra cash for space weapons," *New Scientist* **93,** 644 (March 11, 1982); Clarence A. Robinson, Jr., "New ballistic missile defense proposed," *Aviation Week & Space Technology* **116,** 269-272 (March 8, 1982).
19. Senate Armed Services Committee, Hearings on Department of Defense Authorization for Appropriations for Fiscal Year 1983, Vol. 7, p. 4635.
20. Martha Smith, "Dr. Eberhardt Rechtin: C^3 is the heart of any war in space," (interview) *Military Electronics/Countermeasures* **8** (11), 9-20 (November 1982).
21. Wallop's proposal first appeared in an article, "Opportunities and imperatives of ballistic missile defense," published in the Fall 1979 issue of *Strategic Review.* On December 12, 1979 he described it during hearings on laser technology held at the behest of Senator Howell Heflin (D-Ala) by the Subcommittee on Science & Technology of the Senate Commerce, Science and Transportation Committee. This quote comes from a copy of the testimony supplied by Senator Wallop.
22. Malcolm Wallop, "Opportunities and imperatives of ballistic missile defense," *Strategic Review,* Fall 1979, pp. 13-21.

23. "Defense Dept. experts confirm efficacy of space-based lasers," *Aviation Week & Space Technology,* July 28, 1980, pp. 65--66.
24. Stockholm International Peace Research Institute, *The Arms Race and Arms Control* (Oelgeschlager, Gunn & Hain, Cambridge, Massachusetts, 1982), p. 82.
25. Senate Committee on Commerce, Science and Transportation, *Laser Research and Applications,* November 1980, p. 25.
26. *Ibid.,* p. 30.
27. *Congressional Record-Senate,* May 13, 1981, S4976.
27. Clarence A. Robinson, Jr., "Laser technology demonstration proposed," *Aviation Week & Space Technology,* February 16, 1981, pp. 16-19; figures were given in joules. One joule corresponds to one watt of power delivered for a second, i.e., one joule equals one watt-second.
28. In general, joules per square centimeter is taken to mean vulnerability to repetitively pulsed or continuous lasers, measured in terms of the amount of energy contained in the beam. Watts per square centimeter is taken as an indication of the vulnerability to single laser pulses. Both measures of vulnerability are only approximations, because in practice a variety of other considerations come into play, such as duration of the laser pulses, speed of the target, and whether it is in the air or the vacuum of space.
29. Donald L. Lamberson, plenary talk, Department of Defense Directed Energy Program, presented May 17, 1983, to the conference on Lasers and Electro-Optics, Baltimore. All quotes are taken from a printed version of the talk which he supplied to reporters at the conference, but which is not included in the technical digest of the meeting.
30. Transcript of March 23, 1983, speech: "President's speech on military spending and a new defense," *New York Times,* March 24, 1983, p. A20.
31. Lamberson, *op. cit.*
32. Richard L. Garwin, "Are we on the verge of an arms race in space," *Bulletin of the Atomic Scientists* 37 (5), 48-53 (March 1981).
33. Michael Callaham and Kosta Tsipis, *High-Energy Laser Weapons A Technical Assessment,* Program in Science and Technology for International Security, Massachusetts Institute of Technology, Cambridge, Massachusetts, November 1980; a much briefer but more readily accessible account of the basic findings is Kosta Tsipis, "Laser weapons," *Scientific American* **245** (6), 51-57 (December 1981).
34. One example is laser pioneer Gordon Gould, who I interviewed for *Omni* several days after Reagan's March 23, 1983, speech, and who expressed skepticism about the whole idea.
35. Barry J. Smernoff, "The strategic value of space-based laser weapons," *Air University Review,* March/April 1982, pp. 2-17; this comment appears in a footnote toward the end of the article.
36. My sample is not a scientific one, consisting of a few people I respect, know personally, and have talked with as a friend or colleague, not a reporter in a formal interview. Most of the same people are skeptical that a crash program to develop laser weapons in a few years could be successful.
37. Callaham and Tsipis, *op. cit.,* p. 51.
38. *Ibid.,* pp. 46-47. There are also factual misstatements in this report, such as one that excimer lasers emit visible rather than ultraviolet light.
39. Wallop's *Strategic Review* article was in print about a year before the Callaham-Tsipis report was published in November 1980, but is not even mentioned in the latter. Judging from comments in the introduction, the Callaham-Tsipis report was based at least in part on two workshops held in early 1979. Supporting that assumption is the fact that only three references are to publications later than 1978. Tsipis was known to be working on the report for some time before it was published, but the reasons for the delay were never made clear.
42. Defense Advanced Research Projects Agency, *Fiscal Year 1983 Research and Development Program, A Summary Description,* March 30, 1982, p. III-47.
43. "Washington Roundup," *Aviation Week & Space Technology,* October 25, 1982, p. 15.
44. DARPA, *op. cit.,* p. III-50.
45. Clarence A. Robinson, Jr., "Defense Dept. backs space-based missile defense," *Aviation*

Week & Space Technology, September 27, 1982, p. 14-16.

46. House Committee on Armed Services, *Report on Department of Defense Authorization Act for Fiscal Year 1983,* p. 132.
47. Senate Armed Services Committee, Hearings on Department of Defense Authorization for Appropriations for Fiscal Year 1983, pp. 4909-4916.
48. Jeff Hecht, "House and Senate squabble over laser weapons budget," *Lasers & Applications* **1** (1), 50-54 (September 1982).
49. "House-Senate compromise," *Lasers & Applications* **1** (3), 20 (November 1982).
50. Wallace D. Henderson, "Space-based lasers: Ultimate ABM system?" *Astronautics & Aeronautics* **20** (5), 44-53 (May 1982).
51. Department of Defense, *Soviet Military Power* (U.S. Government Printing Office, Washington, D.C., October 1981), p. 76.
52. Barry J. Smernoff, "The strategic value of space-based laser weapons," *Air University Review,* Mar/Apr 1982, p. 2-17.
53. George C. Wilson, "Soviets reported gaining in space weapons," *Boston Globe,* March 3, 1982, p. 9.
54. House Armed Services Committee, Hearings on Military Posture and Department of Defense Authorization for Appropriations for Fiscal Year 1983, R&D Title II, part 5, p. 571.
55. William A. Barletta, 'The Advanced Test Accelerator," *Military Electronics/Countermeasures* 7 (8), 21-26 (August 1981).
56. G. Bekefi, B. T. Feld, J. Parmentola, and K. Tsipis, *Particle Beam Weapons* (Program in Science and Technology for International Security, Massachusetts Institute of Technology, Cambridge, Massachusetts, December 1978), p. 59.
57. Donald Lamberson, *op. cit.*
58. *Fact Sheet: Particle Beam Technology Program,* Department of Defense, February 1982, p. 3.
59. Clarence A. Robinson, Jr., "Beam-target interaction tested," *Aviation Week & Space Technology,* July 28, 1980, pp. 47-50.
60. *Soviet Military Power* (Department of Defense, Washington, D.C., October 1981), p. 75; *Fact Sheet: Particle Beam Technology Program* Department of Defense, February 1982, p. 7.
61. "Washington Roundup," *Aviation Week & Space Technology,* September 20, 1982, p. 15.
62. "Pentagon spurns Teller's new toy," *New Scientist* 96, December 16, 1982, p. 728.
63. Hans A. Bethe, "Comments on the history of the H-bomb," *Los Alamos Science* 3 (3), 42-53 (Fall 1982). This article was originally written in 1954 but was only recently declassified. Bethe prefaced that comment by saying "Everybody recognized that Teller more than anyone else contributed ideas at every stage of the H-bomb program."
64. "Statement of national-security impact of increased nuclear weapons testing by Concerned Argonne Scientists," *Physics and Society* **12** (1), 7 (Jan 1983); this publication is the newsletter of the American Physical Society's Forum on Physics and Society.
65. Michio Kaku, "Wasting space, countdown to a first strike," *The Progressive* **47** (6), 19-22 (June 1983). The credibility of this article is greatly diminished by a number of serious factual errors in the discussion of ballistic missile defense programs, including statements that the High Frontier proposal calls for beam weapons rather than projectiles; that "it took an entire building to house the 300-watt [sic] power supply for the San Juan Capistrano [laser weapon] tests" when in reality the laser *output* power was around 400,000 W; and a misidentification of the Navy's Chair Heritage ground-based particle-beam program with the supersecret X-ray laser program being conducted by the Lawrence Livermore National Laboratory.

12.
Antisatellite Weapons

Beam weapons for defense against nuclear attack may be many years away, but those for use against satellites seem perilously close. There have been persistent rumors that the Soviet Union is close to testing an antisatellite laser, so close that at times one almost hesitates to open the latest issue of *Aviation Week*. Several years ago there was what evidently was a false alarm: an *Aviation Week* report in late 1975 that an American spy satellite had observed intense infrared radiation from the Soviet Union that some Pentagon sources interpreted as coming from a Soviet laser.[1] Unofficial sources indicate that U.S. officials now believe that the infrared light really came from a natural gas fire within the Soviet Union. American officials are still wary, but now seem less eager to pounce on Soviet lasers to explain malfunctions of U.S. spy satellites. An early 1983 *Aviation Week* report said that United States "analysts doubt hostile [laser] action was to blame for the malfunction [of an Air Force spy satellite] several months ago, but they cannot be sure because the spacecraft was out of radio contact when the failure occurred."[2]

One reason that U.S. officials carefully watch for Soviet antisatellite lasers is that they have plans of their own. Shortly before President Reagan proposed developing missile defense systems, the Air Force was reported to have drafted a formal "statement of need" for antisatellite laser weapons.[3] If that request makes its way through the Air Force and Pentagon bureaucracy, antisatellite lasers will be added to the Air Force's shopping list of new weapon systems it would like to have, a list used in drawing up the federal budget. Meanwhile, the General Accounting Office issued a classified study of antisatellite weapons that sharply criticized the existing antisatellite program and recommended that the Pentagon take a fresh look at other possibilities, specifically mentioning lasers.[4]

The interest in antisatellite lasers is logical, especially given the difficulty of attacking satellites by other means. Although their orbits keep them hundreds of kilometers or more above the ground, satellites are the most vulnerable of targets for high-energy lasers. Their orbits carry them on slow and predictable paths across the sky, making them veritable sitting ducks. Most depend directly or indirectly on optical and infrared sensors, which could be blinded or disabled by laser intensities far below those needed to damage other targets. Even satellites that don't rely on optical or infrared sensors are vulnerable to an upset in heat balance caused by the energy deposited by a laser beam. With satellites having become a critical element in defense strategy, it's no wonder military officials are uneasy.

Military Roles for Satellites

In the quarter of a century since the Soviet Union launched the first artificial earth satellite, space has become a vital part of the strategic equation. Satellites far above the atmosphere have taken over the role of spy planes that flew high above the other side's territory. Where spy planes were once fired upon, spy satellites have become accepted objects in the sky. Space is also cluttered with other military satellites.

The most abundant satellites are those that monitor happenings around the world in a variety of ways. Such surveillance or spy satellites account for roughly half of the 120 military satellites launched each year.[5] Some operate in low orbits, only 150 to 400 km (100 to 240 miles) above the earth's surface, where they collect high-resolution images of the ground. Sometimes these images are recorded on photographic film that is dropped from orbit, but more often they are put into electronic form and relayed home by a network of communication satellites. Objects only 15 to 30 cm across can be spotted on the best spy satellite images,[6] which are widely used by intelligence agencies. For example, satellite photos of a Soviet research facility were used by American intelligence analysts to deduce that Soviet engineers were working on directed-energy weaponry.[7] Other types of optical surveillance satellites survey the oceans to keep track of naval forces. "Ferret" satellites, which monitor the radio spectrum, listen in to signals transmitted in other countries.

One particularly vital group of satellites continually watch for any sign of bomber and missile launches that would be the first warning of an impending nuclear attack. Both the United States and the Soviet Union rely heavily on these early-warning satellites. It was one of these satellites that detected the intense infrared radiation that worried U.S. officials in late 1975. These satellites have largely, but not completely, taken over the functions of the network of radar antennas, which for the United States forms the DEW (Distant Early Warning) line along the northern rim of the North American continent.

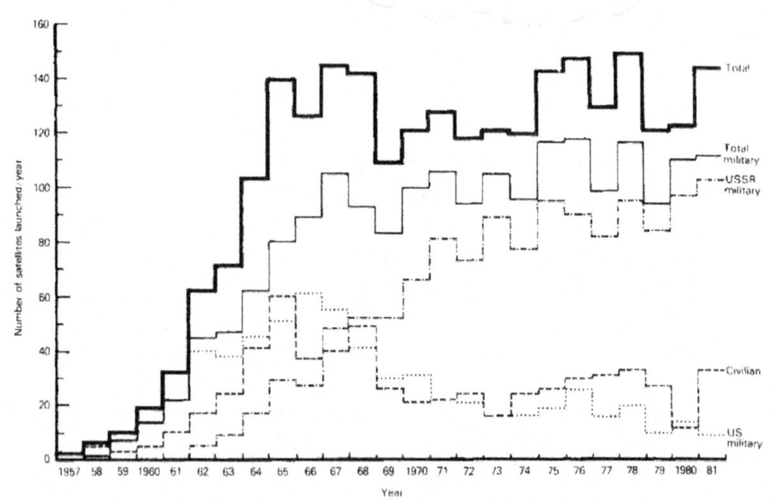

Numbers of military and civilian satellites launched each year between 1957 and 1981. The disparity in number of launches between the United States and Soviet Union reflects the much shorter lifetimes of Soviet surveillance satellites, not a dramatic difference in the number of military satellites operated by the two countries. (Courtesy of Stockholm International Peace Research Institute, from Ref. 6, p. 42.)

A satellite warning is enough to trigger the takeoff of nuclear-armed strategic bombers, which can be called back if the warning proves to be a false alarm rather than an actual attack. There have been proposals that the United States adopt a "launch on warning" policy for its missiles as well. That is an unsettling strategy to many observers because false alarms have occurred, and a massive missile attack cannot be recalled once the missiles have been launched. However, some analysts see that policy as a way around problems that could be created if American land-based intercontinental ballistic missiles should become vulnerable to destruction by a pre-emptive Soviet nuclear attack, a development which would help undermine MAD.[8]

Although best known for their civilian uses, communication satellites also find many military applications. An obvious one is relaying signals from surveillance satellites back to the home country. Satellite communications also link command posts with remote bases, important because both the United States and the Soviet Union have bases scattered around the world.

Military satellites also make measurements and do other jobs related to military needs. Navigation satellites send out signals that make it possible to determine precisely the position of aircraft, submarine, or ship, a vital capability for the guidance of bombs and missiles to their targets. Military satellites collect data on the earth's shape and gravitational fields essential for precise plotting of the courses of missiles. Because weather conditions can be of vital importance to military forces, there are also military weather satellites; around one-third of the $1 billion a year that the government spends on weather forecasting and data collection comes from the Department of Defense.

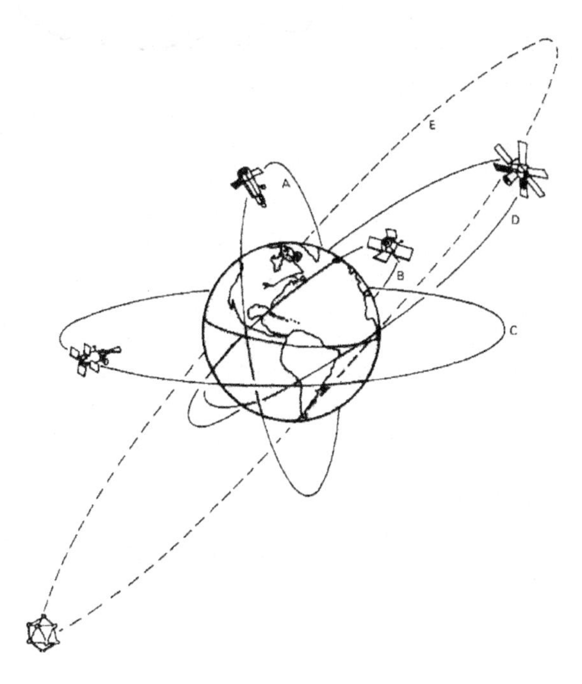

Types of orbits occupied by various military satellites, not to scale because of the different distances involved. Surveillance, navigation, and some weather satellites typically occupy orbits *A* and *B* at altitudes of 180 to about 1400 km (110 to 850 miles). Many early-warning, communications, and weather satellites are in the equator-circling geosynchronous orbit *C* at an altitude of 36,000 km (22,000 miles). Some communications and early-warning satellites are in the eccentric orbit *D*, which allows them to spend most of their time high above part of the earth at an altitude of about 40,000 km (24,000 miles), and a much shorter period low in orbit at altitudes of 250 to 700 km (150 to 420 miles) above another portion of the planet. The distant orbit *E*, at an altitude of 110,000 km (66,000 miles) is occupied by American satellites that watch for nuclear explosions. (Courtesy of Stockholm International Peace Research Institute, from Ref. 6, p. 45.)

These satellites occupy a range of orbits, depending on their functions. Photo reconnaissance satellites typically orbit at altitudes of 130-400 km. Electronic and ocean surveillance satellites typically orbit at 500-650 km. Navigation and some communication satellites orbit at 600-1500 km. Soviet early-warning satellites typically have eccentric orbits ranging in altitude from 600 to about 40,000 km. American early-warning satellites and many communication satellites occupy "geostationary" or "geosynchronous" orbit about 36,000 km up, where their 24-hour orbit makes them appear to stay above the same point on the equator.[9] The reasons for the different orbits are not clear, but it is not obvious that their geosynchronous orbits make U.S. satellites vastly more vulnerable to laser attack because the Soviet satellites also seem to hover over the same area for a long time.

Satellite Vulnerability

Satellites have evolved to be tough yet fragile and vulnerable to attack by lasers or other means. This unusual combination is due both to the peculiar nature of space and to a U.S. military strategy that until 1978 wasn't concerned whether satellites were able to survive a war.

Eberhardt Rechtin, president of the Aerospace Corporation and a former high Pentagon official, explained: "Prior to 1978, investments that had to be made to make satellites survive, even in relatively low-level conflicts, were simply not approved. In some cases, developers of spacecraft systems were told deliberately not even to consider electronic survivability"[10] because strategists had decided the extra cost was not worthwhile. In 1978 the United States shifted its policy, warning that space should be considered a potentially hostile environment. President Reagan has further modified that policy, giving the clear implication that American satellites should be designed to withstand attack. However, Rechtin said that it would take a few years before the new "survivable" satellites make their way into orbit.

Military considerations aside, space is a peculiar environment that can be both benign and hostile to satellite hardware. Space seems benign because there are few material things to get in the way. Above a few hundred kilometers, where there is essentially no air, structures far too flimsy to survive in the atmosphere or under the force of the earth's gravity can last indefinitely. Although the structures must be carefully packaged to survive the rigors of launching into orbit, once they are in space it is possible to unfold a delicate latticework of solar panels and antennas that might topple under their own weight on the ground. The faint wisps of air in the extreme upper atmosphere will eventually slow down satellites in low orbits to the point where they will fall from orbit like meteors. Otherwise, the only normal dangers to satellites are stray micrometeorites and pieces of "space junk," debris from other space launches that remains in orbit.

On the other hand space does put satellites through a kind of torture test. The strong accelerations required to put a spacecraft into orbit can damage delicate components. Once in space, the satellite must be able to operate in a vacuum and to withstand sharp temperature gradients (between its dark and sunlit sides) and rapid temperature changes (when the earth comes between the satellite and the sun). The satellite must be able to withstand short- wavelength radiation, normally blocked by the atmosphere, from the sun and other sources. It also must survive encounters with charged particles trapped by the earth's magnetic field and avoid accumulating damaging levels of electrical charge on insulating surfaces.[11]

Antisatellite Weapons

The environmental problems seem formidable enough that it isn't surprising engineers at first ignored the problems of defending a satellite against a hypothetical attack. In the early 1960s a satellite in orbit a few hundred kilometers above the ground seemed safely above any practical weapon system. Engineers were having enough problems getting satellites into orbit without having to worry about zapping the other side's.

However, as military satellites came to play increasingly important roles, the idea of antisatellite weapons began to attract military planners. The Soviet Union has an active program that may date back as early as 1963. There have been repeated tests of a "killer" satellite, loaded with conventional explosives, that destroys its target by approaching it and blowing up.[12] The Pentagon considers this weapon system "operational," that is, ready to be used when needed. So far, however, it has been used only against Soviet targets.[13] (Use against American satellites would be considered an act of war.)

The Soviet killer satellites are capable of use only in low earth orbits, at altitudes to perhaps several hundred kilometers. Different weapons would be needed to attack satellites in higher orbits, and American officials believe that the Soviets are working on such weapons. Direct physical attack may not be needed; the Joint Chiefs of Staff say that high-altitude satellites could be vulnerable to electronic countermeasures and to electromagnetic pulse effects from nuclear explosions above the atmosphere.[14]

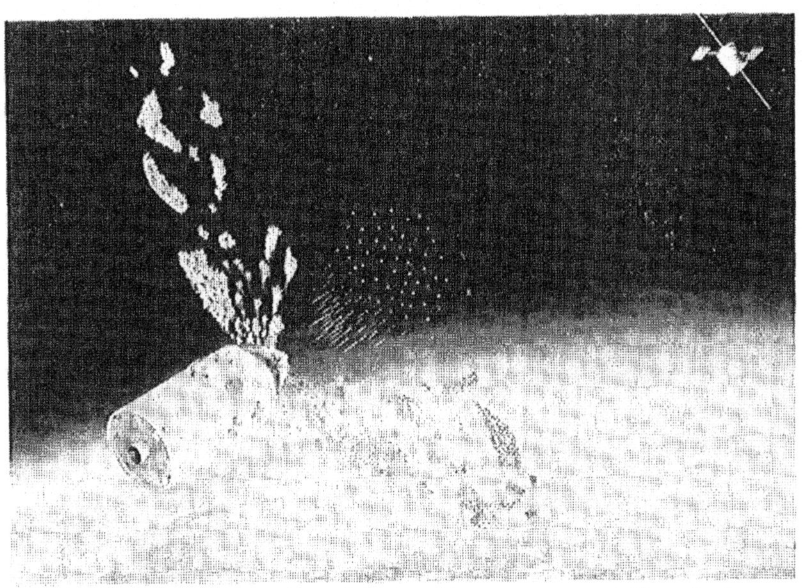

Soviet antisatellite weapon would explode in orbit near a target satellite, releasing many small pellets that would tear through and destroy the target. (Artist's conception courtesy of Department of Defense, from Ref 35, pp. 64-65.)

The United States is also working on antisatellite weapons but has taken a rather different approach. The leading concept is based on a two-stage "miniature homing vehicle" that would be launched from an F-15 interceptor aircraft. In fiscal 1983 the Air Force requested $212 million for the antisatellite program, roughly one-third of the service's budget for strategic-systems research. Plans call for demonstration of the system "in the not too distant future,"[15] and for it to be operational in the mid to late 1980s.[16]

This is the program that was blasted in a report from the General Accounting Office (GAO) in early 1983. According to the unclassified digest,

> When the Air Force selected the miniature vehicle technology as the primary solution to the antisatellite mission, it was envisioned as a relatively cheap, quick way to get an antisatellite system that would meet the mission requirements. This is no longer the case. It will be a more complex and expensive task than originally envisioned, potentially costing in the tens of billions of dollars.[17]

The unclassified version of the GAO report listed four alternative technologies that deserved consideration as antisatellite weapons: missiles, ground-based lasers, airborne lasers, and space-based lasers. All three types of antisatellite laser are on the Air Force's shopping list.[18] Indeed, antisatellite missions are at the head of practically every list of the potential military applications of laser weapons. The main reason is that many satellites are potentially very vulnerable to laser attack. This does not extend to other types of directed-energy weapons. Unlike laser beams, charged particle beams fall apart if they go from the air into space. Neutral particle beams might propagate in space, but the accelerator technology needed to generate them is much farther off than antisatellite lasers. The prospects for microwaves are not yet clear; they might be useful for "electronic warfare" attacks to scramble signals and confuse electronic systems, but their comparatively broad beams might endanger friendly satellites in the same part of the sky as the target.

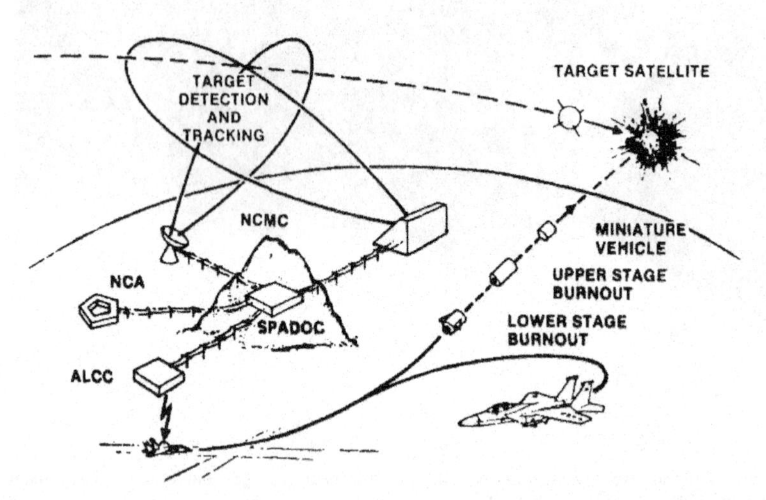

In the U.S. air-launched antisatellite system, target satellites would be tracked from the ground, producing data that would be processed at the Norad Cheyenne Mountain Complex (NCMC), the Space Defense Operations Center (SPADOC). On order from the "national command authority" (labeled NCA in the diagram and shown as the Pentagon), the data would be passed to the air-launch control center (ALCC), which in turn would pass the data to a missile on an F-15. The jet would then take off and launch the missile, which would home in upon and destroy the target satellite. (Courtesy of Department of Defense.)

Satellite vulnerability to laser depends on many factors. One of the simplest to quantify is altitude. The minimum spot size that a laser can produce increases with distance. If a beam can be focused to a 1-m (3-ft) spot size at the

360 km (220-mile) altitude of a surveillance satellite, by the time the beam reaches the 36,000 km (22,000 mile) height of a geosynchronous satellite, it will be 100 m (300 ft) in diameter. The focal spot from a ground-based laser will thus cover 10,000 times more area at the higher altitude. Because the same amount of power could be focused at either distance, this means that the laser intensity at geosynchronous orbit would be only 1/10,000 of that at the lower orbit, unless a larger mirror was used.

The optical and infrared spy satellites that occupy the low orbits where ground-based lasers could produce the highest intensities are particularly vulnerable because their operation depends on sensitive electronic eyes that monitor the ground. These electronic eyes are sensors of optical and infrared radiation, of various degrees of sensitivity and vulnerability to laser attack. Typically, visible light is observed with detectors made of silicon, which are much more rugged than the more exotic types used to observe at some infrared wavelengths. Infrared wavelengths around 10 μm, which are emitted by objects at room temperature, require sensitive detectors that must be cooled to far below room temperature, lest the faint signals they seek be overwhelmed by thermal radiation from objects around them. Heating of such sensors can disrupt their operation. There are also several other ways in which laser illumination can disable a sensor:

- Blinding with power levels much higher than would normally be seen. Like shining a searchlight in a person's face at night, this overwhelms anything else that might be in the field of view, but at lower power levels it won't cause permanent damage.
- Overloading the electronics by illuminating the sensor with so much power that the electronic signal it generates is too high for the electronics which process the signal to handle. This damage could degrade the performance of the system, leaving it operable but not working very well, or it could disable the system altogether.
- Thermal damage to sensors and/or electronics by heating them to temperatures at which the impurity atoms vital to the operation of semiconductor electronics diffuse within the material. The diffusion of these impurities within electronic devices would change their electrical characteristics. This, in turn, could degrade their performance or make them totally inoperative.
- Physical damage to the sensor, optics, and/or electronics, analogous to the physical damage a laser weapon could do to a target such as the skin of an aircraft. For example, a laser might heat a lens enough to shatter it. This type of physical damage probably would tend to follow thermal or overload damage, but it might occur first for some types of optical systems or laser pulses.

Spy satellites are not the only ones with vulnerable optical sensors. Most satellites have to be aligned with respect to the earth, the stars, or the sun to make certain that equipment such as transmitting antennas is pointing in the right direction. A communication satellite is useless if its narrow microwave beam is pointed into empty space rather than towards antennas on earth. One of the most common ways to provide this alignment is by optically sensing the positions of the earth, the sun, or certain stars. Such sensors are not as delicate as the prying eyes of spy satellites, but they nonetheless could be vulnerable to laser attack.

Arrays of solar cells, which generate power for many satellites, are also vulnerable to laser damage because they, too, are optical sensors. Like other optical sensors, they generate an electrical signal when illuminated by light, but solar cells are designed so the current they produce carries a significant amount of power. Because they are designed to withstand full sunlight, solar cells are not exactly delicate, but they can be damaged by intense enough light in much the same way as other sensors. Solar panels are also particularly vulnerable to other types of attack, such as debris from the explosion of a bomb. Such concern is leading military satellite designers to try to avoid them.

Even if sensors can be protected from laser attack, shining a bright laser beam on a satellite could heat it up enough to cause problems. Heat balance is a delicate problem for spacecraft because the only way to get rid of excess heat is to radiate it away. Radiative cooling tends to be slower than convective cooling, when surrounding air molecules absorb heat from an object and carry it away. It also cannot prevent the temperature of a satellite from rising somewhat when a laser beam is depositing energy on the satellite; from the standpoint of thermodynamics, the more energy (in the form of sunlight and laser energy) absorbed by the satellite, the higher its equilibrium temperature. If the laser illumination is intense enough, it could heat the satellite enough to damage electronic components.

The Pentagon is already working on ways to "harden" satellites so they can survive attack by lasers or other types of antisatellite weapons. Most of the major ideas for protection of sensors against laser attack were described in Chapter 9. Ablative coatings, which would evaporate when illuminated by a laser, or highly reflective surfaces might be used to protect against heating effects. New types of alignment systems are also being developed. Shifting to internal power supplies avoids the vulnerability of solar panels, although it tends to limit spacecraft lifetime. Work is also underway on laser warning devices, electronic countermeasures, decoys, and an active defense system intended to destroy the attacker. The Pentagon reportedly was planning to increase funding for satellite laser countermeasures from $5 million in 1983 to $15 million in 1984, with plans for spending around $45 million in fiscal 1988.[19]

Types of Antisatellite Lasers

Three basic approaches to antisatellite laser weapons are under consideration: putting them on the ground, in the air, or in space. If the laser were on the ground, or possibly in the air, an orbiting mirror might be used to redirect the beam. Each approach has advantages and limits.

Stationing the laser on the ground, or more likely on top of a tall mountain, would solve many problems. Current high-energy lasers are huge. They also require power sources that are generally bulky, either electrical power supplies or chemical fuels. Life is a lot simpler for the people building laser weapons if these bulky things can stay on the ground.

Simple accessibility is another advantage of having a high-energy laser on the ground. Today's high-energy lasers are a long way from operating with the reliability sought for a military system. They generally require a small army of PhD-level scientists to fine tune them for proper operation, as well as to repair them if necessary. Military planners hope that actual laser weapons will be easier to operate, but nonetheless they would be far easier to maintain and repair if they were on the ground. It would also be much easier to supply power and/or fuel to the laser if the laser site could be reached by trucks or power lines.

A ground-based laser can also point its beam steadily at a target satellite, avoiding concern over vibrations in an aircraft or motion of the laser itself in an airplane or satellite. In fact, the simplicity of a satellite's trajectory as viewed from the ground makes the satellite something of a sitting duck, simplifying the task of pointing and tracking.

The beam from a laser on the ground would have to travel through the atmosphere, which could interfere with transmission. The degradation of beam quality could be severe, depending on the laser power and on how fast the beam was moving relative to the air. There are compensating factors, however. Because satellites are vulnerable targets, the power that would have to reach them would be lower than for other laser targets, making it possible to spread the beam out in the air to avoid the heating effects that degrade transmission and to reduce the degree of precision required in aiming the beam. Instead of having to hit a spot smaller than 1 m, an antisatellite laser might have its beam spread out to several times the satellite diameter and still do its job. However, the long round-trip distance between laser and satellite, at least 400 km, seems to rule out the use of some adaptive optical techniques that can correct for atmospheric transmission problems over shorter distances, but which require continuous monitoring of atmospheric conditions.

Some problems are inherent in trying to use ground-based lasers against satellites. Foul weather could block the laser beam, and other atmospheric effects could break up the beam. The fixed location of ground-based lasers would also seriously limit their field of view in space, the number of satellites that they could zap, unless orbiting mirrors were used to redirect the beam. Although we think of satellites as being far above the earth, from a viewpoint in space they are actually close to the planet's surface. A satellite 400 km (240 miles) above the earth, for example, is only 7% of the planet's radius above its surface. (Comparatively speaking, that's thinner than the skin of a grapefruit.) Such a satellite is above the horizon at any point on earth only a small fraction of the time. Simple trigonometry shows that a satellite orbiting at 400 km would have to pass within 2300 km (roughly 1400 miles) of a point on the earth's surface to be above the horizon. It isn't reasonable for a laser weapon to shoot at a satellite near the horizon, however; the beam would have to pass through far too much air before reaching its target, and much of the light would be scattered or absorbed, just as sunlight is lost at sunset. If the satellite had to rise halfway up in the sky (that is, 45° above the horizon) to be vulnerable to laser attack, it would have to pass within about 400 km (240 miles) of the laser.

If the satellite passed due overhead at the laser site, it would be within range for only about 2% of its 43,000-km (26,000-mile) orbit, only a couple of minutes. Yet the satellite will not pass due overhead on every orbit because the earth rotates as the satellite revolves around it, so the satellite's path covers different portions of the Earth during each orbit. Most of the time it won't pass within laser range of a particular point on the globe. It might be possible to put certain satellites into orbits that would keep them just out of reach of one or two laser sites, but this wouldn't avoid the threat from a much larger number of lasers. A single ground-based laser could be used to keep spy satellites from looking at the area around it, but that would be taken as a clearly unfriendly act, and an unmistakable indication that something was going on there. (It would also violate provisions of the Anti-Ballistic Missile Treaty that prohibits interference with "national technical means" of verifying compliance–namely, spy satellites.[20]). Keeping spy satellites from looking inside the borders of a country would require placing ground-based laser weapons every several hundred kilometers (a few hundred miles) along the borders. This would require not only many lasers, but putting some of them in places such as the Pacific Northwest, where conditions seem far from ideal for propagation of laser beams.

Higher-altitude satellites may turn out to be easier targets for lasers on the ground. The higher the satellite, the larger is the area on the earth where it will be above the horizon. Thus, there are more places from which a laser could shoot at it; or, conversely, fewer lasers would be needed to provide a comprehensive defense. In addition, the higher the satellite, the more slowly it travels in its orbit; hence, the longer it would linger within range of the laser. From that viewpoint the ultimate sitting duck would be a geostationary satellite, which moves in its orbit at the same rate as the earth turns and hence seems to stay over the same point on the planet. Things aren't that easy, however, and offsetting these advantages are a couple of problems. As was mentioned before, the laser beam spreads out with distance, reducing the intensity with which a distant satellite can be illuminated. And even a geostationary satellite 36,000 km high is not high in the sky all over the globe; above latitudes of 39° North and South, for example, it would be less than 45° above the horizon because such satellites are kept in orbit around the equator to maintain stability of the viewing angle.

One proposed way to avoid the worst of the range limitations with a ground-based antisatellite laser is to use an orbiting mirror to redirect the beam.[21] The mirror in space could be controlled remotely to focus the laser beam over a much larger range. This would avoid some, but not all, of the geometric problems that otherwise would limit the range of a laser on the ground. There would still be some atmospheric absorption, but the space mirror might be able to improve the quality of the beam. However, there are a couple of serious problems. One is getting a large enough mirror into space in the first place. The other is holding it in place above the laser. The mirror would stay put if it was in geosynchronous orbit, but the distance from both the laser and the targets would be large enough to require an unwieldy mirror diameter, about 40 m if the laser divergence was one-millionth of a radian. If the mirror was at a lower altitude, it would keep moving out of range of the laser, so there would have to be many mirrors, one moving into position as the previous one moved out. Neither problem makes the concept impossible, but they do make it harder than it might seem at first.

Lasers in Airplanes

One way to get around the problems inherent in a ground-based laser is to put it into an airplane. Aircraft are mobile; military planes can fly above most of the atmosphere and all of the inclement weather. At a height of 10 km (6 miles or about 33,000 ft), fairly common for long-distance commercial jetliners, an aircraft is above all but one-quarter of the atmosphere. At 15 km (9 miles, or roughly 48,000 ft), only one-eighth of the atmosphere remains between the aircraft and space.[22] That's good enough to get rid of the worst of the atmospheric propagation problems.

Unlike a mountaintop laser, an airplane is mobile. Even if its speed were limited to roughly that of a commercial jetliner, it could fly across the width of the United States in about five hours. Because satellite orbits are generally well known, the laser-equipped plane could fly to the right place *before* the satellite came into range, then zap it. By flying along the satellite's path, it might be able to stretch the illumination time a little beyond what would be possible on the ground, although the plane would travel across the sky much slower than a satellite.

Getting a large laser into an airplane is no trivial task, but it's easier than putting one into orbit. The Air Force has managed to put a 400,000-W gasdynamic carbon dioxide laser into a Boeing KC-135, the military equivalent of a Boeing 707. This system, called the Airborne Laser Laboratory, was designed to test the feasibility of tactical Air Force lasers, as described in the next chapter. Although it indicates that a high-energy laser can be put in an aircraft, the Airborne Laser Laboratory does not have the beam-steering equipment that would be needed for antisatellite tests.

The organization that developed the Airborne Laser Laboratory, the Air Force Weapons Laboratory at Kirtland Air Force Base in Albuquerque, New Mexico, has studied prospects for antisatellite laser aircraft.[23] Air Force officials are said to believe that airborne laser demonstrations have proved the concept viable, although results have been mixed. Costs and time required have been greater than was expected, and the program suffered an embarrassing failure when the Airborne Laser Laboratory failed to shoot down air-to-air missiles in mid-1981 tests, as described in Chapter 2.

Putting an antisatellite laser into a plane will create some problems of its own. First, there's the matter of getting a working laser and power supply into the aircraft. The problem is not merely one of size. The laser, its optics, and the pointing and tracking equipment must be able to withstand the shaking of an airplane without drifting out of alignment. With tolerances in the optical system measured in fractions of a wavelength of light, this requires developing secure mounts for optical components and techniques for damping out vibrations that might affect the laser and optics.

Second, the pointing and tracking system must compensate precisely for various motions of the aircraft: overall motion relative to the ground, longer- term drift, and short-term vibrations and drifts caused by air currents and turbulence. Antisatellite lasers should tolerate minor variations in beam direction better than systems shooting at more difficult targets. However, it is still necessary to keep the beam on the target in order to do the job.

Third, getting above most of the atmosphere does not avoid atmospheric problems altogether. The plane itself creates an envelope of turbulence as it slices through the air. This "boundary-layer" turbulence affects the laser beam at the critical point where it enters the air. Adaptive optics can probably be of some help, but the problem is a complex one.[24]

Antisatellite Laser Satellites

The only way to get a laser above atmospheric problems altogether is to launch it into orbit. That has other advantages. A satellite is much more stable than an plane and not subject to vibrations caused by turbulence or jet engines. Its orbital path is well known, simplifying the task of pointing the beam, especially at a target satellite with a well-known orbit.

The big problem is getting a working laser weapon system into orbit in the first place. The laser and the fuel it requires are heavy. Weight can be reduced, but that comes only at a cost in capabilities: lower laser power and/ or a shorter operating life because the satellite carries less fuel.

Moreover, engineering a laser to survive in space is hard. Unless it was put on a manned satellite, it would have to be designed to operate on demand without any maintenance, a potentially large stumbling block. To pinpoint targets and aim the beam at them, it would require systems for fire and beam control on board the satellite. There would also have to be a jam-proof communication system to permit control from the ground.

The requirements for antisatellite lasers are modest compared with those for ballistic missile defense. Five million watts is the low end of the scale for missile defense, but Pentagon officials are said to consider a 200,000-W hydrogen fluoride laser with a 2.8-m mirror (corresponding to a 1-μrad beam) sufficient for an antisatellite laser satellite. The beam would have to stay on its target with a "jitter" of less than 1 μrad. Five orbiting antisatellite lasers supposedly would give worldwide coverage.[25]

Defending satellites, presumably against Soviet killer satellites, would require a more powerful system. The Pentagon is said to have considered lasers with output of 2-5 million watts, output mirrors 4-8 m in diameter, and jitter of 0.2 mrad or less. Five to ten orbiting lasers would be needed for such a system.[26] Antimissile laser battle stations could also perform the same function.

Plans for Antisatellite Lasers

What is actually being done to turn these antisatellite laser concepts into real hardware tends to be shrouded in secrecy. Unofficial reports indicated that the Pentagon had long-term plans to build two mountaintop antisatellite laser systems by around 1989,[27] even before the Air Force Space Command took the first steps to add antisatellite lasers to the Air Force's shopping list of new weapon systems.[28] The schedule envisioned by the Air Force has not been made public, and in any case would require approval by Congressional budget writers. The Soviet Union is said to have a carbon dioxide laser on the ground that could damage U.S. satellites in low orbits.[29] However, it is not clear exactly what that statement means. It is one thing to be able to damage a satellite that obligingly floats by through the laser beam when beam-transmission conditions are just right, and quite another to be able to zap satellites on demand. That statement probably means that U.S. analysts believe the Soviets have a laser able to produce enough power to damage U.S. satellites in orbit, not necessarily that the laser could do so reliably. For that matter it seems likely that if the output beam from a large American laser such as the 2-million-watt MIRACL was directed upward through a suitable optical system, it could damage Soviet satellites in low orbits.

Soviet Lasers in Space?

There have been repeated reports that the Soviet Union was about to launch a laser weapon into orbit. As far back as 1977, *Aviation Week* reported that American intelligence analysts had detected "ground testing of a small hydrogen-fluoride high-energy laser" and "preparations to launch the device on board a spacecraft." Supposedly the laser was intended for use against satellites, but nothing further was heard in public about any launch or space testing of such a laser satellite.[30]

Operation of the U.S. antisatellite missile system is shown in these two artist's conceptions. A two-stage missile would be carried high into the atmosphere by an F-15 fighter, as shown at top, then the missile would be launched independently to attack an enemy satellite, as shown at bottom. (U.S. Air Force photos.)

In October 1981 there were newspaper reports that "a large Soviet satellite carrying what appears to be a high-powered laser weapon is awaiting launch at a Soviet military space center, according to top-level U.S. defense officials. . . . If a suitable launch booster is available, the giant laser could be test-firing in earth orbit within a matter of weeks."[31] The following year there were a number of reports indicating that high Pentagon officials believed the Soviet Union might put a large laser in space as early as 1983.[32] However, as of October 1983, there had been no public indication that any such laser was in orbit.

Belief that the Soviets may be pushing for an early demonstration of antisatellite laser weapons appears to be widespread in Washington. Such sentiments were expressed by Lt. Gen. Kelly Burke (at the time deputy Air Force chief of staff for research, development, and acquisition) in an unusual early-1981 interview published in the *Wall Street Journal.* He said, "I wouldn't be surprised to see a ground-based laser antisatellite system" in the Soviet Union within five years and added that he "wouldn't rule out" a space-based Soviet laser by the same time. But he stressed that such a Soviet weapon would be very expensive, relatively crude, and have only limited capabilities.

Burke also said that he didn't favor a crash U.S. effort to match the Soviet program and that less exotic types of antisatellite weapons would be more effective. "If we were to build such a laser system today, I don't think we'd be very proud of it." The Soviet Union "might see a lot of political utility in a system that would have very little military utility. I don't think we ought to just throw money into something, just because they're doing it, when we don't know what we're doing."[33]

Similar sentiments were expressed in early 1982 by Richard DeLauer, undersecretary of defense for research and engineering, during testimony before the House Armed Services Committee. He said that if the Soviets could put a chemical laser into orbit within five years it would be more a Sputnik-type demonstration with psychological impact than an indication of militarily useful capabilities.[34] The 1983 edition of the Pentagon's booklet *Soviet Military Power,* which presumably reflects the official Department of Defense position, says: "The Soviets could launch the first prototype of a space-based laser antisatellite system in the late 1980s or very early 1990s. An operational system capable of attacking other satellites within a few thousand kilometers range could be established in the early 1990s."[35]

Laser-weapon enthusiasts have complained that Pentagon officials are dragging their heels on development, so such statements could be viewed as being defensive. However, they also reflect an appreciation of a central fact that was stressed in earlier chapters: it takes much more than a high-energy laser to make an effective laser weapon system. It is entirely possible to orbit a high-energy laser that is functionally useless as a weapon because it is not equipped with fire-control and beam-steering systems good enough to hit the proverbial broad side of a barn.

If the goal was simply a demonstration, it wouldn't hurt too much to skimp on laser fuel, leaving only enough to let it fire a few shots to impress the people on the ground. No one on earth could look inside the fuel tanks to see that they were empty. Nor could they be sure if the laser stopped firing because it was turned off or because it couldn't work any more.

It would even be possible to stage a rather impressive, but nonetheless essentially faked, demonstration of an antisatellite laser weapon by aiming at a friendly target satellite. The friendly satellite could be a "cooperative" one that would intentionally return light to the laser weapon. The light could come from a guiding laser on the target or from a special type of mirror called a "retroreflector," which would return a part of the incident laser beam back along its original path to the weapon system. The feedback would make it much easier to aim a laser weapon at such a "cooperative" target than at an actual enemy satellite, which would not return such a special beam. Such an aiming system was proposed for laser transmission of energy from solar power satellites to the ground, the theory being that reliance on that approach would avoid concern that the powerful laser beam could be used as a weapon. From the ground or from spy satellites it would be hard to be sure that such an aiming system was being used. That would make it possible to stage a fake demonstration of seemingly impressive antisatellite capabilities, without having a working antisatellite laser weapon system. (It also may be possible to stage similar fake demonstrations of other beam weapon missions for the benefit of the other side's spy satellites.)

The possibility of "putting on a show" for the other side's spy satellites points out the uncertainties inherent in intelligence gathering. Information collected from sources ranging from undercover agents to spy satellites has to be interpreted. The more fragmentary the information, the more interpretation is required. The end result is uncertainty, which is expressed in ways such as giving a range of possibilities (e.g., within one to five years the other side should have a particular capability). The problems are greatly magnified when trying to assess future capabilities because it is impossible even for the side doing the work to be sure of the outcome.

A range of possibilities is always difficult to interpret. There is a natural tendency for military officials to assume the "worst case" (i.e., the greatest possible advances by the other side) on the theory that it is better to be over-prepared than underprepared. The press, too, tends to pick up the worst case, transforming a warning that the Soviets could have a laser satellite in orbit in one to five years to a headline warning of a possible launch next year. Sometimes these effects combine to transform a "worst-case" estimate into *the* estimate, particularly if someone is using the information to justify a budget request. The end result can be embarrassment, when critics discover that promised threats have failed to materialize. Such was the case when American military analysts finally got their hands on a Soviet Foxbat fighter that a defecting Russian pilot had flown to Japan, only to find that the much-feared fighter was equipped with vacuum-tube electronics and had a rated top speed far below American estimates.[36]

American Antisatellite Program

Although the United States is clearly working on antisatellite lasers, little has been said about them in public. There have been no public reports of actual demonstrations and some of the major high-energy laser "test beds" apparently are not equipped for antisatellite tests. Experiments with the Air Force's Airborne Laser Laboratory are reportedly only "generally applicable" to studies of aircraft-based antisatellite lasers.[37]

The Air Force reportedly does plan to demonstrate antisatellite laser technology, but that demonstration is to be conducted on the ground. Goals were said to include achieving ranges to 500 km (300 miles), something which could not be done in a straight line on the ground.[38]

The technological barriers to developing antisatellite lasers seem modest compared with those of building antimissile lasers, but they are still significant. There are problems that go beyond illuminating a satellite with a laser beam. A laser weapon has to serve some militarily useful purpose, a criterion that may not be met by a system that can disable only satellites obliging enough to pass directly over a mountaintop laser. Putting lasers in the air or on satellites could solve some problems, but would raise others. There is also the question of where antisatellite weapons would fit into the strategic balance, a question addressed in Chapter 14.

References

1. Phillip J. Klass, "Anti-satellite laser use suspected," *Aviation Week & Space Technology,* December 8, 1975, pp. 12-13. Although I cannot recall seeing any report in *Aviation Week* disclaiming the involvement of laser weapons in this incident, more recent articles in the magazine do not reference it.
2. "Washington roundup," *Aviation Week & Space Technology,* February 21, 1983, p. 13.
3. Craig Covault, "Space Command seeks Asat laser," *Aviation Week & Space Technology,* March 21, 1983, pp. 18-19; see also "Air Force and GAO push antisatellite laser weapons," *Lasers & Applications* 2 (5), 28-30 (May 1983).
4. General Accounting Office, "U.S. antisatellite program needs a fresh look," GAO/C-MA-SAD-83-5 report, issued January 27, 1983 (unclassified digest of classified report); see also Ref. 3.
5. Bhupendra Jasani, "How satellites promote the arms race" *New Scientist* **96,** 346-348 (November 11, 1982).
6. Actual resolution is a highly-guarded secret; these figures are estimates cited in Bhupendra Jasani, ed, *Outer Space-A New Dimension of the Arms Race* (Oelgeschlager, Gunn & Hain, Cambridge, Massachusetts, 1982), pp. 46-47.
7. "Soviets build directed-energy weapon," *Aviation Week & Space Technology,* July 28, 1980, pp. 47-50, an artist's conception based on satellite photos appears on p. 48.
8. This issue has been debated in many forums. See, for example: Senate Committee on Armed Services, *Hearings on Department of Defense Authorization for Appropriations for Fiscal Year 1983, Part 7 Strategic and Theater Nuclear Forces,* pp. 4216-4225.
9. Data taken from Appendix A in Bhupendra Jasani, ed, *Outer Space-A New Dimension of the Arms Race* (Oelgeschlager, Gunn & Hain, Cambridge, Massachusetts, 1982).
10. Martha Smith, "Eberhardt Rechtin: C^3 is the heart of any war in space," (interview) *Military Electronics/Countermeasures* **8** (11), 9-20 (November 1982).
11. Collection of charged particles on certain insulating surfaces can generate a trouble-causing charge differential across the surface. This problem and efforts to deal with it are described in several papers in *IEEE Transactions on Nuclear Science* **NS-29** (6), 1584-1653 (December 1982).

12. Bhupendra Jasani, "How satellites promote the arms race," *New Scientist* **96,** 346-348 (November 11, 1982); for more details see Bhupendra Jasani, ed, *Outer Space-A New Dimension of the Arms Race* (Oelgeschlager, Gunn & Hain, Cambridge, Massachusetts, 1982).
13. Department of Defense, *Soviet Military Power* (U.S. Government Printing Office, Washington, D.C., October 1981), p. 68.
14. Organization of the Joint Chiefs of Staff, *United States Military Posture for Fiscal Year 1983,* p. 77; see Ref. 4 in Chapter 10.
15. House Committee on Armed Services, Hearings on Military Posture and Department of Defense Authorization for Appropriations for FY 1983, R&D Title 2, part 5 of 7 parts, p. 494.
16. Organization of the Joint Chiefs of Staff, *op. cit.,* p. 77.
17. General Accounting Office, "U.S. antisatellite program needs a fresh look," GAO/C-MA-SAD-83-5 report, issued January 27, 1983; this is an unclassified digest of a classified report.
18. Craig Covault, "Space Command seeks Asat laser," *Aviation Week & Space Technology,* March 21, 1983, pp. 18-19.
19. "Washington roundup," *Aviation Week & Space Technology,* November 22, 1982, p. 15.
20. D. L. Hafner, "Antisatellite weapons: the prospects for arms control," in Bhupendra Jasani, ed, *Outer Space-A New Dimension of the Arms Race,* pp. 311-323 (Oelgeschlager, Gunn & Hain, Cambridge, Massachusetts, 1982).
21. See, for example, Wallace D. Henderson, "Space-based lasers: Ultimate ABM system?) *Astronautics and Aeronautics* 20, 44-53 (May 1982).
22. Morris Nieburger, James G. Edinger, and William D. Bonner, *Understanding our Atmospheric Environment,* 2nd ed (W. H. Freeman, San Francisco, 1982), p. 101.
23. "Antisatellite laser weapons planned," *Aviation Week & Space Technology,* June 16, 1980, p. 244.
24. For an unclassified account of the fundamental problems, see Keith G. Gilbert and Leonard J. Otten, eds, *Aero-Optical Phenomena* (American Institute of Aeronautics and Astronautics, New York, 1982).
25. "Pentagon studying laser battle stations in space," *Aviation Week & Space Technology,* July 28, 1980, pp. 57-62.
26. *Ibid.*
27. "Antisatellite laser weapons planned," *Aviation Week & Space Technology,* June 16, 1980, p. 244; these long-term plans may well have changed since this unofficial report. The Reagan administration has accelerated laser weapon development beyond what the Carter administration had planned.
28. Craig Covault, *op. cit.*
29. "Technology eyed to defend ICBMs, spacecraft," *Aviation Week & Space Technology,* July 28, 1980, pp. 32-42.
30. Clarence A. Robinson, Jr., "Soviets push for beam weapon," *Aviation Week & Space Technology,* May 2, 1977, pp. 16-22.
31. One wonders if this system may still be waiting for the massive Saturn-V type booster the Soviets have been developing for a dozen years, described in Chapter 11. This report is from Keith Hindley, "Soviet laser weapon reported," *Boston Globe,* October 13, 1981.
32. Such accounts appeared several places. George C. Wilson, "Soviets reported gaining in space weapons," *Boston Globe,* March 3, 1982, p. 9; "Soviets outspending U.S. on space by $3-4 billion," *Aviation Week & Space Technology,* July 19, 1982, pp. 28-29; Howard Roth, "D.C. Circuit," *Electronic Engineering Times,* August 16, 1982, p. 32.
33. Walter S. Mossberg, "Soviet could build laser weapon to kill satellites in 5 years, Pentagon aide says," *Wall Street Journal,* February 11, 1981, p. 6.
34. House Armed Services Committee, Hearings on Military Posture and Department of Defense Authorization for Appropriations for Fiscal Year 1983, part 5 of 7, R&D Title 2, p. 571.
35. Department of Defense, *Soviet Military Power 1983* (U.S. Government Printing

Office, Washington, D.C., March 1983), p. 68.

36. See, for example, "The Foxbat story," *Astronautics* & *Aeronautics* 15 (1), 20 (January 1977).
37. "Antisatellite laser weapons planned," *Aviation Week* & *Space Technology* June 16, 1980, p. 244.
38. Clarence A. Robinson, Jr., "Defense Dept. prepares its budget for White House talks," *Aviation Week* & *Space Technology,* November 15, 1982, pp. 14-16. This article erroneously indicates that there were plans to spend $87 million on the antisatellite program in fiscal 1984. That figure actually refers to a much larger Air Force program in "Advanced Radiation Technology" which covers the Airborne Laser Laboratory experiments and other work as well as the antisatellite effort. No breakdown of funding for individual elements of the overall program is available.

13. Battlefield Beam Weapons

When the Pentagon first started looking at directed-energy weapons, its planners were looking for a new generation of tactical weapons for battlefield use, whether on the ground, at sea, or in the air. The lion's share of the more than $2 billion the Department of Defense has invested in laser weaponry has gone toward that goal. The recent interest in strategic beam weapons for applications such as missile defense and antisatellite warfare have grabbed the headlines away from tactical systems, but the latter remains a significant program.

Speaking in late 1981, Defense Advanced Research Projects Agency director Robert Cooper said that $2 billion was "an enormous amount of money" for what still remains an exploratory development program. Yet he added that "it's the collective judgment of high officials in the Pentagon that laser weapons present high potential for payoff." There is a very good chance, he added, "that we will put a high-energy laser weapon system on the battlefield," with the Army being the leading candidate to put a system into use first.[1]

It was tactical beam weaponry that was envisioned by science fiction writers such as H. G. Wells. In part, that was because he wrote long before the development of strategic weapons of mass destruction. *The War of the Worlds* was published a decade before the Wright brothers got their first airplane off the ground, in an era when physicists still saw atoms as immutable little balls. In addition, Wells and the writers who followed him had another concern: to earn their keep writing science fiction, they had to tell a story. Stories involve struggles between men (or, in the case of *The War of the Worlds* and much other science fiction, between men and alien beings), and these struggles are carried out with tactical weapons, ray guns and spaceships. With a few exceptions, such as the Death Star of *Star Wars,* fiction writers tend to ignore the use of strategic weapons.

The battlefield is also where the minds of soldiers are molded. The military officials who first became interested in directed-energy weapons in the 1960s had been trained in the classic tradition of war as a series of battles fought between armies. They and their teachers had experienced two such wars, World War II and the Korean War. World War II, in particular, had shown the importance of technology on the battlefield. It was only natural that they should think in such terms.

Moreover, directed-energy weapons could undoubtedly have some very useful capabilities on the battlefield. A key attraction is the high speed of the destructive beam. Light travels at 300,000 km (180,000 miles) per second. That is about 40,000 times the speed of the swiftest rockets and essentially eliminates the need to aim ahead of or "lead" a fast-moving target a few kilometers (or miles) away. Pentagon officials like to say that in the six-millionths of a second that light takes to travel one mile, a jet traveling at twice the speed of sound moves just a little over one-eighth of an inch. Although laser weapons do not have to be aimed ahead of their targets, the laser must track or follow the target for a second or keep the beam on target long enough to do mechanical damage.

Energetic charged or neutral particles can travel close to the speed of light in a vacuum, but boring a path through the air slows a charged particle beam down to about 1000 km/sec. That's still 1000 times the speed of the fastest jets, so little if any leading would be needed. A single shot from a particle-beam generator is expected to do the job, so it shouldn't be necessary to keep the beam on target for long.

Speed is not the only advantage. A highly directional beam should be able to pick out and destroy a single enemy target without damaging friendly forces nearby, an important capability in real-world battles where two air forces may be hopelessly intermingled. Laser beams could be quickly and easily moved from target to target anywhere in the sky simply by moving the output mirror. (Redirecting a particle beam may be a much more serious problem, however.) Because they do not need a missile to carry them, nor a special device called a fuze to trigger the explosion of a warhead at the target, beam weapons are expected to be less vulnerable to electronic countermeasures than conventional tactical missiles.

Beam weapons are expected to pack plenty of firepower if they make it to the battlefield. A laser or particle-beam generator might be bulky, but it could carry enough fuel for many shots and could be refueled to produce many more. A particle-beam generator might be able to produce several shots per second, with a laser able to kill a target every second or two. That's much faster than tactical missiles can be fired on a sustained basis. The rapid-fire capability, combined with rapid retargeting for a laser, could be a big plus on the fast-moving high-technology battlefield, where the air is filled with weaponry and the pace may be too fast for soldiers to do anything but hide until it's all over.

The Pentagon is also looking to beam weapons to save money. That may sound strange, given that a battlefield laser system would probably cost at least $5-$10 million, and that particle-beam systems would probably cost even more. Yet military estimates show that at least for lasers, each shot would be much cheaper than firing a tactical missile. As short-range missiles have increased in sophistication, their prices have soared. Surface-to-air Stinger missiles, which weigh 10 kg (22 lb) and have a 5-km (3-mile) range, cost about $20,000 each. The longer-range surface-to-air Patriot missile, which weighs a ton, costs the Pentagon $300,000 to $500,000 each. In contrast, Pentagon officials estimate that the comparatively expensive deuterium fluoride laser would use about $1000 to $2000 worth of fuel for a shot, which they consider equivalent to the cost of a burst of bullets. Fuel for a single carbon dioxide laser shot might cost only several hundred dollars.[2]

Most of the work on tactical beam weaponry has focused on lasers, but there is some interest in other approaches. The Navy is studying the use of charged particle beams to defend battleships against cruise-missile attack; it was a Navy program for such applications called Chair Heritage that eventually evolved into the Defense Advanced Research Projects Agency's tests of the Advanced Test Accelerator (see Chapter 7). Concepts for tactical microwave weapons remain largely in the speculative "blue sky" stage, with better definition awaiting an improved understanding of microwave effects. Because the laser program is much broader and seems somewhat better defined, I will concentrate mainly on laser weapons.

Types of Missions

In theory, beam weapons could perform varied missions on the ground, at sea, or in the air. As one defense official noted informally, "lasers can shoot anything that moves or flies, but they can't kill four inches of tank armor." That exception is a noteworthy one, though. It arises from the fact that it takes momentum to penetrate armor, and for all practical purposes laser light does not have momentum. Projectile weapons such as missiles and artillery shells get their penetrating power from their momentum, and they will continue to be used against armored targets even on battlefields filled with beam weapons. Beam weapons might be used against armored targets in cases where there was an Achilles heel, such as a window for sensors that was vulnerable to laser attack.

Military developers seem to have looked seriously at the prospects of using lasers to zap just about everything on a battlefield *except* heavily armored vehicles. All three armed services have been conducting their own programs, reflecting in part genuinely different needs. Different types of lasers will probably be needed to attack different targets. And the different services also need to put weapons in different environments. A ship, for example, is a very different environment from the inside of an aircraft, and both are quite different from what would be found inside a tank.

Yet the three separate programs also reflect the traditional bureaucratic rivalry of the armed services. Critics have said that it's wasteful to have three separate programs so early in development of the technology. One prominent laser enthusiast, Senator Howell Heflin (D-Ala) has complained that dividing the Pentagon's laser weapon program among the three services and the Defense Advanced Research Projects Agency has caused resources to be devoted to comparatively few technology demonstrations rather than on more-needed general research. He urged forming a strong central office to coordinate the Pentagon's laser weapon program, and forming a National Laser Institute to coordinate all government-sponsored civilian and military laser research.[3] Providing such high-level coordination was one evident goal of an early 1982 reorganization which put the head of the directed-energy weapon program directly under the undersecretary of defense for research and engineering. That slot, initially filled by Major General Donald L. Lamberson, is now occupied by Brigadier General Robert R. Rankine, Jr.

Despite the fundamental differences in the nature of laser weapon missions, there are some important common features to tactical laser weapons. The best way to look at them is by taking the nature of the target as a starting point. There are three main potential targets: soldiers, sensors, and other hardware.

Antipersonnel Lasers

The ways in which beam weapons might actually be used against soldiers bear no resemblance to the near-instantaneous incineration envisioned by H. G. Wells in *The War of the Worlds* or to the instantly fatal death rays of pulp science fiction. It is theoretically possible to cook soldiers with intense microwave beams, but the large antennas needed would be impractical on the battlefield. Charged particle beams, if they could make their way through the air, could to kill soldiers, but it would be rather like shooting flies with a bazooka. A tightly focused laser beam could burn the skin but would not be an efficient way to bum a soldier to death except at near point-blank range, where no rational soldier would be intentionally. However, one part of the body is very vulnerable to damage by laser light: the eyes.

The problem with the eye is similar to that with other types of optical sensors: it is extremely sensitive to light. This sensitivity varies widely with wavelength and is highest at visible wavelengths. Prolonged staring directly at the sun, or directly into a laser beam carrying only several thousandths of a watt of visible light can cause permanent damage to the retina, the thin layer at the back of the eyeball that senses light and relays images to the brain. This can happen because the lens of the eye focuses visible and near- infrared light, concentrating its power to high enough levels to bum the retina. Higher powers take less time to cause damage, with short, fairly intense pulses being particularly dangerous. The result is not total blindness but rather partial obstruction of vision, which may be permanent or temporary depending on the laser power.

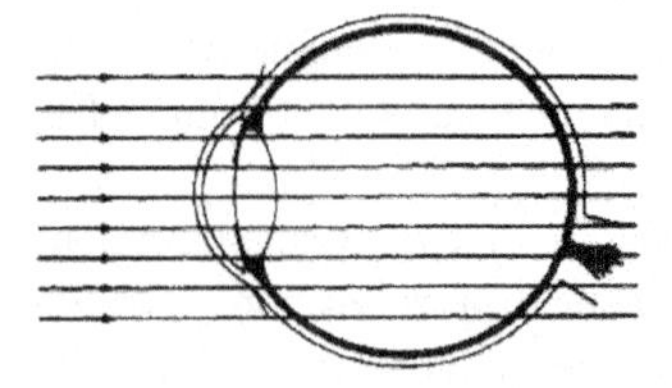

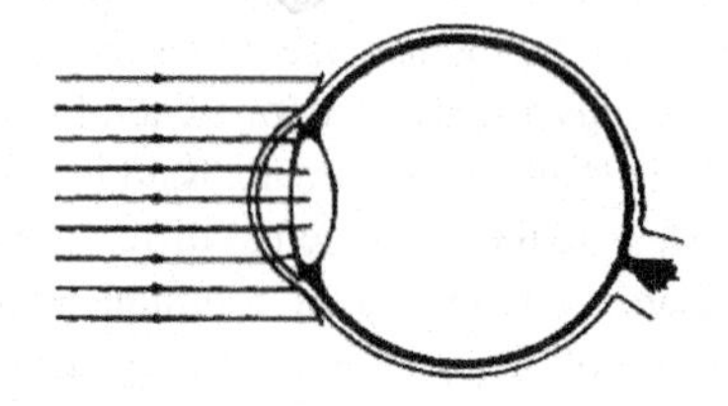

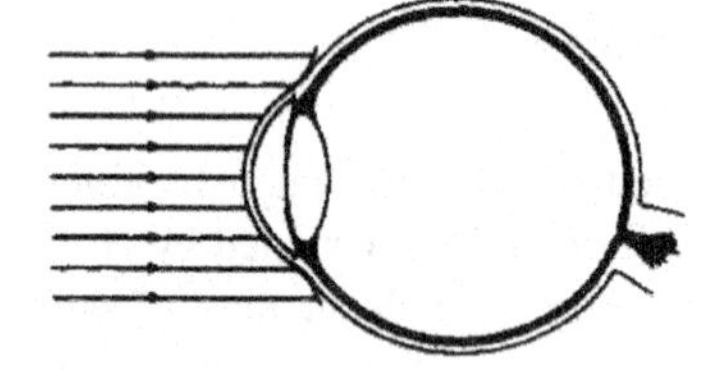

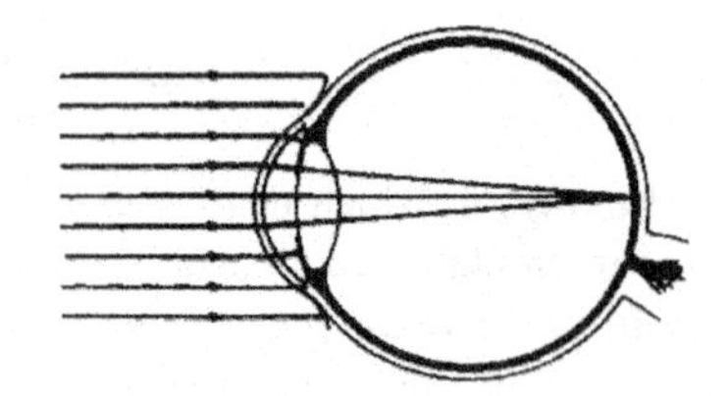

The interaction of electromagnetic radiation of various wavelengths with the eye. Microwaves, gamma rays, X rays, and some other types of radiation pass right through without much interaction. Near-ultraviolet radiation, which for practical purposes is that transmitted well by the atmosphere, is absorbed by the lens of the eye. Infrared radiation longer than a few micrometers is absorbed by the eye's outer layer, the cornea, where intense laser beams can cause burns. Visible and near-infrared laser beams are focused to a point on the retina at the back of the eyeball and can cause damage if they are intense enough. (Courtesy of David Sliney and Myron Wolbarsht. From Ref. 6, p. 3.)

Military organizations have long been concerned about the effects of laser beams on the eye. Their original interest came not from laser weaponry but from the use of low-power visible and near-infrared lasers to find the range to targets and to mark spots for "smart bombs" to home in on. Such laser systems produce short pulses too weak to cause serious skin bums but that could permanently damage the eye.

Needless to say, that's not a concern if the laser is aimed at enemy soldiers during a battle. However, it does present difficulties if friendly troops are involved in "war games" or training exercises. Such training is vital to military preparedness and should involve the use of realistic weapons. In the United States concern over eye hazards to soldiers has led to research on new laser materials that emit "eye-safe" wavelengths that can't penetrate the eyeball. The problem is even worse in Europe, where civilians often are so close to military training grounds that they, too, could be endangered by stray laser pulses even if no live ammunition is fired.

Eye damage causing partial blindness would be only an incidental effect of the use of laser rangefinders and designators in a war where more deadly weapons were being guided by the laser systems. Yet it is also possible, at least in theory, to use a high-power laser to try to blind enemy soldiers totally, either permanently or temporarily, depending on laser power and wavelength. If the United States is working on this concept, it is being kept extremely quiet. There have been unofficial reports that Soviet chemical lasers were used to blind Chinese soldiers during the war between China and Vietnam in the late 1970s.[4] One official Pentagon publication, the *Soviet Military Power* brochure published in October 1981, mentioned in passing that the Soviet Union was interested in antipersonnel laser weapons.[5]

The type of physical injury that a laser can cause depends on the laser power, pulse duration, and wavelength. The wavelength is particularly critical in determining what type of eye damage, if any, will occur. Light with wavelengths between about 0.4 and 1.4 μm, in the visible and what is called the "near" infrared regions, can penetrate the eyeball. The lens focuses this light to a point on the retina, the light-detecting layer at the rear of the eyeball. A short, intense pulse can severely burn the retina, causing bleeding and a permanent blind spot. Longer-term exposure to lower laser powers can also impair vision in some spots on the retina.[6] The eye damage may be serious, and may permanently interfere with vision, but is not enough to blind a soldier completely. It is unclear how effective such eye damage would be in putting the enemy out of action. An eye-damaging laser pulse might leave him able to fight and madder than ever.

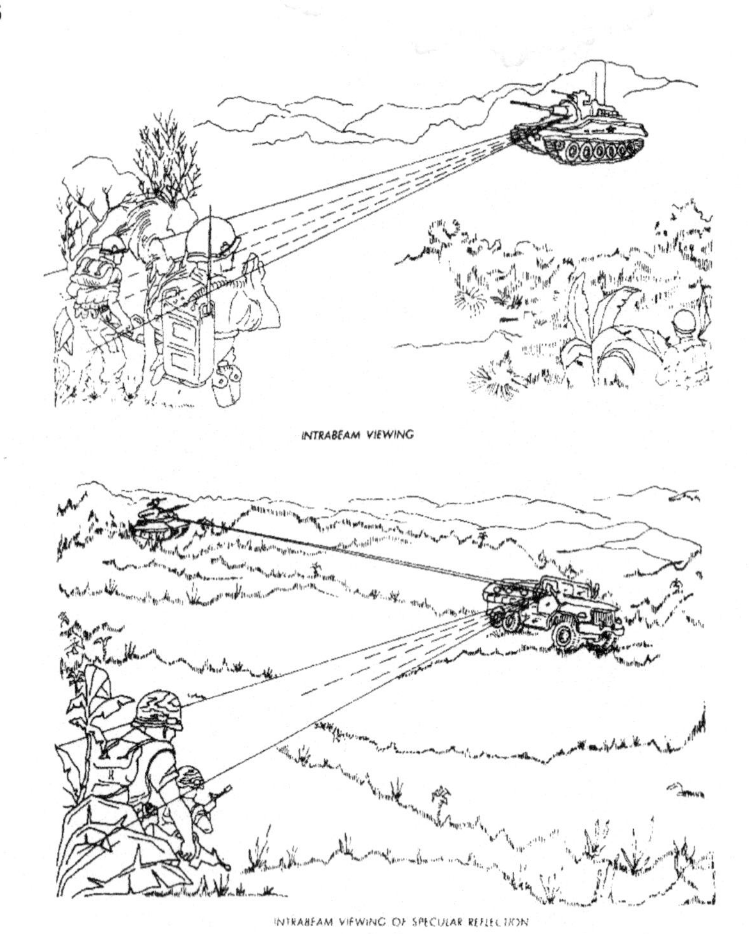

Soldiers can be zapped in the eye with pulses from a laser rangefinder either by standing directly in the way of the beam (top) or by being hit by reflections of the beam from mirror-like surfaces, such as the rear-view mirror of a truck (bottom). Either type of incident can cause eye damage, even though that is not the specific purpose of the laser system. (Courtesy of David Sliney and Myron Wolbarsht. From Ref. 6, p. 658.)

Light at slightly shorter and longer wavelengths can penetrate the eye slightly, but much shorter and longer wavelengths cannot. Light that can't penetrate the eyeball can still cause damage. Intense ultraviolet light can cause a variety of problems, including temporary blindness and a form of damage to the cornea (the outer surface of the eye) that is similar in nature to sunburn, an effect also caused by ultraviolet radiation. The corneal bum seems to depend on total exposure, with little sensitivity to how fast or slow the exposure occurred. The thresholds are fairly low, about 0.005 J/cm^2, equivalent to 0.005 W/cm^2 for 1 sec or 0.0001 W/cm^2 for 1 min.[7] However, like sunburn, the effect typically takes a few hours to show up, a delay that probably makes it unsuitable for weapon use.

Short, intense pulses from infrared lasers, lasting one ten-millionth to one billionth of a second, can be quite dangerous to the eye. They can bum the cornea in a way not yet fully understood, but different than ultraviolet light. Experiments have shown that injury can occur at 0.005-0.01 J/cm^2.[8]

Long exposures to long-wavelength infrared light such as that produced by a deuterium fluoride or carbon dioxide laser can also bum the cornea, which absorbs nearly all the incident light at those wavelengths. The eye's natural blink reflex provides a safety mechanism because infrared intensities high enough to damage the cornea also cause pain. Continuous laser powers of more than about 10 W/cm^2 would be needed to damage the cornea by delivering about 0.5-10 J/cm^2 before the eyelid could shut.[9] Once the eyelids were closed, the strong absorption of skin would prevent long infrared wave- lengths from reaching the eye. The soldier could also move to hide his eyes from the laser light, by turning away or burying his face in his hands or on the ground.

Infrared laser powers comparable to those needed to burn the cornea could also bum exposed skin, which strongly absorbs infrared radiation. Indeed, the skin may be more sensitive to laser light at some infrared wavelengths than the eye; there have been reports of intense infrared light caused by explosions being powerful enough to bum the skin, but not harming the eyes of people exposed.[10] The power levels required to bum the skin are high, but within the range of laser weapons. Burns can be caused by a continuous beam delivering about 10 W/cm^2–roughly the intensity of a 100,000-W beam focused to a 1-m (3-ft) spot, something which should be possible on the battlefield. The skin absorbs similar percentages of light at carbon dioxide and chemical laser wavelengths but absorbs much less in the visible and near ultraviolet spectral regions.[11] (The seemingly obvious differences between light absorption of fair and dark skins exist only at visible and near-infrared wavelengths; at longer infrared wavelengths there is no significant difference in light absorption.)

A bright flash of visible light that isn't intense enough to cause permanent eye damage can still cause a phenomenon called "flash blindness," which is familiar to anyone who has ever been on the receiving end of a flashbulb. Afterimages from the flash can obscure vision until they fade away with time. How long the effect lasts depends both on the intensity of the flash and the amount of illumination on the scene. Military simulations of the effects of a nuclear explosion flash far from the blast show that a person outside on a clear day might be blinded for 10 sec, while a pilot flying an airplane at night might be unable to see for around 100 sec.[12] Lasers might be used to cause such flash-blindness effects.

Some simple measures could provide soldiers with considerable protection against laser effects. Even ordinary clothing would provide some protection, and a suit of aluminum-foil armor would do better. Special safety goggles have been developed that strongly absorb light at certain laser wavelengths; they are used in laboratories to keep the wearer's eyes from being zapped by a stray laser beam. As mentioned in Chapter 9, military researchers concerned with flash blindness have developed optical materials that darken in a few thousandths to a few millionths of a second–not quick enough for some applications, but much faster than the eye's blink reflex.

Nature has been kind in providing some simple protection against the infrared and ultraviolet. Ordinary lime glass absorbs strongly at ultraviolet wavelengths shorter than about 0.3 μm. Pyrex glass absorbs strongly infrared wavelengths longer than about 4 μm, soaking up about 80% of the 3.8-μm output of a deuterium-fluoride chemical laser and virtually all of the 10-μm output of a carbon dioxide laser.[13] Thus, simple spectacles, or the window of a tank or airplane cockpit, can protect against many present laser threats, unless intensities are high enough to shatter the glass. One wonders if any of the Chinese soldiers who reportedly encountered the Soviet laser were wearing glasses.

Near-infrared and visible lasers are not as powerful as those emitting long-wavelength infrared light, but they are harder to defend against and operate at wavelengths where the eye is especially sensitive. Virtually all common optical materials transmit light from visible and near-infrared lasers as readily as glass transmits visible light. Not only do optical instruments fail to protect soldiers, but they can be downright hazardous. For example, a pair of binoculars can concentrate a diffuse pulse of laser light enough to endanger the eyes of a soldier staring through the binoculars. Special safety goggles are available that can block selected laser wavelengths, but the use of such goggles is practical only if the laser wavelength is known in advance. Even then, they often obstruct vision, and even laboratory researchers have found them uncomfortable to wear. Most rangefinders and designators now operate at 1.06 μm in the near infrared, but researchers are working on other types of lasers that would operate at other wavelengths, some of which are potentially dangerous to the eye. Others would be "eye safe" because they could not penetrate into the eyeball.

None of the U.S. military publications I have seen give any indication that laser rangefinders and designators are considered as weapons for use against enemy soldiers–at least on an official level. Unofficially, in the heat of battle soldiers are liable to shoot anything they can at the enemy. That could include the beams from laser rangefinders and designators. While such beams could cause eye damage to anyone close enough, it is unclear how useful they would be militarily in disabling enemy soldiers.

In practice, two simple logistic problems may mitigate against the use of lasers as antipersonnel weapons. Both stem from the fact that a laser beam travels in a straight line. Although that is an advantage for many types of laser weapons, it may be a serious disadvantage for use against soldiers. Armies teach their soldiers to stay out of sight, and that is one lesson that soldiers learn very well because the penalty for failure on the battlefield can be death. Conventional projectile weapons can go after soldiers in their hiding places. A shell or grenade can be lobbed into the air so it drops into a trench and kills the enemy soldiers hiding there. A nearby explosion of a shell can kill a soldier hiding behind a tree. However, a laser beam can zap only soldiers who are standing in the line of fire. Matters are made worse by the fact that this is a two-edged sword. The soldier *firing* the laser has to be standing in a straight-line path to his target, putting *himself* in the line of fire. Most soldiers have more sense than that, a fact that may present problems because some guidance systems require the soldier steering a missile to be standing in an exposed position (leading to the interest in "fire-and-forget" weapons described in Chapter 10).

There is little doubt that a high-energy laser beam sweeping across the battlefield could be a potent psychological weapon, particularly if it could ignite some dried brush along the way. It might be militarily useful against soldiers with little protective equipment. But it would not be able to bum holes through trees to get at soldiers hiding behind them. Given the practical problems of laser bulk and cost, and the comparative ease of protecting against or hiding from the beam, it is hard to see the current generation of high- energy infrared lasers as cost-effective antipersonnel weapons. New types of high-power lasers in the visible or near-infrared spectral regions might be useful in blinding soldiers, but not for killing them. And bullets are still cheap.

Attacks on Sensors

The priorities of the modem battlefield make electronic eyes more inviting targets than human ones. As mentioned in Chapter 10, military forces are relying increasingly on electro-optical sensors, which are part of a broad range of electronic equipment used in command, control, communications, and intelligence–C^{3}I ("C-cubed I") in the jargon of military planners. Because electro-optical sensors are designed to be sensitive to infrared light and/or visible radiation, they are inherently vulnerable to laser attack. Electronic systems and the sensors themselves are also vulnerable to damage from laser heating, electromagnetic noise or other damage from high-intensity microwaves, and to effects of particle beams at intensities much lower than needed to knock a hole in the skin of a missile.

Attacks on battlefield sensors would involve mechanisms similar to those used in the attacks on satellite sensors described in Chapter 12. However, the roles of battlefield sensors generally are quite different, as are the missions of the weapon systems they help to control. There are several types of attack on battlefield sensors:

- Blinding sensors with modest-power laser beams, causing them to lose track of what they were observing. If the sensor is guiding a weapon to its target, such blinding could cause it to miss.
- Confusing the sensors that trigger the explosion of a warhead on a missile or bomb. This could either cause a premature explosion that does not harm the intended target or prevent the warhead from exploding at all.
- Overloading sensors and/or the associated electronics to damage the components, thus causing the weapon to malfunction and miss its target.
- Causing thermal or physical damage to the sensor itself or to the optics that focus light onto it, again causing a malfunction leading to a

miss.

- Generating intense electromagnetic interference with an intense microwave beam or a burst of charged particles, confusing or damaging the internal electronics, and causing the weapon to miss its target or detonate prematurely or not at all.

The emphasis of anti-sensor warfare is not so much to destroy the target as to put its most vulnerable components–its electronic eyes–out of commission or to otherwise disable it. Different sensors have different degrees of vulnerability. Most sensors are designed to operate over a limited range of wavelengths and light intensities. In general, the longer the wavelength and the greater the sensitivity, the more vulnerable the sensors are to laser attack. Sensors of visible light are usually made of silicon and tend to be rugged. The most vulnerable sensors are those designed to detect "thermal" radiation at 8 to 12 μm wavelengths (the name comes from the fact that these wavelengths are emitted by ordinary objects at room temperature). Military "night vision" systems operate in this part of the spectrum because warm objects such as soldiers and the engines of jeeps and tanks veritably glow in the dark at these wavelengths, making them easy to spot.

Other infrared sensors, typically operating at somewhat shorter wavelengths, are used in heat-seeking missiles. Such missiles are used mainly against aircraft, with the idea being that the infrared sensor would direct the missile toward the hottest thing in sight, the exhaust of the target jet aircraft. An infrared laser could be used as a decoy to steer the missile along the wrong path, or could preferably burn out the sensor altogether, blinding the missile completely.

One particularly important type of sensing-system target is what military engineers call the "fuze" or "fuse." It is the fuze, rather than actual impact, that triggers the detonation of a missile warhead. The fuze is preprogrammed to set off the warhead when the weapon comes close enough to its target to destroy it, or in some cases when the missile is attacked and disabled. Many fuzes rely on optical sensors. For example, an optical sensor used in many antiaircraft missiles includes a small, low-power semiconductor laser that emits short pulses of light. A detector watches to see when light from the laser is reflected back. If the reflection comes very soon after the pulse is sent out, it means that the target is very close; when the target is close enough, the fuze detonates the warhead. The occupants of an aircraft being pursued by such a missile would like nothing better than to trigger the fuze and blow up the warhead when the missile is too far away to do any damage, or to disable the fuze altogether. That is a goal now pursued in electronic warfare. Military planners hope that lasers could do a better job of fooling missiles equipped with optical sensors. The same approach could be used on the battlefield to prevent warheads from doing damage to other targets.

With sensors playing an ever-increasing role on the battlefield, and being particularly vulnerable to laser attack, they are likely to be the first targets for tactical laser weapons. There are some potential drawbacks, however. A major one is that many of the "kill" mechanisms don't produce any obvious immediate evidence of their success. A pilot being pursued by a heat-seeking missile would feel much more comfortable if he had a weapon that could blow the missile up than if his weapon merely made the missile appear to drift off course. Premature detonation of the warhead by the fuze can produce a satisfactory explosion, but other types of sensor "kills" may not be obvious enough for use on the battlefield.

Causing Mechanical Damage

At powers higher than needed to disable sensors, directed-energy weapons can cause physical damage to other components of military targets. The few laser weapon demonstrations that have been described in public have involved producing some type of physical damage to the target, perhaps reflecting a need to impress military brass. In general, a high-energy laser is expected to take somewhere from a second to several seconds to do enough physical damage to "kill" a target (actual times are classified, and would depend on the target type). An intense charged particle beam could do the job in a single short pulse, if the beam could make it through the atmosphere. There are many types of physical or mechanical damage that could be lethal to a military target. Missiles and aircraft could be destroyed by rupturing fuel tanks and causing explosions of the fuel. The rotor blade core of a helicopter could be heated enough to weaken it, causing it to fail and send the copter plummeting to the ground. Heating or mechanical damage could cause the wing of an aircraft to fail, or damage some of the critical components that control a plane's aerodynamic characteristics during flight. Aircraft windshields could be shattered by thermal and/or mechanical shock from a laser pulse, making it virtually impossible to fly today's high-speed military planes. Control wires could be cut or broken. Triggering premature explosion of a warhead is also sometimes put in this category.

Most major mechanical damage produces results that are immediately obvious: a fuel tank blowing up, an aircraft veering out of control, or a helicopter dropping out of the sky. Such visible results are important in the real world of battle, where it's vital to be sure enemy equipment is really dead.

Because much higher laser intensities are needed to cause mechanical damage than to zap human or electronic eyes, visible damage is harder to produce. More is needed than simply cranking up the laser power. As laser beam intensity in the atmosphere increases, the harmful atmospheric effects described in Chapter 5 begin to manifest themselves. High-energy laser beams are liable to be bent away from their targets, or to be dispersed by thermal blooming effects in air. It appears at least theoretically possible to correct for some of these effects using the adaptive optical techniques described in Chapter 5 to change the output characteristics of the laser. The problem is that adaptive optics technology is far from perfection, probably farther than that needed to build big enough lasers.

Battlefield Missions

To the armed services the special promise of beam weaponry is the potential to rapidly zap many enemy targets with a highly directional beam that can single out enemy targets in a field of view that includes friendly forces. That sort of capability seems increasingly attractive as new hardware is developed that promises to fill the air over the battlefield with an array of potentially hostile hardware in rapid flight. The military is looking at possible roles for beam weapons on land, at sea, and in the air.

The Army considers laser weapons extremely important because of their effectiveness against electro-optical equipment, which has become a very important part of modem warfare.[14] Military planners have talked about the Army using high-power lasers either as front-line weapons or to protect "high-value" targets in the rear such as command headquarters. The main difference between lasers designed for the two roles would probably be in mobility, which would be more important on the front lines but still necessary to some degree anywhere on the battlefield. (The difference between "offense" and "defense" in such battlefield roles is largely an academic one, although it can impact the range of conditions under which a weapon system must be able to function.)

Any type of laser for use on the battlefield probably would be carried in a tank-like vehicle. In the mid-1970s, as will be described later, the Army put a moderate-power laser in a vehicle for laser weapon tests. One vision of a battlefield laser weapon it the illustration (on page 38) from the Pentagon's brochure *Soviet Military Power* showing an artist's conception of a Soviet laser weapon: a pair of tank-like vehicles. One has a beam-directing turret on top and houses the laser itself; it is connected by a thick cable to the second, which includes two large exhaust fans and presumably is the power supply.[15]

The Army shoehorned a carbon dioxide laser capable of producing 30 to 40 kW into this armored personnel carrier (which looks like a tank but has lighter armor and normally carries less firepower) called the Mobile Test Unit. The laser beam emerged from the turret at top; the fans at left rear worked furiously to try to cool the laser. Targets had to be tracked by hand. (Courtesy of Department of Defense.)

In theory, a high-energy laser could shoot at anything that flies, moves, or even lies around the battlefield. The most likely targets are either fast moving or particularly vulnerable to laser attack, such as:

- Enemy fighters, bombers, helicopters, and other aircraft
- Short-, medium-, and long-range missiles
- Remotely piloted vehicles, Pentagonese for small, remotely controlled aircraft designed for battlefield surveillance or to carry small weapons
- Sensing and communications equipment

Soldiers are possible targets, as is such "soft" military equipment as jeeps and trucks, as long as they're in exposed positions. Military planners do not consider them the *prime* targets, however, because they are quite vulnerable to existing types of weapons. Heavily armored tanks are a different matter altogether because laser beams cannot realistically be expected to penetrate a few inches of metal armor; instead, the Army is working on new types of armor-piercing artillery shells.

The extent to which high-energy lasers are used on the battlefield depends both on overcoming remaining technical obstacles and on cost. A price tag of several million dollars would not seem unreasonable to the Pentagon; a single new MI tank costs the Army over $2 million. Once the laser is purchased, individual shots should cost much less than sophisticated missiles. If laser weapons prove feasible but very costly, they might defend only such critical targets as field command centers. If the cost is moderate and the capabilities attractive, laser weapons might wind up in the thick of battle.

Portability is a key requirement for any tactical battlefield system. Although it seems feasible to put a high-energy laser into a large tank-like vehicle, it would not be easy. It would be even harder to carry particle-beam generators or large antennas for microwave weapons. Thus, with many technological questions unanswered, there is so far little evidence of serious army interest in putting such weapons on the battlefield.

Beam Weapons at Sea

Some military analysts have said that large surface ships are little more than sitting ducks for modem weapons. There is little doubt that a real problem exists, as pointed out by the losses of British ships during the Falkland Islands War. Some observers even say that surface battleships and aircraft carriers have become obsolete because of their high cost and vulnerability to air attack. Predictably, naval officers dispute such claims, but they do recognize the need for better defenses for surface ships. One of their hopes is that beam weapons will be able to fight off attacks by aircraft, cruise missiles, or other missiles that pack a wallop big enough to sink a ship.

Navy officials are most concerned about a large-scale air assault in which a barrage of fast-moving missiles close in on the target ship. Conventional weapons can knock out individual missiles, and there are Gatling-gun-type devices that can rapidly spew forth a rain of bullets in an effort to blunt a larger-scale attack. But Navy officials are not confident that they can fight off massive attacks. They have studied beam weapons because they offer the prospect of rapidly zapping many missiles attacking from different directions without endangering friendly aircraft or missiles that might be involved in the battle.

The beam generator would probably be fixed somewhere inside the ship, with the output beam emerging from a point high enough to have a clear view from horizon to horizon. The ability to hit missiles coming from any direction is vital and might even justify the use of two beam generators if one couldn't cover the whole sky. Space on even a large ship is limited, but the bulk of a laser or particle-beam weapon would be far less of a problem on a ship than on the battlefield, in the air, or in space. Both high-energy chemical lasers and charged particle beams have been studied by the Navy.

Lasers in the Air

Aircraft can move much faster than surface ships, so they are far from the sitting ducks that battleships may seem. Nonetheless, they can face a similar array of threats in the air. To counter those threats, the Air Force would like to put beam weapons into aircraft, ideally into fast and maneuverable fighters. The bulk of present beam weapons poses a serious problem, however, and for the time being the Air Force is working with a high-energy laser in a much larger transport-size aircraft. (The bulk and power requirements of a particle-beam generator, along with probable beam-propagation problems, seem to rule out particle-beam weapons. Bulk seems a fatal problem for microwave antennas too.)

The difference between fighters and cargo planes is an important one, comparable to that between a sports car and a large truck. Typical U.S. fighters and attack planes (two different categories for military purposes) are 12-20 m (40-60 ft) long and can carry 10,000-30,000 kg (20,000-60,000 lb) fully loaded. Half or more is usually the empty weight of the airplane, and much of the rest is fuel. They are fast, capable of flying at twice the speed of sound, and maneuverable.

Cargo planes are bigger and slower, flying below the speed of sound. The giant of the military fleet is the 76-m (250-ft) C5A, able to get 350,000 kg (770,000 lb) off the ground, 43% of which is the aircraft's empty weight. The Air Force has put a 400,000-W gasdynamic carbon dioxide laser into the smaller KC-135 aircraft, the military version of a Boeing 707. This 41-m (136-ft) long plane can get almost 136,000 kg (300,000 lb) off the ground, three times its weight when empty.[16]

As early as 1978, the Air Force had its Airborne Laser Laboratory off the ground for this photo. However, firing the 400,000-W gasdynamic carbon dioxide laser from the plane in the air with enough accuracy to destroy real military targets proved a much more difficult job. The laser beam emerges from the turret at the top of the aircraft, toward the left in this photo. (Courtesy of U.S. Air Force.)

The problems of compressing a high-energy laser enough so it could fit into a fighter are formidable, but the concept is attractive. Laser-equipped fighters could zap enemy aircraft and missiles out of the sky in missions including self-defense, bomber escort, or attack. The ability to defend against antiaircraft missiles is particularly important because improvements in guidance have made such missiles increasingly lethal. They are also hard to fend off with conventional weapons.

Bombers are considered much better candidates to be equipped with laser weapons than the fighters that would escort them. That's because bombers are comparable in size to cargo aircraft, and hence seem better able to carry a big laser. A bomber could use the laser to zap antiaircraft missiles. For antiaircraft use the beam could be aimed into the cockpit of the enemy plane. One device proposed for installation on B-52 bombers "would be devastating against interceptors. That device would turn the windshield of an aircraft incandescent and bum up everything in the cockpit," according to an unnamed Pentagon official quoted in *Aviation Week*.[17] The Air Force reportedly has tested the concept against a fighter aircraft. Another possibility cited in the same article is shattering the windshield with an intense laser pulse delivering about 10 J of energy to each square centimeter of the windshield surface. That energy density requirement is apparently a rough estimate. It seems to be somewhat smaller than the energy delivered to a windshield used in a standard test for measuring impact resistance: firing a dead chicken from an air cannon at 300 miles per hour, to simulate the effect of a bird hitting the plane in flight. However, the energy requirement is undoubtedly affected by the way in which the laser (and the chicken) deliver their energy. In any case shattering the windshield seems sure to put an aircraft out of action; one simply does not fly a plane with wind velocities of a few hundred miles an hour hitting his face.

In addition to the usual technical limitations on building high-energy lasers and getting the beam through the atmosphere to the target, the missions envisioned by the armed services present their own special problems. For the Army the problem is dust, dirt, and smoke on the battlefield, which could block the laser beam or prevent adequate fire control. For the Navy the problem is moisture over the ocean, which could increase absorption of the laser beam and speed degradation of components of a laser or particle-beam generator. For the Air Force the problem is turbulence in the air generated by the plane itself, which complicates pointing, tracking, and the beam-propagation problem. These problems are among those being addressed in the Pentagon's laser and particle-beam weapon programs.

Department of Defense Programs

The Pentagon has been working on concepts for tactical laser weapons for over a dozen years. The current focus is on demonstrating "lethality," the capability to kill militarily important targets such as helicopters and tactical missiles in realistic situations. Only after this capability is demonstrated, probably around 1985 if all goes well and according to current schedules, will military engineers set to work developing actual prototypes of laser weapons. Demonstrations of the prototypes would not begin until the early 1990s according to current schedules. Production of laser weapons for selected offensive and defensive applications would not begin until the prototypes had been tested to see what modifications would be needed. Pentagon estimates indicate that laser weapons wouldn't reach the battlefield until the late 1990s.[18]

This long timetable has drawn criticism from many observers who feel that it should be possible to put lasers on the battlefield earlier and worry that the Soviet Union will get there first. Pentagon officials have repeatedly said they consider it "premature" to attempt to develop specific weapon systems before they are sure that laser weapons will work under realistic conditions, not just the friendly conditions of the laboratory. They cite potential problems such as the inability to kill targets fast enough, uncertainty if the target has been killed, and the need to be able to overcome countermeasures that could help targets withstand laser attack.

Serious efforts to study laser weapons began after the gasdynamic laser was developed in the late 1960s. By the early 1970s, the Air Force was trying to shoot down targets at the Sandia Optical Test Range at Kirtland Air Force Base in Albuquerque, New Mexico. Visitors to the site describe charred areas where the beam had missed its target–vivid evidence of atmospheric transmission problems. Nonetheless, Air Force scientists kept at it, and on November 14, 1973, shot down a winged drone using what the Pentagon calls a "high-energy gasdynamic laser of moderate power,"[19] evidently a carbon dioxide laser emitting over 100,000 W.

Films of these tests, called Project Delta, were kept secret for many years, but at least some have recently been declassified. One movie was shown in an open session at the Conference on Lasers and Electro-Optics held in Phoenix during April 1982 by the Optical Society of America and the Institute of Electrical and Electronics Engineers. I was part of a large audience, including at least one visiting Soviet physicist, who saw that film. Although its quality wasn't good, it was adequate to show the drone flying through the air, being illuminated by the laser, and eventually catching fire as the fuel tank ruptured. In a second test the laser cut the control wires and caused the drone to crash out of control. These are not the type of tests that would be faked.

The next major military demonstration was of the Army's Mobile Test Unit in which a laser was mounted in a tank-like vehicle. Shoehorned in would be a more accurate term. Army engineers managed to get an electrically powered carbon dioxide laser built by Avco Everett Research Laboratory to produce about 30,000 W into a Marine Corps LTVP-7 armored personnel carrier. However, the laser was fitted so tightly inside that the vehicle had to be partially dissembled in order to service the laser, a frequent task given the comparatively primitive technology used. The tight squeeze is also said to have caused serious heat-dissipation problems, which presumably limited the running time of what was essentially a laboratory system.

The Mobile Test Unit did manage to shoot down helicopters at a reasonable distance in 1975 tests,[20] but that didn't convince Army brass that laser weapons would be useful on the battlefield. The targets were slow moving and easy to kill, and the laser's pointing system had to be hand aimed. The laser was eventually removed from the vehicle and, because it was too low in power and primitive in technology for further tests, became the first Army-surplus laser weapon. It was eventually donated to NASA's Marshall Space Flight Center in Huntsville, Alabama, for tests of laser propulsion of rockets. The Navy was the next to demonstrate a laser weapon, using a 400,000-W deuterium fluoride chemical laser built by TRW Inc. during 1978 tests in San Juan Capistrano, California. The Navy-ARPA Chemical Laser (NACL) system occupied a fair-sized building. It used a pointer-tracker built by Hughes Aircraft in tests against TOW antitank missiles (also built by Hughes), which travel at roughly the speed of sound. In one case the laser disabled the missile's guidance system; in another, it detonated the missile's fuze.[21] Roughly a half foot (15 cm) in diameter and 3.8 ft (1.2 m) long, the TOW missile is far from a sitting duck, although in a battle a soldier steering the wire-guided missile to its target may find *himself* a sitting duck.

The San Juan Capistrano (Calif.) site of the Unified Navy Field Test Program, where TOW missiles were shot down with a 400,000-W chemical laser. The complex of buildings at center house the laser, pointing and tracking system, fuel, and power-generating equipment. (Courtesy of Department of Defense.)

The most recent tests were performed with the Air Force's Airborne Laser Laboratory, a 400,000-W gasdynamic carbon dioxide laser mounted in a KC-135 military transport plane. The laser itself occupies the front third of the interior of the plane; the middle third houses the power supply, and the rear third the computers and other equipment needed to control the system. The program began in 1974, and by early 1981 the laser was being demonstrated on the ground.

Apparently overconfident, Air Force officials then made a tactical mistake, announcing plans to put the laser into the air and have it shoot down an AIM-9 Sidewinder air-to-air missile fired from beneath the aircraft. The tests were performed at the Naval Weapons Center in China Lake, California. In the first test the laser illuminated the target but failed to destroy it, although reportedly causing some damage to infrared sensors.[22] The laser also failed to shoot down another missile in a second test a week later, reportedly because of a "mechanical failure."[23] Pentagon officials insisted "there were no significant failures associated with these tests," and that they had provided some useful data,[24] but they didn't want to talk about it further.

The tests definitely served to remind the Air Force of the serious problems involved in trying to get a laser weapon to work in the air. They also reminded Pentagon officials of the virtues of keeping their mouths closed about weapon-related experiments. Confronted with headlines such as "Laser weapon flunks a test," the Air Force eventually overhauled the Airborne Laser Laboratory, and quietly began a new series of tests in early 1983. That May, the Airborne Laser Lab finally did its job, destroying five Sidewinders in a row at China Lake, after first taking eight shots to adjust the equipment. The results were announced in July, while Congress was pondering the fate of the program.[25]

Shortly before the results were disclosed, Major General Donald Lamberson had said, "the current test series will be completed in fiscal year 1984 and the program brought to a planned and disciplined termination."[26] That would be none too soon for Congressional critics, who tried to cut the program out of the fiscal 1984 budget. It has come in for some sharp attacks. Senator Malcolm Wallop (R-Wyo), one of the leading advocates of space-based laser weapons, has labeled the Airborne Laser Lab "the worst boondoggle in this field . . . a weapon without a mission, and a testbed wholly inappropriate for working out the engineering problems of space lasers."[27]

The biggest feasibility demonstration planned was the "Sea Lite" program, which the Navy was to conduct at the High-Energy Laser National Test Range nearing completion at the White Sands Missile Range. Tests were to be performed with a 2.2-million-watt deuterium fluoride chemical laser known as MIRACL, for Mid-InfraRed Advanced Chemical Laser. However, Congress deleted the Sea Lite program from the fiscal 1984 budget, reassigning the laser to DARPA. Exact plans were uncertain as this book went to press.

Part of MIRACL, the Mid-InfraRed Advanced Chemical Laser, built by TRW Inc. to produce a 2.2-million-watt beam for tests at the National High-Energy Laser Systems Test Facility at White Sands, New Mexico. This photo evidently shows some of the gas-flow equipment and perhaps part of the laser cavity while the system was initially assembled at San Juan Capistrano, California, but the Department of Defense identifies the photo only as showing MIRACL. (Courtesy of Department of Defense.)

At White Sands MIRACL was to be tested with an advanced pointer-tracker developed by Hughes Aircraft against a wide range of targets. The Navy's original goal was to collect data on lethality of laser weapons against realistic military targets and on the capability of laser weapons to fend off saturation attacks.[28] Although the immediate emphasis was on defense of ships against missile attack, Gen. Lamberson said that the new White Sands facility "provides us with a national capability that will be used by all three services, DARPA [the Defense Advanced Research Projects Agency], DNA [the Defense Nuclear Agency, which is responsible for military nuclear programs], and others to obtain badly needed vulnerability and lethality information and data on experimental system performance."[29]

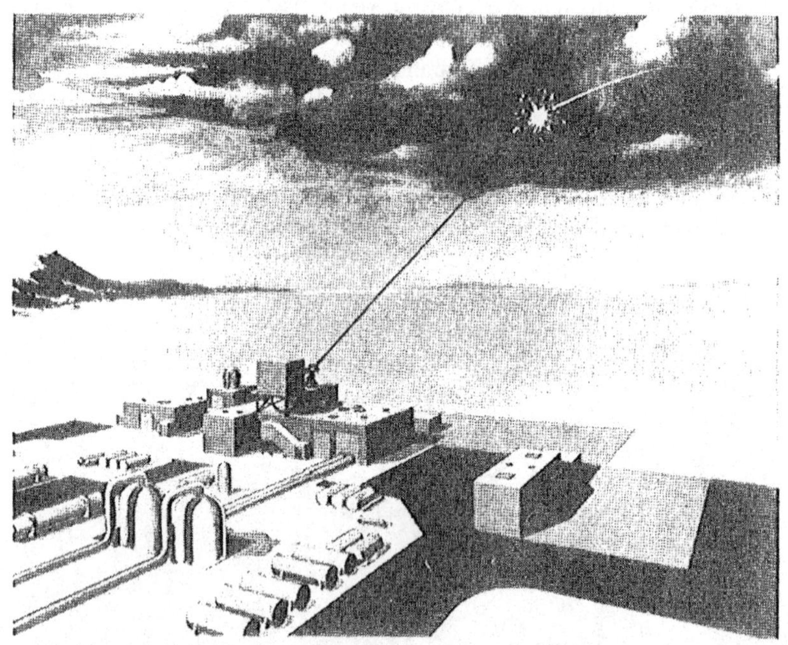

Target practice at the National High-Energy Laser Systems Test Facility is shown in this artist's conception. The laser, on the ground, has just blown up a missile. The two large tanks at lower left are presumably for the fuels that power the chemical laser that will be used in the tests. Scale is not explicitly indicated, but if the smaller opening in the main building is intended as a door for people, the main portion of that building (not counting the beam-pointing turret) is probably about 25 ft high. (Courtesy of Department of Defense.)

As this book was going to press, the fate of MIRACL and the Navy program were unresolved. Having been deleted from the defense authorization bill for fiscal 1984, the Navy program looks dead, a decision which Navy officials are said not to be fighting. However, MIRACL is too big to just throw away, and there are plans to use it for other tests, probably to be handled by the Defense Advanced Research Projects Agency. The money available for the tests will depend on the fiscal 1984 defense appropriations bill, which had not been passed by the start of the fiscal year. The management shift follows a physical one for MIRACL; the laser was first assembled at TRW's San Juan Capistrano site on the California coast, then moved to White Sands for a combination of reasons.[30]

The Army is working on its own laser weapon demonstration, called "Roadrunner." According to Lamberson, this program "will use relatively mature technology to demonstrate a capability for supporting forward area forces by engaging battlefield targets vulnerable to laser radiation." The field demonstration "will give us an understanding of the technical requirements of an operational weapon system."[31] Few details have been described in public. Plans for another test, called FALW-D for Forward Area Laser Weapon Demonstration, were scratched from the fiscal 1983 budget by Congress.

The only major particle-beam experiment in the works that is relevant to tactical warfare is the Advanced Test Accelerator at the Lawrence Livermore National Laboratory. The program, which started as a Navy program called Chair Heritage, was intended to evaluate particle beams for warship defense; this remains the only major tactical application envisioned for particle beams. The program was transferred to the Defense Advanced Research Projects Agency at the start of fiscal 1979. With the transfer came a broadening of scope to consideration of the entire problem of charged-particle defense of "point targets" such as warships or ballistic missile launch sites (a role described in Chapter 11).

Scheduled for completion in 1983, the $42-million accelerator will generate a beam of electrons, each carrying 50 million eV of energy. The immediate goal is not to see what happens when targets are zapped, but simply to show that a powerful enough electron beam can punch its way through the air. If that works as well as theorists hope, there will be efforts to adapt the technology to defend large ships against massive air assaults. However, much work remains; physicists know that charged particle beams can do massive damage to slabs of metal, but they have yet to demonstrate that they can hit even a "sitting duck" target a reasonable distance from the beam generator.

The Soviet Program

The Soviet Union appears to have its own program to develop tactical beam weapons, but its nature is something of a mystery. Although most weapons-related work remains under a predictable heavy layer of security, many papers on high-energy lasers and their effects appear in openly circulated Russian-language scientific journals. The idea of weaponry is never mentioned explicitly, but the strong interest in high powers is evident. In the past such papers have seemed more common in Russian laser journals than in unclassified English publications, a fact which may reflect differences in classification policy rather than in magnitude of effort.

Publicly cited estimates by U.S. intelligence sources say that the Soviet laser weapon effort is three to five times the magnitude of that in the United States.[32] Exactly what that statement means is subject to considerable interpretation. When it first surfaced around 1981, the Pentagon's laser weapon budget was running $200 million a year. It was still being used in 1983, when the Reagan administration had boosted spending to nearly double that amount. It has been used repeatedly by advocates of laser weaponry who warn that the United States is being outspent. It is unclear, however, whether the Soviet program is growing at the same pace as the American effort, or whether the estimate refers to a particular year.

In fairness, it should be pointed out that comparing program sizes is an extremely difficult job. Simple ruble-to-dollar conversions would understate the size of the Soviet program because wages are much lower there than in the United States. Simple counts of personnel working on the program might misstate the size the other way. In fact, most information on the Soviet program seems to be shrouded in secrecy, making the job of intelligence analysts even harder. Given those problems, and the need to protect clandestine information sources, it is not surprising that estimates are vague.

There seem to be some differences of opinion within the Pentagon. The management of the directed-energy program gives this analysis in its fact sheet on high-energy laser development:

> The Soviet Union is apparently concentrating large resources on high-energy laser technology. In particular, they may be beginning the development of specific laser weapon systems. We, on the other hand, have decided to keep our high-energy laser program as a technology program for the next few years. We believe that we understand the technical issues basic to translating high-energy laser technology into weapon systems, that our decision is correct, and that the Soviets may be moving prematurely to weapon systems. However, we are continually conducting a careful review of our program, as well as watching Soviet progress with great interest, in a continuing re-evaluation of this decision.[33]

A somewhat different view–less skeptical about Soviet prospects for success–emerges from the booklet *Soviet Military Power,* approved at higher levels in the Pentagon and first issued in October 1981 at the request of Secretary of Defense Caspar W. Weinberger, who included it in his annual report to Congress submitted the following February:

> [The Soviet Union's] development of moderate-power [laser] weapons capable of short-range ground-based applications, such as tactical air defense and antipersonnel weapons, may well be far enough along for such systems to be fielded by the mid-1980s. In the latter half of this decade, it is possible that the Soviets could demonstrate laser weapons in a wide variety of ground, ship, and aerospace applications.[34]

Aviation Week has gone even farther, reporting that the Soviet Union was planning to install a high-energy chemical laser on a Kirov-class battle cruiser that, in mid-1982, was being built at a Leningrad shipyard. Reportedly the system would be used to destroy ocean-skimming missiles perhaps at ranges to 16km (10 miles) by attacking sensors. The warship is said to weigh in at 23,000 tons and to be the largest type of warship other than aircraft carriers built by any nation since World War II.[35] However, a year later nothing more had been heard of those plans for a laser. If not totally spurious, such reports may reflect confusion between experimental demonstrations such as MIRACL and the Airborne Laser Laboratory and an actual operational laser weapon ready to fight a battle–something which is probably still years away.

Similar uncertainty exists over the status of particle-beam weaponry in the Soviet Union, although there seems to be a general agreement that for tactical missions, lasers are more advanced than particle-beam sources.

What's really happening is hard to ascertain. The problems encountered by the Airborne Laser Laboratory point out that there are serious technical obstacles to be overcome before laser weapons are ready for the battlefield. These obstacles have to be overcome by both the Soviet Union and the United States, and there's no real likelihood that the two sides will intentionally help each other build tactical weapons in the current international climate, regardless of President Reagan's grand gesture of offering to share with the Soviet Union any beam weapon technology the U.S. might eventually develop for missile defense.[36]

Different observers tend to see different aspects of the situation. Some point to the feasibility of building big lasers and say that the time for weapon systems is here, neglecting the serious problems in getting the beam to hit the target. Others see only the problems, not the promises, of the technology. Although there is no sign of unanimity within the Department of Defense, Pentagon officials seem to have reached a working compromise: a moderately paced program intended to consider both the problems and promises of beam weapon technology for the battlefield. They insist they are not in a race with the Soviet Union but rather are setting the pace of the laser program according to "the advance of the technology and . . . budget priorities."[37] Those are reasonable-sounding sentiments, but they are hard to live by when the politicians who ultimately determine the budgets are more likely to be influenced by leaked warnings of Soviet capabilities than by sober technological assessments.

References

1. Robert Cooper, talk at Electro-Optical Systems and Technology Conference, October 1981, Boston, sponsored by American Institute of Aeronautics and Astronautics, quoted in *Laser Report,* October 12, 1981, p. 2.
2. Cost estimates came from a talk by Col. Frederick Holmes, staff officer in the Pentagon's directed-energy office, at a November 1980 Electro-Optical Systems and Technology Conference in Boston, sponsored by American Institute of Aeronautics and Astronautics; quoted in "Go-ahead to develop laser weapons awaits demonstrations of lethality," *Laser Focus* **17** (2), 24-26 (February 1981).
3. Howell Heflin, talk at Laser Systems and Technology Conference, July 9, 1981, Washington, sponsored by American Institute of Aeronautics and Astronautics; see also *Congressional Record-Senate* March 12, 1981, pp. S2072-S2075.
4. This information came from two independent sources, one of whom told me he spoke with a Chinese physician who had treated the laser burns.
5. Department of Defense, *Soviet Military Power* (U.S. Government Printing Office, Washington, D.C., October 1981), p. 76.
6. David Sliney and Myron Wolbarsht, *Safety with Lasers and Other Optical Sources* (Plenum Press, New York, 1980), pp. 116-144.
7. *Ibid.,* pp. 108-116.
8. *Ibid.,* p. 149.
9. *Ibid.,* pp. 144-149.
10. *Ibid.,* p. 146.
11. *Ibid.,* p. 165.
12. *Ibid.,* p. 138-141.
13. The glass data comes from William L. Wolfe and George J. Zissis, eds, *The Infrared Handbook* (Office of Naval Research, Washington, D.C., 1978), pp. 7-38. It should be comparatively easy to fabricate special infrared-absorbing glasses, although

they would run the danger of damage from intense infrared laser beams. Another possibility would be application of coatings that selectively reflect a broad range of infrared wavelengths; such coatings are easier to make than those that reflect only a narrow range of laser wavelengths.

14. J. Richard Airey, talk at Laser Systems and Technology Conference, Boston, Massachusetts, July 1981, sponsored by American Institute of Aeronautics and Astronautics.
15. Department of Defense, *Soviet Military Power* (U.S. Government Printing Office, Washington, D.C., October 1981), p. 75.
16. Dimensions and other characteristics of aircraft are taken from a table of specifications of U.S. military aircraft in *Aviation Week & Space Technology* 116(10), 116 (March 8, 1982).
17. "U.S. nears laser weapon decisions," *Aviation Week & Space Technology,* August 4, 1980, pp. 48-54.
18. Arms Control and Disarmament Agency *Fiscal Year 1983 Arms Control Impact Statements* (U.S. Government Printing Office, Washington, D.C., March 1982), p. 300.
19. Department of Defense, "Fact Sheet: High-Energy Laser Program," February 1982, p. 4.
20. *Aviation Week & Space Technology,* September 8, 1975, p. 53.
21. Philip J. Klass, "Laser destroys missile in test," *Aviation Week & Space Technology,* August 7, 1978, pp. 14-16.
22. "Laser weapon fails to destroy missile," *Aviation Week & Space Technology,* June 8, 1981, p. 63.
23. "News notes," *Aviation Week & Space Technology,* June 15, 1981, p. 34.
24. J. Richard Airey, talk at Laser Systems & Technology Conference, Boston, July 1981, sponsored by American Institute of Aeronautics and Astronautics.
25. "Airborne laser lab finally kills missiles," *Lasers & Applications* 2(9), 240, September 1983.
26. Donald L. Lamberson, plenary talk on Department of Defense Directed Energy Technology given May 17, 1983 at Conference on Lasers and Electro-Optics, Baltimore. Quotes are taken from a written version of his talk distributed to reporters; the technical digest of the meeting contains only a cursory five-line summary.
27. Letter from Malcolm Wallop, quoted in William H. Gregory, "The great laser battle--continued," (editorial) *Aviation Week & Space Technology,* June 14, 1982, p. 13.
28. "Navy schedules laser lethality tests," *Aviation Week & Space Technology,* August 4, 1980, p. 55.
29. Donald L. Lamberson, *op. cit.*
30. Philip J. Klass, "Laser destroys missile in test," *Aviation Week & Space Technology,* August 7, 1978, pp. 14-16.
31. Donald L. Lamberson, *op. cit.*
32. Department of Defense, *Soviet Military Power 1983* (U.S. Government Printing Office, Washington, D.C., March 1983), p. 75.
33. Department of Defense, "Fact Sheet: High-Energy Laser Program," February 1982, p. 6.
34. Department of Defense, *Soviet Military Power* (U.S. Government Printing Office, Washington, D.C., October 1981), p. 76; this same statement appears word-for-word on p. 75 of *Soviet Military Power 1983* (Ref. 32), a revised edition of this booklet.
35. "Washington roundup," *Aviation Week & Space Technology,* June 7, 1982, p. 13.
36. Benjamin Taylor, "U.S. could offer ABM to Soviets, Reagan says," *Boston Globe,* March 30, 1983, p. 1.
37. Department of Defense, "Fact Sheet: High-Energy Laser Program," February 1982, p. 6.

14. A Revolution in Defense Strategy

Beam weapons hold out the promise of realizing two long-sought capabilities in strategic warfare: ballistic missile defense and antisatellite weapons (BMD and ASAT in the acronym-laden jargon of the Pentagon). Both the United States and the Soviet Union have been working on ways to achieve those two capabilities for the past two decades. Billions of dollars (and rubles) have been poured into efforts to adapt various types of projectiles and explosives (both nuclear and conventional) to the job, as described briefly in Chapters 11 and 12. Systems have been built, some only to be taken apart later as being useless. Other systems are in research and development. None of them have come close to meeting the ideal capabilities that military planners both seek for their side and fear that the other will develop. Beam weapons are important because they represent a whole new approach in trying to achieve these capabilities, which, if realized, would bring about a veritable revolution in defense strategy.

By far the most dramatic impact on defense strategy would come from the successful development of a weapon system that could zap nuclear-armed ballistic missiles long before they approached their targets. This is the capability that President Reagan hopes to achieve. From the standpoint of the president and top military planners, the nature of the technology is less important than the capabilities. If such a system can be developed, it would dramatically change the rules that underlie the game of military strategy, no matter what technology is involved. For better or for worse, it would push us away from the current strategy of mutual assured destruction (MAD). The result would probably be a period of uncertainty, perhaps culminating in a shift of power from offensive to defensive technology. There are, however, many unknowns.

The current strategy of MAD is based on the premise that there is no effective defense against a massive attack by nuclear-armed missiles. The United States and Soviet Union have lived with that uneasy nuclear balance of terror since the mid 1960s. The world's single array of antiballistic missiles surrounding Moscow probably wouldn't be able to stop nuclear devastation from raining down on the city. The United States has no operating system, having dismantled its Safeguard system shortly after it was built in the mid-1970s because it seemed unlikely that it would be effective. Construction of more missile defense systems is prohibited by the 1972 ABM Treaty.

The result has been a sort of standoff. Neither side dares to attack the other because both know that even a massive first strike would leave the other side with a nuclear arsenal large enough to devastate the attacker. Manned bombers, nuclear-armed missiles launched from submarines, and even intercontinental ballistic missiles in underground silos would remain to counterattack. If either side triggered a nuclear war, both would be devastated, wiping out vital military and civilian facilities. Or so the theory goes.

The MAD strategy is one that evolved rather than was picked intentionally by military strategists. The notion that keeping the peace requires maintaining a balance of terror–the threat of nuclear attack–bothers President Reagan and many other observers, some of whom have remarked how apt the acronym MAD seems as a description of the strategy. Yet MAD does make a certain kind of perverse sense. It is basically a consequence of offensive technology in the form of nuclear bombs having gotten far ahead of defensive technology. With no credible way to stop a nuclear attack, the only defense becomes threatening nuclear retaliation.

There have been attempts to build missile defense systems, which would rely on missiles launched from the ground to destroy nuclear re-entry vehicles approaching their targets. This is the approach the Soviets use around Moscow and the United States used in its now-dissembled Safeguard system. However, as mentioned in Chapter 11, this approach has serious limitations. It is inherently a "point" defense system because it would be outrageously expensive to pepper the entire countryside with defensive missiles and the huge radars they would require. It also is limited to defending "hardened" targets that could withstand a nearby nuclear explosion, although not a direct hit. Cities are much too vulnerable to survive such explosions well, although the Soviet Union has developed elaborate civil defense systems for Moscow to protect its residents from the fallout and other side effects of an antimissile missile system. But even if many of the residents did survive, they would be greeted by widespread devastation when they emerged from their shelters.

Because of the inherent limits on what they can defend, such antimissile systems can fill only a limited strategic role. In the past United States officials have viewed such ballistic missile defense capabilities primarily as a way to assure survival of American land-based ballistic missiles in the event of a Soviet nuclear attack. As recently as 1982, long before President Reagan's plea for broader forms of missile defense, Secretary of Defense Caspar Weinberger said that "ground-based deployment of MX ultimately may require a [ballistic missile defense system] for survivability."[1] Such ballistic missile defense systems would really protect the offensive weapons needed for MAD, and thus fit into the MAD way of strategic thought.

Pentagon officials do not believe that present ballistic missile defense technology is adequate to defend against Soviet missiles.[2] They are, however, working on other concepts, notably the Low Altitude Defense (LoAD) program mentioned in Chapter 11. Ground-based charged particle-beam systems are also being studied for point defense, although they would be farther off. Even if they do offer adequate defense against incoming Soviet warheads, such systems would be able to defend only a limited area containing "hardened" targets. They could be strategic bargaining points in establishing the balance of power, but they would still be part of the MAD game.

Suppose, however, that there was an effective way to defend the whole country against nuclear attack. Suppose that one side could intercept most of the other side's ballistic missiles and manned bombers long before they approached their targets. This is the sort of system President Reagan is seeking, and it would indeed change the ground rules of the strategy game.

Ideally, the defense should be perfect to prevent any losses. But military planners are realistic enough to know that in the real world there is no such thing as a perfect system, and some enemy missiles would almost certainly leak through any defensive shield. To try to minimize the leakage, they have developed a concept called a "layered" defense in which enemy missiles would have to run a gauntlet of defensive systems. Any missiles that leaked through the first layer would be subject to attack by succeeding layers. For example, the first layer might be a group of orbiting battle stations that zapped missiles in their boost phase, before they had a chance to launch multiple warheads. This layer would thin the ranks of the attackers, perhaps destroying 90% of them, leaving 100 out of an original attack force of 1000. Each of these missiles might then launch five separate warheads (as "re-entry vehicles") toward separate targets. A second layer (another set of orbiting battle stations) could attack these warheads during their flight through space, perhaps destroying another 90%, leaving 50 warheads. A third and final layer (point defenses around strategically vital sites such as missile launching fields) would destroy some of the remaining warheads, with the numbers destroyed depending on how many of the surviving warheads were targeted at defended areas.

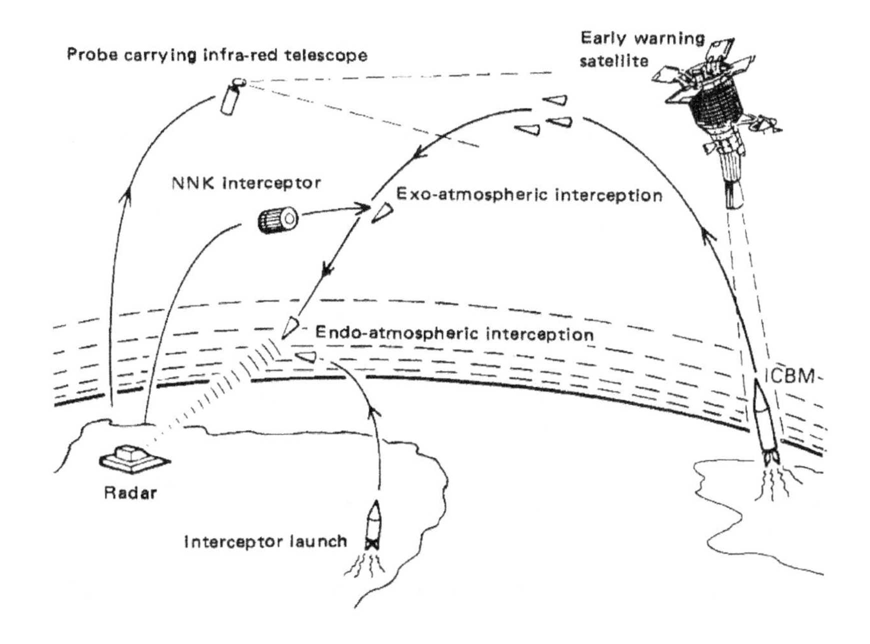

An early concept of a "layered" ballistic missile defense, similar to that envisioned by the U.S. Department of Defense before beam weapon concepts hit the headlines. Launch of ICBMs (intercontinental ballistic missiles) would be detected by early warning satellites. Warheads released by the ICBM would have to survive two layers of defense–nonnuclear kill vehicles (NNK interceptors in the drawing) above the atmosphere and ground-based interceptors in the atmosphere (endo-atmospheric interception). Beam weapon concepts are similar but would add a space-based beam weapon to zap ICBMs as they rose out of the atmosphere, and would substitute a beam weapon for the nonnuclear space interceptors. (Courtesy of Stockholm International Peace Research Institute, from *Outer Space-A New Dimension of the Arms Race,* edited by Bhupendra Jasani, p. 101.)

New Questions for Strategists

Development of such a defense system would raise a host of major new questions for which military strategists have no ready answers:

- How good would such a system have to be? That is, what would be an "acceptable" level of damage from nuclear weapons that leaked through the defensive shield?
- How much would the strategic value of offensive nuclear weapons be reduced?
- Could one side put a defensive system into operation fast enough to gain a decisive advantage on the other? Would that advantage be temporary or could it be made permanent?
- If the other side was about to put a missile defense system into use, should you attack before it went into operation, while your missiles still had the capability to do extensive damage?
- If you had just put a missile defense system into operation, should you attack the other side before it could get its own defense system working?
- How effective are missile defense systems, in practice as opposed to in theory?
- How can you ascertain the effectiveness of the other side's missile defense system short of launching an attack?
- How can you test the effectiveness of your own missile defense system without the other side's spy satellites observing the results?
- How can you convince the other side that your system works?
- If you had a defense system, how far could you go in provoking or pushing the other side? Would the other side be impressed enough by your defense system and offensive arsenal to give in readily, or would they be willing to risk a war?
- If the other side had a defense system, how far should you let yourself be pushed around?
- What ways are there to circumvent the missile defense system using other weapons in your arsenal? What about new technologies for new types of weapons, or countermeasures that would defeat the defensive system?

What these questions add up to, in a word, is uncertainty; a state military planners like to see–on the other side. The underlying theory is that the less certain a potential opponent is about the results of launching an attack, the less likely he is to launch the attack. With a layered missile defense system an attacker could not be sure of the damage he could cause by attempting a nuclear first strike. From a probabilistic standpoint he might be able to estimate that perhaps 25 warheads out of the 5000 he launched would hit their targets, but he would not know in advance *which* 25 would hit. Indeed, the most critical strategic targets would probably be surrounded by ground-based point defense systems, making them especially hard to destroy. Maj. Gen. Donald L. Lamberson said when he headed the Pentagon's directed-energy program that deployment of a layered missile defense system "could vastly increase the uncertainty of the outcome of even a mass attack. Such a [missile defense] capability would present any rational political-military planning system with uncertainties too large to permit such attacks to be a viable option. In essence, such a capability would strengthen the deterrent to war."[3]

Most discussions of nuclear strategy implicitly focus on the possibility of a full-scale nuclear war in which the United States and Soviet Union each launch massive nuclear attacks on the other. However, strategists have been considering other alternatives besides a continuing uneasy peace and an all- out nuclear war, and the development of missile defense systems could impact these alternatives as well. A number of accounts have suggested that the Reagan Administration is seeking the capability to fight an "enduring" nuclear war, one that might continue for a matter of months. Fred. C. Ikle, under secretary of defense for policy, told the Senate Armed Services Committee's subcommittee on strategic and nuclear forces that the real interest was in being able to survive what might be considered a partial first strike. According to Ikle, this would occur if the Soviet Union launched only part of its nuclear arsenal against the United States and kept the rest in reserve to threaten further devastation if the United States should counterattack.[4]

There are other questions and possibilities as well, some of them quite subtle. Defense strategy is something like a poker game, full of bluff and counterbluff, with the cards in each player's hand carefully hidden. Unlike poker, however, the goal of the military strategy game is to avoid a "showdown," where both sides lay down their hands on the table to see which is superior. In the real world of military strategy the ultimate showdown would be a full-scale war, and in that event there would probably be only losers.

In the nuclear strategy game the cards are arsenals of bombs and missiles. The impact of a system that could defend against nuclear attack would be to change the values of the cards and the rules of the game. However uneasy the status quo may be, any changes in the rules that determine the balance of power would tend to make both sides uneasy. Yet, despite the likelihood that it would cause instability, the idea of having an effective defense against nuclear attack is in many ways alluring. In making his plea for development of missile defense technology, President Reagan was able to invoke some noble-sounding sentiments, such as "the human spirit must be capable of rising above dealing with other nations and human beings by threatening their existence," and "Wouldn't it be better to save lives than to avenge them?"[5] By shifting the balance of power from offensive weapons to defensive weapons, he and other missile defense advocates say they want to remove the continual threat of nuclear devastation.

The Case for Beam Weapons

The cornerstone of the argument for any type of new missile defense technology is the notion that "assured protection" is better than "assured destruction." From that viewpoint it is foolish to continue relying on the MAD doctrine to protect the United States because it is a tenuous and unstable balance, a slippery handhold to which some military strategists cling from fear that nothing else could save them from sinking into the bottomless swamp of nuclear war. Some observers have begun questioning the morality of a defense strategy based on a threat to carry out a devastating nuclear attack against another country, a concern echoed in some of President Reagan's remarks.

There is also an important political element to the debate over beam weapons, one which is often glossed over. By and large, the most vocal advocates of developing beam weapons are conservatives who believe that the Soviet Union is trying to circumvent the present balance of power and gain a strategic superiority. For example, Reagan prefaced his advocacy of missile defense technology with a long account of evidence of a Soviet military buildup: "For 20 years, the Soviet Union has been accumulating enormous military might. They didn't stop when their forces exceeded all requirements of a legitimate defensive capability. And they haven't stopped now."[6] This fundamental distrust of Soviet motives has helped to stimulate interest in new defenses against the Soviet military machine.

Reagan's interest centers on the capability for ballistic missile defense, not on the technology used to realize that goal. He is not alone. Indeed, from a strategic viewpoint the specific technology chosen does not matter–what is important is the defense capability it offers. Beam weapons have been mentioned often as a dramatic and powerful-seeming new technology; however, as illustrated by the High Frontier proposal described in Chapter 11, there may be other ways to achieve missile defense.

Senator Malcolm Wallop (R-Wyo) made his first public plea for the development of space-based antimissile laser weapons more than three years before President Reagan's speech, and he can legitimately be considered the first major public advocate of the idea. His vision was expressed in an article in the Fall 1979 *Strategic Review:* "technology is rendering the 'balance of terror' obsolete. Technology now promises a considerable measure of safety from the threat of ballistic missiles. To be more precise, it offers that safety to whichever advanced nation is willing to grasp it."[7]

Wallop and other advocates of space-based beam weapons generally believe that the MAD strategy is bankrupt. They see the United States as dithering aimlessly, while the Soviet Union builds up massive military power aimed in the long run at strategic superiority. Wallop has said that while "the goal of space defense is within our grasp . . . it is also within the Soviet

Union's grasp, and we know that while we are not doing our utmost to seize it, they are."[8]

Defensive weapons tend to sound better than offensive ones, and Wallop, like Reagan, has been invoked some noble-sounding rhetoric. Advocating faster development of space-based laser weapons (he doubts that particle beams or X-ray lasers will work) in a 1981 Senate speech, Wallop said:

> We are at a crossroads in this country. We have spent money, dollars after dollars and billions and billions, for weapons whose only consequence is to kill people. Now we have within our capability the possibility of developing weapons whose only real role in the world is to kill the things that kill people. . . . Here we have the chance to make a commitment to the American people at last to try to defend them against some of the horrors of the nuclear holocaust.[9]

Similar arguments have been made by other advocates of faster development of space-based laser weapons. Senator Howell Heflin (D-Ala) told a 1981 meeting on laser technology, which I moderated, that:

> Space laser weapons . . . should bring an end to the fear-filled era in which we have lived for the last few decades. Technology offers us, and the Soviet Union I hasten to add, the chance to begin to overcome this vulnerable mass destruction strategy–it, conversely, will enable us to employ a new and stronger defensive stance. . . . High-energy laser development and weaponization programs could give us a needed edge, but we must not drag our feet.[10]

Much the same philosophy is behind other proposals for orbital ballistic missile defense, including those that rely on particle beams, X-ray lasers, or projectile technology. (Probably the most comprehensive explanation of this viewpoint is in the High Frontier report, which advocates projectiles for a "first-generation" system.[11])

Some statements made by Wallop and other beam weapon advocates have been sharply disputed by others. For example, in his original proposal Wallop wrote that he did not believe space-based laser weapons would be provocative because they would be defensive weapons incapable of mass destruction.[12] Others inside and outside of the United States seem to see them as very provocative. When President Reagan put himself behind efforts to develop antimissile technology, Soviet leader Yuri Andropov responded by accusing him of "attempting to disarm the Soviet Union in the face of the U.S. nuclear threat."[13]

Wallop also made a sweeping statement that the worst a space-based laser system could do in peacetime would be to abort accidentally the launch of a Soviet weather satellite.[14] That is probably something of an overstatement. The Soviets would undoubtedly be upset about having a would-be weather satellite shot down, but they would be much unhappier if the inadvertent target had been carrying a crew of cosmonauts.

Even advocates of beam weapons for missile defense do concede that they will have limitations. It may be possible to develop countermeasures to keep them from destroying their targets, although advocates of beam weapons say that could be done only at the cost of building a new generation of offensive weapons, or by reducing the effectiveness of existing weapons. Beam weapon battle stations themselves will not be invulnerable to attack (a point that critics seize upon). And for all the high hopes, I have yet to see or hear anyone label beam weapons as the "ultimate" weapons.

Some beam weapon advocates appear driven by the hope of "stealing a march" on the Soviet Union. Evidence for Soviet programs to develop beam weapons is also stimulating some Americans to push for more work on antimissile beam systems, lest the Soviets get there first. For many of these people, staying ahead of the Soviets is an urgent priority, for they believe that the Soviet Union seeks to develop their military might in order to dominate the world.

Some of the most forceful statements of this conservative viewpoint have appeared on the editorial pages of *Aviation Week*. When the magazine first publicized charges that the Soviets were developing particle-beam weapons, now-retired editor Robert Hotz compared the development of particle-beam weapons with "a chess game, which ends in the early 1980s with the triumphant Soviet shout of 'check and mate.'"[15] That statement was made in 1977, and while there are still occasional warnings that Soviet beam weapons of some sort are likely to appear in the skies soon, there are no obvious signs those early prophecies of doom are going to be proved right.

Warnings of Danger

Other observers, mostly outside the Department of Defense, hold very different views, saying that beam weapons are unlikely to be feasible, that they might dangerously destabilize the strategic balance, or both. Kosta Tsipis, a physicist at the Massachusetts Institute of Technology, states flatly that "lasers have little or no chance of succeeding as practical, cost-effective defensive weapons."[16] In a study co-authored with Michael Callaham, now on the electrical engineering faculty at Carnegie-Mellon University, Tsipis warns that even though space-based laser weapons wouldn't work, they would make the balance of power unstable.[17]

Callaham and Tsipis paint a disturbing picture of the consequences of trying to build a space laser system. If one side tried to build such a battle station, they argue, the other would be tempted to destroy it before it could be finished. Such a pre-emptive strike against a defensive satellite system could trigger an escalation of hostilities, perhaps culminating in an all-out nuclear war. This could happen even if the other side doubted the effectiveness of beam weapons, they warn, because there would be no way to be *sure* the system wouldn't work.

A report prepared by Tsipis and three other MIT physicists–George Bekefi, Bernard T. Feld, and J. Parmentola–is even more pessimistic about the prospects for particle-beam weapons in space. That group concluded that "it is not unfounded to expect that eventually accelerators with performance characteristics required for a beam weapon system could be built. On the other hand, the operational difficulties as a particle beam weapon as a system... appear insurmountable."[18] They say the prospects for building laser weapons look good in comparison.

Similar pessimism comes from Richard L. Garwin, a strategic analyst based at IBM's Watson Research Center in Yorktown Heights, New York. He maintains that

> the nemesis of a satellite-based system is the "space mine," a small weapon which follows it around in orbit and can be detonated by the opponent by radio command. Even on the sea, it is certainly hallowed by international law and custom that every nation has the right to maintain the vessels of another nation within range of its guns (except in domestic water). To attempt to establish the opposite will lead to war in space--and that as a prelude to war on earth, not as a substitute for it.[19]

Critics raise a number of other points which they say would make beam weapons technically unfeasible, including the large size of the devices that would have to be put into space and the ease of developing countermeasures against them. Such concerns are discussed elsewhere in this book because they relate more to hardware than to strategy.

A central strategic issue raised by critics is the relative costs of nuclear bombs and beam weapons. If beam weapons turn out to be much more expensive than bombs and missiles, the simplest strategy against them may be to build enough bombs to overwhelm the defense system. Thus, if the goal is to have 1000 warheads hit their targets, a 90%-effective ballistic missile defense system could be countered by building 10,000 warheads. Military planners might even want more warheads, on the grounds that they couldn't be sure *which* 1000 warheads would get through. The actual strategy would depend on the way in which the defense system worked, that is, whether it knocked out a certain percentage of targets or whether it had a fixed maximum capability. Nonetheless, the basic temptation would remain to simply build more bombs, throwing more fuel onto the fires of the arms race.

Skeptics have raised some important points in expressing concern about instabilities caused by the deployment of missile defense systems and possible contributions to acceleration of the arms race. But, like the conservatives who want to build defense systems because they don't trust the Soviets, the opponents of beam weapons have their own biases. Some of them seem to have a prejudice against *any* new types of weaponry. Others seem to be thoroughly trained in the MAD strategy, and reluctant to look at alternatives. The reactions to President Reagan's proposal to develop missile defense systems were illustrative; politicians lined up "for" or "against" the idea on the basis of their party affiliation and agreements or disagreements with Reagan, not on the merits of the idea *per se*.

Even before President Reagan got into the act, a few skeptics had produced some analyses that seem deliberately pessimistic. As mentioned in Chapter 11, estimates of incredibly large masses for battle stations have been based on the assumption that the satellites would be in geosynchronous orbit 36,000 km (22,000 miles) above the earth, an idea not advocated by beam weapon proponents because it would put the weapon too far from the target. Warnings that countermeasures could be developed tend to ignore the time it takes to develop them and put them into service.

Ironically, the most vocal critics of space-based beam weapons seem to have largely ignored, at least until recently, a potentially serious problem raised by cruise missiles. Manned bombers, and ballistic missiles launched from land or submarines, all travel through space or the upper atmosphere where they would be exposed to attack by high-energy conventional lasers. (X-ray lasers and neutral particle beams may not be able to penetrate far enough into the atmosphere to zap bombers.) However, once it is launched, the cruise missile follows a path much closer to the ground. X- ray lasers and neutral particle beams simply could not penetrate that deeply into the atmosphere. Space-based conventional lasers would also have trouble because of atmospheric effects. The adaptive optical techniques described in Chapter 5 can provide some compensation for atmospheric effects, but targets deep in the atmosphere are inevitably harder to hit than those in space.[20]

Beam Weapons and a "First Strike"

At least one critic of Reagan administration policy has charged that efforts to develop beam weapons are part of a larger strategy to develop the capability for the United States to win a nuclear war by scoring a first strike against the Soviet Union. The rationale, attributed to "nuclear warfighters" within the Pentagon, is that an American first strike would be unable to destroy *all* Soviet missiles, with perhaps 10 to 20% remaining to launch a counterattack against the United States. A beam weapon defense system might not be able to cope with a full-scale Soviet attack, but it probably could knock out virtually all of the much smaller force of missiles remaining after an American first strike.

Those charges were made in an article in the June 1983 issue of *The Progressive,*[21] which unfortunately is deeply flawed by factual errors.[22] The strategy is one that I have never heard espoused by anyone. It is one which I believe the majority of Americans would consider amoral, to say the least. Yet, perversely, it makes some sense if viewed through the eyes of a hard-line military strategist who believes that the Soviet Union is evil incarnate, patiently waiting to pounce on the United States and enslave its populace. There is a faction of "first strikers" in the military, but this is nothing new. As long as the United States has had nuclear bombs, there have been people suggesting that the bombs be dropped on the Soviet Union.[23] Fortunately, cooler heads have always prevailed.

Most of these unsettling possibilities were mentioned before there was serious consideration of "pop-up" systems-X-ray lasers or battle mirrors for ground-based lasers-which would be put into space only when needed. So far beam weapon opponents have made few comments about pop-up systems *per se,* but the strategic implications could be disturbing. The critical problem is reaction time. The technical challenge of launching a defense system upon warning of an enemy attack is formidable because it takes an intercontinental ballistic missile only about half an hour to reach its target. If a half-hour warning was not sufficient–and it is hard to see how it would allow enough time to get the satellites into orbit and stabilized–the satellites would have to be launched *before* warning of an enemy launch. Presumably that would be done in times of tension, when one or both sides felt there was a real danger of nuclear war. However, the very act of launching the defensive system could have serious consequences:

- A further escalation of international tensions, an indication that one side thought nuclear war was likely.
- It could be interpreted as an actively hostile action, perhaps as a sign of an impending first strike. (Orbiting the battle stations could be seen as a defensive maneuver against the other side's missiles that survived the first strike, as in the scenario in *The Progressive* article.)
- The simultaneous launch of many boosters might be misinterpreted as the first stage in a nuclear attack by early warning satellites. triggering a nuclear attack by the other side.

The Pentagon Position

The somewhat unlikely occupant of the middle ground in the debate over the feasibility of beam weapons for missile defense is the Pentagon hierarchy. The official position is a mixture of interest in the theoretical capabilities of beam weapons and reluctance to make a major commitment to the technology without having seen a convincing demonstration of its effectiveness. It is basically a cautious response, typical of engineers and bureaucrats who, for different reasons, tend to distrust paper predictions of impressive capabilities. In part, that distrust is only rational. The Pentagon and other government agencies have many times been burned by impressive proposals that turn out to be impossible to implement. There has been a common tendency to overpromise results and capabilities. The scars linger in Washington as ghosts of boondoggles past, which come to haunt military officials evaluating new proposals or testifying before Congress.

Yet in part, some people in the Pentagon seem reluctant to accept the possibility that new technology might revolutionize the concept of strategic balance of power. Although the Pentagon now professes to support research on space-based beam weapons, it took Congressional pressure to get the ball rolling. The armed forces have tended to have more interest in laser weapons that would have less revolutionary capabilities on land, at sea, or in the air.

When Congress's investigative agency, the General Accounting Office, took a look at the space-based laser program at the request of Congress, it concluded that caution was justified. Donald E. Day, senior associate director of GAO's mission analysis and systems acquisition division, told the Senate Armed Services Committee's subcommittee on strategic and theater nuclear forces: "It is premature to believe that the effectiveness and affordability issues can be resolved before important technical uncertainties are favorably resolved. Resolving these uncertainties is necessary before even a limited first-generation weapon system is possible."[24]

However, GAO warned that as of the beginning of 1982, the program was limited by funding rather than by development of the technology. The report said, "This approach risks keeping the potentially revolutionary technology in component development for the foreseeable future." It stressed the need for "a well structured, funded and managed effort," but questioned if such a program existed at the time. It recommended that the Department of Defense establish a plan for space-based laser development with "clear and specific milestones and objectives which recognize the relative priority of laser missile defense within DoD," and that enough money be committed to the program to meet these objectives.[25]

Pentagon officials clearly recognize that successful development of space-based beam weapons could provide some important new capabilities. Some of these are listed in Chapter 11. They recognize that successful development of space-based beam weapons could have an impact on the strategic balance and could permit the use of new military strategies. According to H. Alan Pike, then deputy director of the directed-energy office at the Defense Advanced Research Projects Agency, if space-based laser weapons prove truly revolutionary, we can skip the costs of more gradual, "evolutionary" program development and avoid the danger of technological surprise without the high costs of catching up.[26]

Other Visions

There are almost as many scenarios for the impact of space-based beam weapons as there are observers. Most fall between the rosy projections of a beam-weapon umbrella to protect the country from nuclear attack, and the gloomy warnings of a multi-billion-dollar boondoggle that would trigger World War III. It is impossible to go into them all here, but a few deserve special mention because they avoid the extremes of easy answers–be they bright-eyed optimism or gloom-and-doom.

One observer who sees a long-term hope that development of space-based beam weapons could lead us out of the present nuclear standoff is Barry J. Smernoff, formerly an independent arms-control consultant and now a senior fellow at the Strategic Concepts Development Center of the National Defense University in Washington. He believes laser weapons will eventually be developed, although not as soon as their most enthusiastic advocates hope. He shares concerns that the deployment of a new type of weaponry could lead to temporary instability. But he also brings an optimistic message, that the development of space-based beam weapons may help to create a way out of the nuclear balance of terror.

The importance of laser ballistic missile defense, he writes, is that it is "the single most obvious and credible technological innovation that could facilitate the initial steps of a gradual transition away from a world in which the recognized currency of strategic power is the nuclear weapon," a *status quo* which he believes has "very limited (if any) meaningful political utility and quite negative moral implications."[27] He does not believe the transition will be easy, but hopes that it will be possible. Although he does not see a space-based laser system, at least in its initial form, as a perfectly effective defense against ballistic missile attack, he says it should provide some defense that could help to reduce the value that the United States and the Soviet Union put on their offensive nuclear arsenals. The long-term hope is that once each side had erected its own defensive shield, they would be willing to make negotiated reductions in offensive weapons. Such a defense-dominated world, Smernoff hopes, will be more stable than one dominated by offense. It also would eliminate the moral dilemma that the United States would face in having to rely on the threat of a nuclear holocaust to maintain national security,[28] a concern he expressed over a year before President Reagan cited it in his proposal for new missile defense technology.[29]

A less optimistic vision comes from A. M. Din of the University of Lausanne in Switzerland. He writes that even if the United States and the Soviet Union reached a rough balance of beam weapon deployment, the appearance of strategic balance would be illusionary. He explains: "the characteristic technological time scale of warfare would change by several orders of magnitude. Today this time scale is, strategically, of the order of minutes, but in [the case of beam weapons] split-second decisions would have to be taken." That would put the strategic balance "at the mercy of computers."[30] Some might take issue with the forcefulness of that statement, but if the goal was to destroy ballistic missiles during their vulnerable boost phase, clearly action would have to be taken very rapidly.

Interestingly, Din's "educated guess" based on the past performance of Soviet and American military engineers is that the United States probably holds a lead in the beam weapon race. He writes that the two powers probably have a near parity in the basic physics and technological development of beam weapons. However, Din believes that "actual deployment of effective beam weapons will require the solution of a hitherto unseen collection of 'packaging' and logistics problems in which the U.S. military-industrial complex should have a significant edge."

An intriguingly innovative suggestion of what to do with antimissile beam weapons was made by Paul J. Nahin, associate professor of electrical engineering at the University of New Hampshire in Durham. He proposes "what might be the world's best military use of space–the deployment of appropriate [antiballistic missile] weapons in orbit under the control of an international space patrol."[31] The battle stations would be built using expertise from the United States, Soviet Union, and other countries and staffed by "elite crews selected from all the nations of the world." Its mission would be to stay on perpetual alert, ready to destroy any nuclear-armed ballistic missiles that might be launched and to attack warplanes in flight. These international patrol stations would be the only weapon facilities allowed in orbit; individual nations would be allowed to maintain their tactical armed forces on the ground. In addition to protecting the United States and Soviet Union from each other, Nahin says, it would protect them from "the one threat neither can now defeat–an unsophisticated missile attack from a third country." In a way Nahin's concept brings to mind his avocation–he is a published writer of science fiction (so, for that matter, am I). The idea would require a truly radical departure from the way that nations have traditionally managed their military affairs. Yet it comes at a time when, at least technologically speaking, some radical departures from traditional strategy are inevitable. It also is one of the few proposed scenarios that leads to a way out of our current strategic crises.

A wide range of other scenarios can also be envisioned for the deployment of antimissile beam weapons, which might be deployed wholly or partly in space, or deployed on the ground or in submarines for launch into space when needed. It is impossible to go through them all here. Ultimately there can be as many possible scenarios as there are people to envision them. With the shape of the technology to come still largely undefined, it is very hard to sort them out.

Impact of Antisatellite Weapons

The development of beam weapons, most probably lasers, for use against satellites would not have as great a strategic impact as the development of beam weapons for use against missiles. Although satellites play an important military role, they are not as highly visible (and terrifying) as nuclear bombs are, both to the public at large and to military strategists. Yet as indicated earlier, antisatellite laser technology is much closer--something recognized even by such laser weapon critics as Michael Callaham[32]–and its impact will be felt first.

The roles played by antisatellite and antimissile weapons would be rather different. Antimissile weapons would be used to blunt a hostile attack. With the exception of "killer" satellites intended to destroy other satellites, however, satellites themselves do not shoot at anything. Disabling a satellite would in itself be an aggressive act (except if it was a killer satellite approaching its target).

As was mentioned in Chapter 12, satellites have come to play a variety of militarily important roles. Most of them are essentially passive support roles that have come to be tolerated in peacetime, such as communications, surveillance, and navigation. Zapping one of these satellites in peacetime would clearly be a provocative act, much like shooting at one of the other side's ships on the high seas, which is the closest equivalent in international law. Such incidents are taken very seriously indeed, and can be interpreted as an act of war.

The stakes are highest for early-warning satellites, whose function it is to monitor for signs of a nuclear attack by the other side. Destruction of an early-warning satellite could be considered a prelude to such an attack. How- ever, the early-warning system has some important safety features, such as the use of multiple observing satellites and multiple communication paths. Several satellites would have to be destroyed in order to knock out an early- warning capability, reducing the chance of a false alarm being triggered by an ordinary failure of hardware. Nearly simultaneous failure of several satellites would probably be interpreted as a sure sign of enemy action, but the failure of a single satellite, although sure to be checked carefully, would be unlikely to trigger a nuclear confrontation. For example, Pentagon officials carefully analyzed data on the late-1982 failure of a single spy satellite, rather than immediately jumping to the conclusion it had been zapped by a Soviet laser.[33]

Some observers are concerned that the deployment of laser or other antisatellite weapons could enhance the danger of war during periods of international tension. Michael Callaham of Carnegie-Mellon University warns that one side's imminent capability of degrading the other's command, control, communications, and intelligence network by using antisatellite weapons "may, in a period of heightened tension and suspicion, provide an opponent with a strong incentive to attack pre-emptively in mistaken anticipation of an imminent counterforce attack."[34] How real that incentive would be is hard to assess; it certainly would not be the only incentive.

A more important point is made by D. L. Hafner of Boston College's political science department: "Many of the surveillance satellites that the USA and USSR would regard as prime ASAT targets in the event of war actually serve a very desirable stabilizing function in crises. . . . A satellite which can warn that an attack is under way can also confirm that an attack is not under way or is not coming from the other great power."[35]

The existence of antisatellite lasers would inevitably heighten the level of anxiety caused by the failure of any military satellite. However, an understanding that the other side shared the same fears would tend to prevent the use of antisatellite weapons except as part of a massive attack. That is particularly true in the present environment, where the U.S. Government has refused to rule out the possibility of firing the first nuclear shot at the Soviet Union, or to believe Soviet claims that they would never make a first strike. The best that an attacker could hope for in knocking out the other side's satellites would be a temporary paralyzed paranoia in which the other side's leaders waited with fingers poised uneasily over the ATTACK button. The worst would be an all-out nuclear attack in quick response. Those high stakes–and large uncertainty–would tend to limit the strategic value of antisatellite weapons.

On the other hand antisatellite lasers would undoubtedly be of use in the case of a full-scale nuclear war. Once the nuclear shooting has started on a large scale, each side will want to knock out the other's communications, spy, and navigation satellites. Where in the case of peace or lower-level conventional war, it pays to reassure the other side that no nuclear attack is under way, during a war there's no reason to let the other side watch. It's also desirable to break the communication channels that link enemy forces around the world and to zap navigation satellites to make it easy for their ships to get lost. The value of such capabilities would be small if a nuclear war ended with a single massive exchange of bombs, but it could be high if the war lasted longer. This points up a central irony: while moderate-power laser weapons for use against satellites would make their main contribution to war-waging capabilities, higher-power beam weapons could–at least in some scenarios–help to keep the peace by threatening to destroy nuclear bombs.

Long-Term Potential

In the long term antimissile weapons can't help but have a greater impact than antisatellite weapons, assuming that the technological barriers to the former can be overcome. Indeed, an antimissile beam weapon could also zap satellites, although the high power of the beam wouldn't be needed to do the job. If the technology is feasible, the actual deployment of beam weapons may be inevitable unless there are breakthroughs in arms control and inter- national understanding.

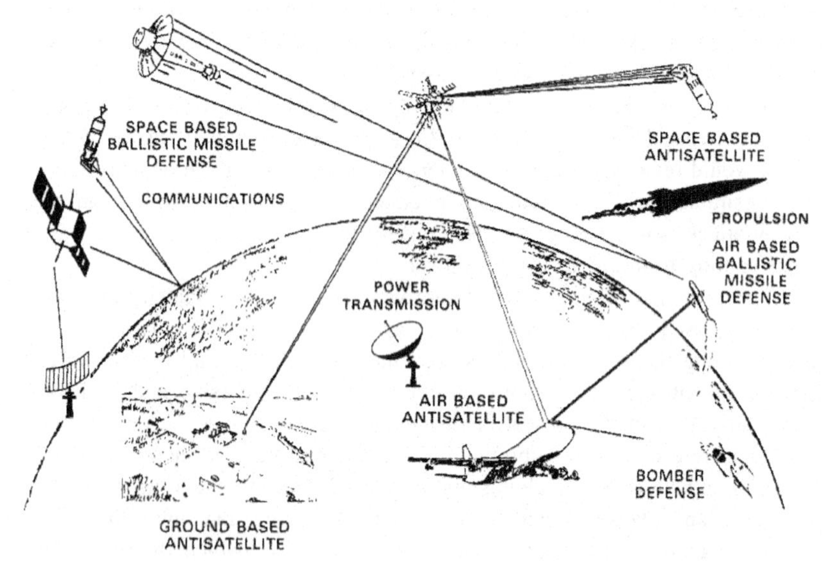

Potential applications of lasers in space envisioned by the Department of Defense range from weaponry to such comparatively innocuous tasks as communication and propulsion of spacecraft. High-power lasers would be needed for all of the applications shown except for communications. (Courtesy of Department of Defense.)

Further in the future, new technologies could add still further complications. Successful efforts to "colonize" or "industrialize" space, whether with factories, solar power satellites, or cities, would raise the stakes in space. The international political climate could change.

The factors affecting the global balance of power will probably grow increasingly complex. The independent-minded Chinese are thought by some to be pointing the way to a "multi-polar" world with more than two great powers. Their government has shown a strong interest in laser technology,[36] although there has been no indication that they are developing laser weapons *per se*. The United States has bilateral agreements to exchange information on tactical laser weaponry with the United Kingdom, Germany, and Australia, but these agreements do not extend to strategic weapons,[37] and none of those countries is known to have a large laser weapon program.

Former Pentagon official Allan D. Simon envisions that by the year 2032 the world will have at least a dozen "greatly powerful nations" with some mixture of assets including large population, natural resources, strong technology, and/or military might. He believes these nations will be able to build nuclear weapons, launch ballistic missiles, and put satellites into orbit. He sees beam weapons as part of that future, along with "intelligent" computerized weapons and nearly invincible communications.[38]

The real future may not look like any of the scenarios described above. However, if beam weapons work, they will clearly be an important issue to be addressed in establishing a balance of power.

References

1. Casper W. Weinberger, *Secretary of Defense Annual Report to Congress Fiscal Year 1983* (U.S. Government Printing Office, Washington, D.C., February 8, 1982), p. III-65.
2. *Ibid.*
3. Donald L. Lamberson, plenary talk titled "Department of Defense Directed Energy Program," presented May 17, 1983 at the Conference on Lasers and Electro-Optics, Baltimore; all quotes are based on a printed version of his talk given to reporters at the meeting. The technical digest includes only a five-line summary.
4. Senate Armed Services Committee, *Strategic Force Modernization Programs* (Hearings before the Subcommittee on Strategic and Theater Nuclear Forces, conducted October 26-30 and November 3-10, 1981) (U.S. Government Printing Office, Washington, D.C., 1981), p. 73.
5. A transcript of the speech appears as "President's speech on military spending and a new defense," *New York Times,* March 24, 1983, p. A20.
6. *Ibid.*
7. Malcolm Wallop, "Opportunities and imperatives of ballistic missile defense," *Strategic Review,* Fall 1979, pp. 13-21.
8. *Congressional Record-Senate* May 13, 1981, p. S4975. 9.
9. *Ibid.,* p. S4976.
10. Howell Heflin, talk given at Laser Systems & Technology Conference, July 1981, Washington, sponsored by the American Institute of Aeronautics and Astronautics; this quote is taken from his prepared text.
11. Daniel O. Graham, *High Frontier A New National Strategy* (High Frontier Inc., Washington, D.C., 1982).
12. Malcolm Wallop, "Opportunities and imperatives of ballistic missile defense," *Strategic Review,* Fall 1979, pp. 13-21.
13. Dusko Doder, "Andropov calls ABM proposal 'insane,' " *Boston Globe,* March 27, 1983, p. 1.
14. Wallop, *op. cit.*
15. Robert Hotz, "Beam weapon threat," *Aviation Week & Space Technology,* May 2, 1977, p. 11.
16. Kosta Tsipis, "Laser weapons," *Scientific American* 245 (6), 51-57, (December 1981).
17. Michael Callaham and Kosta Tsipis, *High-Energy Laser Weapons: A Technical Assessment* (Program in Science and Technology for International Security, Massachusetts Institute of Technology, Cambridge, Massachusetts, November 1980).
18. George Bekefi, B.T. Feld, J. Parmentola, and K. Tsipis, *Particle Beam Weapons* (Program in Science and Technology for International Security, Massachusetts Institute of Technology, Cambridge, Massachusetts, December 1978), p. 59.
19. Richard L. Garwin, "Ballistic missile defense: silos and space," author's population version of paper presented at American Physical Society meeting March 1982 in Dallas, p. 3. Garwin mentions the possibility of putting antisatellite lasers in low orbit or geosynchronous orbit but only gives the characteristics for the more difficult geosynchronous orbit case in Richard L. Garwin, "Are we on the verge of an arms race in space?" *Bulletin of the Atomic Scientists* 37 (5), 48-53 (May 1981).
20. The first observer I saw point out the problem in destroying a low-flying cruise missile in independent flight was A. M. Din, "The prospects for beam weapons," pp. 229-239 in

Bhupendra Jasani, ed, *Outer Space-A New Dimension of the Arms Race* (Oelgeschlager, Gunn & Hain, Cambridge, Massachusetts, 1982).
21. Michio Kaku, "Wasting space, countdown to a first strike," *The Progressive,* 47 (6), 19-22 (June 1983).
22. On a quick reading I caught several glaring errors. The High Frontier proposal was said to involve beam weapons; their initial system would instead rely on projectiles launched from orbiting battle stations. The total cost of the U.S. laser weapon program was quoted as half a billion dollars; the actual total is over $2 billion. The chemical laser used in the 1978 Navy tests at San Juan Capistrano was said to require a 300-W power supply that laser generated a 400,000 W beam; 300-W lasers are available commercially. The charged particle-beam program at the Lawrence Livermore National Laboratory was confused with the lab's X-ray laser program. Combined, these leave the overall impression that the author did not do enough homework.
23. For a sampling of some of the bizarre opinions held by the political extremists of the early 1960s, see: Mike Newberry, *The Yahoos* (Marzani & Mansell, New York, 1964). At that time there were a number of groups and some retired military officers advocating that the United States declare war against the Soviet Union. A flyer for one such group, with the headline "We want war!" is reprinted on the book's back cover.
24. Donald E. Day, statement before the Senate Armed Services Committee, Subcommittee on Strategic and Theater Nuclear Forces, March 16, 1982; copy supplied by the General Accounting Office.
25. *Ibid*
26. .H. Alan Pike, talk at Laser Systems and Technology Conference, July 1981, Washington, D.C., sponsored by American Institute of Aeronautics and Astronautics.
27. Barry J. Smernoff, "The strategic value of space-based laser weapons," *Air University Review,* Mar/Apr 1982, pp. 2-17.
28. *Ibid.*
29. For a transcript of the speech, see: "President's speech on military spending and a new defense," *New York Times,* March 24, 1983, p. A20.
30. A. M. Din, *op, cit.,* p. 237.
31. Paul J. Nahin, "Orbital BMD and the space patrol," in Bhupendra Jasani, ed, *Outer Space- A New Dimension of the Arms Race,* pp. 241-255 (Oelgeschlager, Gunn & Hain, Cambridge, Massachusetts, 1982). The same type of idea had been kicking around in my head before I read Nahin's article, but not as well formed.
32. Michael Callaham, private communication.
33. "Washington roundup," *Aviation Week & Space Technology,* February 21, 1983.
34. Michael Callaham, "Strategic laser weapons: arming for a shoot-out on the high frontier," to be published in proceedings of a 1982 symposium conducted by International Student Pugwash, Washington.
35. D. L. Hafner, "Anti-satellite weapons: the prospects for arms control," in Bhupendra Jasani, *Outer Space-A New Dimension of the Arms Race,* pp. 311-323 (Oelgeschlager, Gunn & Hain, Cambridge, Massachusetts, 1982).
36. For an indication of the emphasis the Chinese have placed on laser technology in general, see: "China emphasizes applications," *Laser Focus* 16 (8), 12-24 (August 1980).
37. Senate Armed Services Committee, Hearings on Department of Defense Authorization for Appropriations for Fiscal Year 1983, Part 7, Strategic and Theater Nuclear Forces, p. 4632, testimony of J. Richard Airey. An agreement with France cited in this testimony expired in 1982 and was not renewed.
38. Allen D. Simon, "The coming weapons," *Astronautics & Aeronautics* 20 (11), 40-44 (November 1982).

15.
Beams on the Battlefield:
A New Generation of Tactical Weapons

Beam weapons would undoubtedly have a serious impact on tactical warfare if they were brought to the battlefield. Yet the changes they would cause in military tactics would be much less dramatic than the changes that strategic beam weapons could cause in defense strategy. The reason is that the two types of beam weapons would do fundamentally different things. As described in the last chapter, strategic beam weapons would offer capabilities that for the most practical purposes would be new, notably ballistic missile defense. Tactical beam weapons would serve more to replace existing military equipment that doesn't perform as well as would be desired.

To understand that point, consider the similarities between the battles in World War II and those depicted in such science fiction movies as *Star Wars* and *The Empire Strikes Back.* Luke Skywalker may fly a rocket plane and fire some kind of directed-energy weapon, but his role in the action is still similar to that of a World War II fighter pilot. The Imperial Walkers in *The Empire Strikes Back* look like the result of crossing a tank with a long-legged Martian invader from *The War of the Worlds,* and they fire what look like laser bolts. Yet in battle they still play much the same role as tanks. Similarly, the laser ray guns and cannons fired by both sides are simply taking the place of rifles and artillery, respectively. The beam weapons would presumably be more deadly (they wouldn't be used in a real battle unless they were), but they are still performing the same function as the weapons used on today's battlefields.

Tank-mounted laser weapons defend an Army field command center against air attack in this artist's conception of tactical laser warfare. (Courtesy of Department of Defense.)

Zapping More Targets

From the standpoint of the Pentagon the central attraction of beam weapons is their potential for killing battlefield targets more effectively than conventional weapons. In theory, laser weapons could be used against anything that flies or moves, except heavily armored tanks. The list of potential targets ranges from individual soldiers to highly sophisticated missiles and aircraft. In practice, the interest seems to center on fast-moving or otherwise hard to kill targets with conventional weapons, or targets particularly sensitive to beam weapon attack.

At the top of the list of highly vulnerable targets are electro-optical sensors, which have become increasingly important in modem warfare. Because such devices are designed to sense visible or infrared light, they are especially vulnerable to laser attack, as described in Chapter 13. There are other ways to knock out sensors, including the use of electronic warfare techniques to scramble the signals they generate. However, the results can be hard to measure, and multiple layers of countermeasures may make it difficult to be sure what happened until the dust has settled. Lasers can attack such sensors more directly than electronic warfare, and in some cases it may be possible to monitor the attack by watching for reflection of part of the laser beam from the target.

That capability complements one of the Pentagon's biggest interests: shooting down tactical missiles before they reach their targets. An amazing variety of missiles have been developed around the world, including types for launch from the air, sea, or ground at targets in any of those places. The United States armed services have some four dozen types in development, production, and/or service, and other countries have their own types.[1] When modem armies meet, there are going to be many of these highly sophisticated missiles flying through the air.

These missiles are small, fast, and deadly. Their effectiveness against targets such as fast-moving fighter planes and heavily armored tanks makes them important to disable. However, their small size and fast speed make them hard to kill with conventional weapons. Some special-purpose missiles have proven useful against such missiles, but beam weapons promise to be even faster and more precisely targetable. Laser weapons are particularly attractive because many sophisticated antiaircraft missiles are guided to their targets by infrared seekers, which could be blinded or disabled by lasers.

Military planners thus see a major role for laser weapons in ridding the sky of enemy missiles. The same weapon could be used against hostile fighters, helicopters, or other aircraft or to attack vulnerable ground targets. Because of its tightly focused beam, a directed-energy weapon would be able to zap enemy targets in the midst of a field containing friendly targets as well, a vital capability in trying to shoot down an antiaircraft missile before it could hit a friendly plane. High speed and rapid retargetability, other characteristics of beam weapons, would also be important.

Preserving the Warship

The capabilities discussed above may offer a way to protect what some observers see as the biggest sitting ducks on today's battlefield: surface-going warships and aircraft carriers. Even before the Falkland Islands War, naval commanders were concerned about the vulnerability of their warships to missile attack. The damage the comparatively weak Argentine air force was able to inflict upon the British navy reinforced this view in the minds of some observers. Others remained unconvinced, such as *Aviation Week* editor William H. Gregory, who wrote: "The Falklands campaign did not prove anything conclusively about ships versus missiles . . . . While Exocets [French-built missiles fired by Argentine aircraft] did sink two British ships, four others were lost to iron bombs dropped by obsolete aircraft without modem countermeasures equipment. It was no test of large-carrier battle group survivability against missile swarms."[2] Later reports have indicated that the HMS *Sheffield,* Britain's most serious loss, was sunk because its computer system had been programmed to think of the Exocet as friendly and did not fire at the missile approaching the ship.[3]

Laser weapons zap aircraft and missiles attacking a fleet of warships in this artist's conception. The attacker at left is passing by its target on fire and out of control. Navy officials have looked at laser weapons as one way to keep the surface ship from evolving into an extinct species of sitting duck. (Courtesy of Department of Defense.)

Still, there are genuine reasons for concern that hostile forces able to fire swarms of missiles from the air could attack in enough force to send even ultrasophisticated warships to the bottom of the sea. The problem is that such a swarm could provide more targets than existing defense systems, based on Gatling-gun-like equipment, could cope with. Either particle beams or high-energy lasers might provide the rapid firing and retargeting capabilities needed to fend off such a massive missile attack. Such a defense system would probably require sophisticated automated controls, with reaction time much faster than a human soldier. (However, beam weapons show no immediate prospects for the other major menace to surface-going ships: submarines, which are effectively shielded from beam weapons by the water.)

The Emerging Automated Battlefield

The development of directed-energy weapons would fit in with the larger- scale development of the automated battlefield described earlier. There are no near-term plans to phase out human soldiers, but in the long term there is a continuing trend toward giving important front-line jobs to hardware.

The first generation of such weapons are called "precision-guided munitions." These include systems like "smart bombs," which automatically home in upon a spot on the target marked by a soldier on the sidelines holding a laser. They also include other systems that are remotely controlled by human operators at the rear or on the sidelines. As part of this trend, large sums are being spent on perfecting C^3I–the military code for command, control, communications, and intelligence–equipment that will let a commander run a battle more effectively from the rear.

The next generation under development are "fire-and-forget" missiles and other projectile weapons, where the only role for a soldier is to pick out the target and fire the weapon. The missile will then automatically home in on its target. As was mentioned in Chapter 10, there is even talk of completely automating the control of weapon systems so that the weapon will pick out its target, decide when to fire, and guide itself to the target. (Smart bombs can only guide themselves to targets identified by a soldier.) This leads to discussion of "intelligent" weapons[4] because, in some sense of the word, weapons with enough internal computer power to control their own firing would have to have some form of "intelligence." (This is one of the reasons why the Pentagon has been heavily supporting research in such esoteric-sounding areas of electronics and computer science as artificial intelligence.)

Beam weapons fit naturally into this trend. Those intended to fend off massive attacks would require some degree of computerized intelligence, if only because human reaction times wouldn't be fast enough. Even slower systems designed for human control would require sophisticated pointing and tracking capabilities, to say nothing of special equipment to control atmospheric effects on the beam.

In some cases beam weapons might be used to terrify enemy soldiers. This would probably be most effective against fairly unsophisticated troops without any protective equipment, who nonetheless had some idea of the damage a beam weapon could do. However, a more sophisticated army, with access to protective gear, would soon realize what protective measures were needed and take them. In many cases it would seem that chemical or biological warfare would be more terrifying to the troops under attack.

In this context it is probably worth noting that, although there have been public claims that the Soviet Union has used chemical and biological warfare against Afghan rebels, there have been no public reports of the use of laser weapons. In practice, it may be that the toxic chemicals used in and produced by chemical lasers are so hard to handle on the battlefield that they prove to be effective chemical warfare agents against friendly troops.

The Battlefield Arms Race

The real significance of tactical beam weaponry appears to lie not in the concept itself but in what it represents–the continuing tactical arms race. Strategic nuclear missiles generally are considered the most visible part of the arms race because they're the most threatening. We in the United States do not expect our front yards to turn into battlefields, but we are afraid that nuclear missiles might land there.

That attitude is understandable. Yet in terms of dollars, in terms of numbers of programs, and in terms of technological effort and complexity, the main thrust of the arms race is aimed at tactical weapons. While the United States and Soviet Union at least devote lip service to strategic arms control (see Chapter 16), there is very little effort to control the development of tactical weapons.

According to the Pentagon, about 85% of our defense budget goes to nonnuclear forces.[5] Much of that money goes to military pay, retirement benefits, and other items far from the tactical arms race. But billions more go into an imposing arsenal of sophisticated weapons.

Comparatively small programs weigh heavily in specialized technologies. During 1982, armed forces in the noncommunist world spent some $124 million on low-power lasers for use in tactical rangefinding and target designation. That is over one-third of all sales of lasers during that year.[6] Yet it's just a tiny fraction of the billions that are being spent on research, development, production, deployment, and maintenance of high technology weaponry.

There's no end in sight to this tactical arms race. The Pentagon is pouring billions of dollars into new battlefield hardware and concepts. Engineers are playing the conceptual chess game of countermeasures, counter-countermeasures, *ad infinitum.* If tactical beam weapons work, they will be absorbed into that continuing game without changing its essential course.

References

1. *Aviation Week & Space Technology* publishes a tabulation of missiles used by all nations of the world in its annual "Aerospace Forecast & Inventory" issue in early March. These figures are based on the listing on pp. 142-147 of the March 8, 1982 edition.
2. William H. Gregory, "New lessons from combat," (editorial) *Aviation Week & Space*

Technology **117** (3), 13 (July 19, 1982).

3. "HMS Sheffield thought Exocet was friendly," *New Scientist* 97, February JO, 1983, p. 353.
4. Allan D. Simon, "The coming weapons," *Astronautics & Aeronautics* 20 (11), 40-44 (November 1982).
5. Caspar W. Weinberger, *Secretary of Defense Annual Report to Congress Fiscal Year 1983* (U.S. Government Printing Office, Washington, D.C., February 8, 1982), p. 1-28.
6. C. Breck Hitz, "The laser marketplace–1983," *Lasers & Applications* 2 (1), 45-53 (January 1983).

16. Beam Weapons and Strategic Arms Control

Arms control has received increasing attention over the past two decades, but progress has been excruciatingly slow. The basic goal of arms control–limiting the deployment of new weaponry that would increase the danger of war without increasing anyone's national security–is as hard to argue against as the goal of peace on earth. Unfortunately, both arms control and peace are exceedingly difficult goals to attain in practice.

Diplomacy and treaty making are basic tools of arms-control negotiations, inevitably making the process delicate. The result should be an agreement that makes sure neither side gains an unfair advantage in quantity or quality of arms. With vital national interests at stake, negotiators tend to quibble over every definition and interpretation in a proposed treaty. Even such a seemingly simple task as counting the numbers of weapons possessed by each side can prove to be very complicated. That issue, in particular whether nuclear weapons controlled by European members of the North Atlantic Treaty Organization should be counted in the total of U.S. weapons, stalled 1983 talks between the United States and Soviet Union on limiting the deployment of nuclear weapons in Europe.

Over the past couple of decades arms-control negotiations have focused on strategic and nuclear armaments, which both the United States and the Soviet Union recognize as the most threatening types of weapons. Because development of missile defense beam weapons would have a dramatic impact on the strategic balance of power, such systems are emerging as a major arms-control concern. The issue is a complicated one, particularly because existing treaties mention ballistic missile defense but do not consider the possibility of beam weapons–leaving a hazy ground of wording that either side may try to interpret to its own advantage.

Both the United States and the Soviet Union clearly take the beam weapon issue very seriously. Soviet leader Yuri Andropov reacted strongly to President Reagan's suggestions that the United States develop a missile defense system, saying that it would violate existing arms-control treaties.[1] President Reagan evidently took Andropov's criticisms seriously and responded by saying that the United States might eventually offer any missile defense technology it developed to the Soviets "to prove to them that there was no longer any need for keeping" offensive nuclear missiles.[2]

Arms-control specialists were interested in beam weapons long before President Reagan's March 23, 1983, speech. That interest is evident to anyone willing to plod through the ponderous bureaucratic prose of the annual *Arms Control Impact Statements,* which the Arms Control and Disarmament Agency submits to Congress. As far back as 1979, the Carter Administration devoted 10 pages of its 273-page report to directed-energy weapon programs.[3] The attention devoted to beam weapons has grown since then. Of the 389 pages in the edition prepared to accompany President Reagan's fiscal 1983 budget, 35 are devoted to directed-energy programs.[4] That is slightly over 1/12 the space in the book, yet the nearly $500 million President Reagan requested for beam weapons that year was a much smaller fraction (1/40) of the $20 billion requested for military research and development. (Congress eventually cut about $60 million from the Administration's beam-weapon request, and the final fiscal 1983 budget figure is closer to $400 million.) The directed-energy program was miniscule compared with the $258 billion the Reagan Administration requested for all military activities in fiscal 1983.

There is another indication of sensitivity: security restrictions. Two versions of the *Arms Control Impact Statements* are prepared, one classified for distribution to officials with security clearances and one unclassified for distribution to the general public. Lacking a security clearance, I receive the general distribution version in which it seems that "[deleted]" may be the most common word. Outside of the directed-energy section, it is possible to find many unexcised pages. However, only two pages in the beam weapon section don't carry the censor's "[deleted]" trademark, and one of these is the final page that contains a mere two lines. On one other page the only word is "[deleted]," another includes two "[deleted]s" and the word "systems," and a third contains three "[deleted]s" and the subheading "Program status." Judging from the context, some of the missing material has been publically disclosed in other unclassified publications. For example, the *Arms Control* book deletes identification of the most expensive part of the particle-beam research program,[5] which is identified as the Advanced Test Accelerator at the Lawrence Livermore National Laboratory in the Pentagon's fact sheet on particle-beam technology.[6] Similarly, the goal of the accelerator experiments is deleted in the arms-control book but described as the demonstration of propagation of an intense electron beam through the atmosphere in the fact sheet.

The heavy-handed security policy presumably reflects the sensitivity of one of the major goals of the beam weapon program: space-based defense against ballistic missiles. Development of such a system would have serious implications for existing treaties, described below, that prohibit all but sharply limited types of defense against ballistic missiles.

The ABM Treaty

On May 26, 1972, the United States and Soviet Union signed a pair or treaties collectively known as SALT-I (for Strategic Arms Limitation Treaties).

One of them is the ABM (for Anti-Ballistic Missile) Treaty, which at the time limited each country to building only two missile defense systems, one around the national capitol and the other around a missile-launching area. This was later reduced to one system per country around either site. The Soviet Union picked Moscow; the United States picked a Minuteman missile site in North Dakota but scrapped its system in 1976.

The ABM Treaty also bans "the development, testing, and deployment of all ABM systems and components that are sea-based, air-based, space- based, or mobile land-based."[7] In this case the word "development" has been subject to some special interpretation. The United States government defines it as: "The obligation not to develop such systems, devices or warheads would be applicable only to that stage of development which follows laboratory development and testing. The prohibitions on development contained in the ABM Treaty would start at that part of the development process where field testing is initiated on either a prototype of a bread-board model."[8] In short, the ban on development doesn't apply to development, but rather to testing of an already-built system. Admittedly, a stringent ban on development in the usual sense of the word would probably be impossible to enforce. Nonetheless, it is interesting to note that government officials couldn't find a different word in trying to explain how "development" had been defined in the treaty.

The treaty contains another loophole allowing "development and testing of fixed, land-based ABM systems and components based on other physical principles" that could substitute for the antimissile missiles used in the missile defense systems being developed at the time. That mention of "other physical principles" might be interpreted to cover lasers and particle beams, but the exclusion does not include systems for use in space. It also prohibits the actual deployment of such fixed, land-based systems and components unless the treaty is modified.[9] The U.S. government also interprets the treaty as constraining the field testing of directed-energy weapons intended for certain "[deleted]" applications.[10]

Not surprisingly, the U.S. government's official position is that its beam weapons program complies with the ABM Treaty. The Arms Control and Disarmament Agency concluded that "the directed-energy related research and development efforts funded in this fiscal year 1983 budget have not more than marginal arms control effects now, [but] this technology deserves continuing attention in the future."[11] This opinion is based on plans to test the three elements of DARPA's Space Laser Triad separately, with the Alpha chemical laser and the large optics demonstration experiment staying on the ground. Tests of an integrated system in space probably would be considered a field test–at least by the Soviet Union.

The arms-control book mentions plans for mid-1980s demonstrations intended to allow a decision on whether to go ahead with "advanced development prototypes," which would be demonstrated by the early 1990s. If those prototype demonstrations work, the report continues, "laser weapons could be available in the late 1990s for selected offensive and defensive applications such as defense of ships, aircraft, high-value ground targets or satellites, destruction of ground and airborne sensor systems, and ballistic missile defense."[12]

The report does not mention what would happen to the existing treaties at that point. Actual deployment of an antimissile weapon system would require either modification or abrogation of the ABM Treaty. Proponents of the High Frontier space-based missile defense system have urged just such a course.[13] Another possibility would be trying to find a loophole big enough to accommodate a laser or particle-beam battle station.

The *Arms Control Impact Statements* book generally uses the generic term "directed-energy weapon" but at one point indicates[14] that the main concern is laser weaponry. This is consistent with the Pentagon's public position that it is too early to assess the feasibility of particle-beam weapons.

The Soviet Union seems to think that particle beams might be usable as weapons of mass destruction and has raised the issue both in public and in discussions of the Committee on Disarmament Working Group on Radiological Weapons and New Mass Destruction Weapons. One Soviet proposal would ban weapons using "charged and neutral particles to affect biological targets"[15]—that is, people. The American arms-control report does not identify any specific U.S. response to that proposal but does say that most proposed particle-beam weapons focus their energy onto a point and hence could not be weapons of mass destruction.[16] The Soviets appear to be concerned about the interaction between an intense particle beam and the air producing what is called "secondary radiation."[17] The same thing occurs on a much smaller scale when a cosmic ray hits the atmosphere. As the energetic particle strikes atoms in the air, it can break up atoms to generate other particles or cause the atoms to emit short-wavelength electromagnetic radiation: gamma rays and X rays. At high doses this secondary radiation can be harmful to people, and the theory is that an intense enough particle beam could generate a cone-shaped shower of secondary radiation powerful enough to kill or disable any people in the way. It is far from clear that the concept could be militarily useful in practice. A powerful enough particle-beam generator might be too big to haul onto the battlefield or put into orbit. And power would be important because without enough power such a weapon might kill soldiers very slowly–hardly a useful way of fighting off an enemy.

Weapons in Space

The idea of stationing weapons in space has become a controversial one in itself. The Outer Space Treaty, which went into effect in 1967, includes some high-minded statements about the goals of space exploration to maintain international peace and security and to promote cooperation and understanding among nations.[18] Yet it puts only a few limitations on weapons in space. A careful reading shows that nuclear bombs are expressly banned from space, along with other weapons of "mass destruction." There are no restrictions that would prohibit artificial satellites from housing military bases or other types of weapons. Thus, the treaty appears to allow particle-beam and most types of laser battle stations. One exception is the X-ray laser battle station described in Chapter 6, because the nuclear bomb that would power it is expressly forbidden from space. The X-ray laser proposal would also run afoul of the 1963 Limited Test Ban Treaty, which prohibits nuclear tests in the atmosphere or space,[19] tests needed to verify the X-ray laser weapon concept.

Concern over orbiting weapons goes beyond battle stations for missile defense to include antisatellite weapons that require putting something in orbit, such as the Soviet killer satellites. The United States and Soviet Union talked about banning space-based weapons altogether in 1978 and 1979, and in 1981 the Soviets proposed a draft treaty that would ban space weapons.[20] The next year the United Nations held a conference on the exploration and peaceful uses of outer space (UNISPACE 82), which many of the participating countries hoped would make a similar proposal. However, strong opposition from the United States led to a mere mention of "general concern" in the final report. Talks held the same year by the United Nations Committee on Disarmament also failed to produce significant action on space-based weapons.[21]

Some observers outside of the two major powers are disturbed not just by weapons in space but by all military equipment in orbit.[22] There has even been some talk of trying to keep military equipment out of space altogether. That's a noble sentiment, but as long as there are large nuclear arsenals, spy satellites can help keep the peace by making sure that those arsenals are where they belong.

Antisatellite Weapons

The role of spy satellites is recognized implicitly in the ABM Treaty, the SALT-I Interim Agreement, and the unratified SALT-II agreement. All three ban the use of weapons that might interfere with "national technical means" of verifying that the other side is complying with arms-control treaties. That strange-sounding term has not been defined formally in any international agreement but is generally taken as covering surveillance and early-warning satellites.[23] In short, "national technical means" are electronic spies, recognized in somewhat prettier words by treaties that protect them from being shot at by the spied upon.

Although the three agreements prohibit *use* of antisatellite weapons against at least some types of satellites, they do not outlaw the development, testing, or even deployment of such weapons. As far as the treaties go, laser (or other types of) satellite killers can be built, tested against target satellites intended for that purpose, and deployed ready for use. They can even track the other side's satellites. What is forbidden is to pull the trigger.[24] The United States and Soviet Union have talked about limiting or banning antisatellite weapons, but no talks are under way at this writing. The *Arms Control Impact Statements* state only that the United States "is reviewing its policy in this area."[25]

More Push for the Arms Race

As in most parts of the arms race, research on beam weapons serves to stimulate other military research. Despite the denials from the Pentagon, it is clear that the United States and the Soviet Union are engaged in a race to develop beam weapons, a race President Reagan's interest can't help but stimulate. Each side believes (with good reason) that the other side is working on beam weapon technology. Each side seems to believe that at least some beam weapon technologies offer the potential for important military capabilities, most importantly ballistic missile defense. Thus, both sides have their own programs in directed-energy technology, with each apparently following its own development strategy.

It is hard to see anything on the horizon that might stop that race, short of a "show-stopper" type of problem that can convince both sides the whole idea is ridiculous. In theory, arms-control agreements might be used to limit development programs, but in practice that has yet to be done. Spy satellites can count missile silos and other large weapon systems, but they can't peer inside research laboratories. That would require on-site inspections of military laboratories, something neither the United States nor the Soviet Union have been willing to accept. Without some way of verifying compliance with arms-control treaties, there is little incentive to sign them in the present atmosphere of mutual distrust.

The problem with most beam weapon concepts, from an arms-control standpoint, is that development programs are not likely to produce any effects that the other side could monitor to verify compliance. The X-ray laser is an exception because it can't work without the explosion of a small nuclear bomb, which the other side *can* detect. But as long as other types of beam weapons are being tested in a closed laboratory, it is impossible to monitor development programs if security is tight enough. Without some way to verify compliance, there are no incentives for countries that don't trust each other to sign treaties limiting development programs–and then to comply with them. The same type of reasoning has led to treaties that limit the deployment, rather than the development, of missile defense systems in general.

What about prospects for limiting the actual deployment of beam weapons for missile defense? Those are hard to call because they will depend on political and technical factors yet to be defined. Neither Ronald Reagan nor Yuri Andropov are likely to be in office when the technology is ready; the views of their successors are as unknown as their identities. The technological nature of a beam weapon system for missile defense is also undefined. If it proves to be as bulky, expensive, and restricted in capabilities as critics predict, both sides might be willing to negotiate limitations, especially if they are at a rough parity in development. If, as seems likely, construction of orbiting hardware is a highly visible and lengthy process, that might add to the incentives for negotiating limitations. On the other hand it might prove very hard to negotiate meaningful and effective limitations if the technology promises to be highly effective and not horrifyingly expensive, if one side feels that it is decisively ahead of the other, or if one or both sides are not in a mood to negotiate, for their own political reasons.

Will other countries try to get into the game? The high stakes and inherent secrecy cloud the issue, but no other countries seem as willing to make the necessary investments of money and technological resources as the United States and the Soviet Union. China might be interested. The country has been putting high priorities on both laser and space development. Yet China is still recovering from the scars of the Cultural Revolution, which caused a major hiatus in scientific and technological development.[26] China is also the one major country to have cut its military budget recently, citing more urgent needs for economic development.[27] Thus, it seems unlikely that China is an active participant in the beam weapons race.

The United States has had agreements to share classified technical information on high-energy lasers for tactical uses with four of its allies: Britain, France, West Germany, and Australia.[28] Those agreements do not cover information on strategic laser weapons, that is, those for missile defense. There are also no obvious signs that any of those four nations have strong high-energy laser programs, apparently preferring to let the United States lead the way and pay the bills. Japan has never been very active in high-energy laser development and only recently has begun building industrial lasers with continuous power levels of more than 1000 W. The Soviet Union rarely shares its advanced military technologies with its Warsaw Pact allies, and there are no signs that high-energy lasers are an exception to that pattern. The U.S. *Arms Control Impact Statements* does warn that other nations might become interested in laser weaponry once prototypes become available, but that day has yet to come, and the report notes that cost would still be a problem.[29]

What of the long-term impact on arms-control efforts? In the eyes of many observers the deployment of beam weapons for missile defense would simply be another step in the continuous arms race, with no sign of change in an overall pattern they find distressing. In some ways it might be even more disturbing because it would mark a failure of the 1972 ABM Treaty to effectively stop deployment of missile defense systems. (The construction of a pop-up system of X-ray laser weapons would be even more unsettling because it would mark the demise of two other major arms-control treaties dating from the 1960s: the Limited Nuclear Test Ban Treaty and the Outer Space Treaty. That appears to be the reason why government censors delete official indications of interest in X-ray laser weapons from virtually all unclassified government publications.) Pessimists warn these moves would push us further toward a dangerous instability in the balance of power.

Outside of the most vocal advocates of beam weaponry, few observers see the development of beam weapon missile defense systems as a quick cure to the threat of nuclear war. However, some observers besides President Reagan have expressed hope that strategic beam weapons might help ease a transition away from national defense policies based on threats of nuclear doom. Barry J. Smernoff of National Defense University suggests that the development of effective defensive weapons, and the shift in power from offense to defense, might help convince the great powers and their allies to reduce the sizes of their nuclear arsenals.[30] As mentioned in Chapter 14, Paul J. Nahin of the University of New Hampshire has gone a step farther and suggested that orbiting antimissile battle stations be operated by an international peacekeeping force, as a step toward dismantling strategic nuclear arsenals.[31]

Realization of those visionary-sounding hopes will require real international cooperation, not merely new technology. The alternative is a continuation of the arms race to ever-higher technological levels. As one of the staunchest of laser weapon advocates, Malcolm Wallop, has noted, "Like all other weapons in history, they would be neither 'absolute' nor 'ultimate.' "[32]

References

1. Dusko Doder, "Andropov calls ABM proposal 'insane,' "*Boston Globe,* March 27, 1983, pp. 1, 8.
2. Benjamin Taylor, "US could offer ABM to Soviets, Reagan says," *Boston Globe,* March 30, 1983, pp. 1, 7.
3. Arms Control and Disarmament Agency, *Fiscal Year 1980 Arms Control Impact Statements* (U.S. Government Printing Office, Washington, D.C., March 1979), pp. 94-103. This is the earliest of these statements which I have seen.
4. Arms Control and Disarmament Agency, *Fiscal Year 1983 Arms Control Impact Statements* (U.S. Government Printing Office, Washington, D.C., March 1982). A new edition of this publication is prepared each year.
5. *Ibid.,* p. 313.
6. Department of Defense, "Fact Sheet: Particle Beam Technology Program," February 1982, p. 5.
7. Arms Control and Disarmament Agency, *op. cit.,* p. 321.
8. *Ibid.,* p. 321, footnote to testimony by Ambassador Gerard Smith before the Senate Armed Services Committee in 1972.
9. *Ibid.,* p. 32
10. *Ibid.,* p. 322.
11. *Ibid.,* pp. 331-332.
12. *Ibid.,* p. 300.
13. Daniel O. Graham, *High Frontier, A New National Strategy* (High Frontier, Washington, D.C., 1982), p. 106.
14. Arms Control and Disarmament Agency, *op. cit.,* pp. 323-324.
15. *Ibid.,* p. 324.
16. *Ibid.*
17. Potential use of ground-based charged particle beams to generate a lethal shower of secondary radiation is mentioned in Bhupendra Jasani, *Outer Space-A New Dimension of the Arms Race* (Oelgeschlager, Gunn & Hain, Cambridge, Massachusetts, 1982), pp. 82, 235. There has also been some discussion of shooting a particle beam downwards from a satellite to irradiate an area on the ground, but this appears to require unreasonably high powers. See: Peter Laurie, "Exploding the beam weapon myth," *New Scientist,* April 26, 1979, pp. 248-250.
18. The full treaty is reproduced in Bhupendra Jasani, ed, *Outer Space-A New Dimension of the Arms Race* (Oelgeschlager, Gunn & Hain, Cambridge, Massachusetts, 1982), pp. 370-374.
19. *Ibid.,* pp. 368-369.
20. O. V. Bogdanov, "Banning all weapons in outer space," in Bhupendra Jasani, ed, *Outer Space-A New Dimension of the Arms Race*, pp. 325-329 (Oelgeschlager, Gunn & Hain, Cambridge, Massachusetts, 1982); the proposed text appears on pp. 401-403.
21. Bhupendra Jasani, "How satellites promote the arms race," *New Scientist* 96, 346-348 (November 11, 1982).
22. *Ibid.*

23. Arms Control and Disarmament Agency, *op. cit.,* p. 322.
24. D. L. Hafner, "Antisatellite weapons: The prospects for arms control," in Bhupendra Jasani, ed, *Outer Space-A New Dimension of the Arms Race* (Oelgeschlager, Gunn & Hain, Cambridge, Massachusetts, 1982), pp. 311-323.
25. Arms Control and Disarmament Agency, *op. cit.,* p. 323.
26. The scope of the Chinese laser program is indicated in "China emphasizes applications," *Laser Focus* 16 (8), 12-24 (August 1980), but laser weapons are not mentioned.
27. Stockholm International Peace Research Institute, *The Arms Race and Arms Control* (Oelgeschlager, Gunn & Hain, Cambridge, Massachusetts, 1982), p. 4.
28. Senate Armed Services Committee, *Hearings on Department of Defense Authorization for Appropriations Fiscal Year 1983,* Volume 7, p. 4632; testimony of J. Richard Airey. The agreement with France expired and was not renewed.
29. Arms Control and Disarmament Agency, *op. cit.,* p. 325.
30. Barry J. Smernoff, "The strategic value of space-based laser weapons," *Air University Review,* Mar/Apr 1982, pp. 2-17.
31. Paul J. Nahin, "Orbital BMD and the space patrol," pp. 241-247 in Bhupendra Jasani, ed, *Outer Space-A New Dimension of the Arms Race* (Oelgeschlager, Gunn & Hain, Cambridge, Massachusetts, 1982).
32. Malcolm Wallop, "Opportunities and imperatives of ballistic missile defense," *Strategic Review,* Fall 1979, pp. 13-21.

17. Conclusion: Where are Beam Weapons Going?

In many ways it is premature to try to write a conclusion to the story of beam weapons. Critical tests of feasibility have yet to be performed. Vast differences of opinion remain both on the possibility of building beam weapons and on the effect such weapons could have on military strategy and tactics. President Reagan's call for an effort to develop beam weapons for missile defense is just the opening of a debate over whether the United States should try to do so. Yet those same differences of opinion demand that some effort be made to sort them out.

Although I have never been involved in beam weapon development on an official level, I have spent enough time watching the field to fancy that I can fill in some of the "deleted" portions of government publications. I've learned which sources to trust and which ones to question. I have tried to keep my objectivity and to avoid letting personal prejudices color my judgments. I don't claim to have all the answers (no one does), but I do feel that I have a general idea of what is happening.

The answers are not clear and unambiguous; neither the world nor the technology are that simple. Beam weapons are very immature, but they do show signs of promise. Pencil-and-paper projections can be made of their characteristics, but the results differ widely. The fact is that until some critical experiments are performed, no one will be *certain* how well the technology will work. There clearly is much to be learned, and that is the job of a research and development program. It would be surprising if there were no surprises around the corner.

Will beam weapons work? I've seen statements ranging from a flat "no" to a definite "yes, and soon." These conclusions come from respectable sounding sources who have studied the question carefully. To understand something of how such widely divergent answers were reached, let's first take a look at some historical lessons.

The story of Archimedes and the mirrors in Chapter 2 is well worth remembering. Generations of eminent scholars thought the idea was ridiculous, and some of them disproved it–they thought–with pencil and paper. They were wrong. An experiment in the mid-1700s showed that the idea would work, but it was conveniently forgotten. Another experiment a decade ago again showed that the idea could work, as mentioned in Chapter 2. We will never know if Archimedes was able to overcome the practical problems of logistics and to get an adequate number of mirrors with good enough surfaces to bum the Roman ships. But we certainly know that eminent scholars can say some foolish things.

History is full of such failures of imagination. Science fact and fiction writer Arthur C. Clarke noted that and went on to formulate what he called "Clarke's law":

> When a distinguished but elderly scientist states that something is possible, he is almost certainly right. When he states that something is impossible, he is very probably wrong. [1]

In this context Clarke defines an "elderly" physicist as one over 30.

In the 1940s and 1950s a number of prominent physicists declared that intercontinental ballistic missiles and landing a man on the moon were technologically impossible. They were wrong, of course. The ICBM example was pointed out to me by Maxwell W. Hunter II during a break in a mid-1981 seminar on high-energy lasers, which I moderated for the American Institute of Aeronautics and Astronautics. As a young engineer, he had been one of the people who developed the ICBM. Now a high-level engineer at Lockheed, he was instrumental in shaping Senator Wallop's proposals for orbiting laser battle stations for use against ICBMs. There are major obstacles to be overcome with high-energy lasers, he said, but analogous barriers were overcome in the ICBM program. He predicted that high-energy laser weapons could be made to work, if the country was willing to make the needed commitment.

On the other hand more than one brainchild of Pentagon planners has succeeded only in consuming large quantities of the taxpayers' dollars. Over $1 billion was spent between 1945 and 1961 in an effort to develop a nuclear-powered aircraft, which literally never got off the ground. There was even an apparently erroneous report in *Aviation Week* that the Soviet Union was flight-testing a nuclear-powered plane. [2] It's an example I point out to friends in the laser community, as I wonder aloud which of today's programs will in two decades be the boondoggles we knew and loved.

The commercial world has made some expensive wrong guesses of its own. One example is the Bell System's unsuccessful attempt to convince the general public that a Picturephone belonged in every home. Another was the supersonic transport, aviation's great white hope of 1970, which evolved into the white elephant of 1980. Browse through technological magazines of the late 1960s and you can find many more examples. There was, for example, a holographic tape system, which in late 1969, officials of the RCA Corp. predicted would find a $1-billion market for video playback in 10 years;[3] by 1973, the holographic system had vanished from sight, leaving only the name "Selecta-Vision," which the company would later apply to the videodisk system it finally put on the market using completely different technology nearly a decade later.

The problem is hype. Anyone trying to sell an idea is liable to get carried away in translating a bright idea into a program and promise more than he can possibly deliver. The developers of the nuclear-powered aircraft finally couldn't find a mission for the plane. The holographic videotape ran into technical problems. The same story is written over and over again in program slippages in government and industry. I would half-seriously propose Hecht's Law of Technology Marketing: "Any sales projection is an overestimate, and any marketing timetable is overoptimistic." That applies not just to industry, but also to programs sold to the government. Indeed, writing proposals urging military agencies to support research programs was once described as a special type of science fiction writing by Ben Bova,[4] a person certainly qualified to make such judgments because he was responsible for marketing the Avco Everett Research Laboratory's laser development programs to the government before leaving to become a full-time science-fiction and -fact writer, and editor of *Analog Science Fiction* and *Omni.*

Both Clarke's law and Hecht's law apply to beam weaponry. At least some of the technology–most likely infrared, visible, and ultraviolet lasers– has a reasonable shot at working. Not working according to the most optimistic timetables, but working sometime within the next couple of decades.

The pessimists seem to have their hearts in the right place in their desire to avoid further escalation of the arms race. But their analyses are so determinedly pessimistic they can become unrealistic. Perhaps inadvertently, they have stacked the deck to come out with the answers they want to hear. Although the most pessimistic studies, such as those of Kosta Tsipis of the Massachusetts Institute of Technology, have gotten widespread attention in the general press,[5] they are largely shrugged off by most people active in laser weapon development. Scientists who I have come to respect complain of flawed analyses and an unwillingness to listen to critical questions and say that information in the classified domain basically disproves the most pessimistic conclusions. There are obvious problems with some of the assumptions made in the more pessimistic studies, including selection of unrealistically small mirror diameters[6] and stationing of the hypothetical battle stations in geosynchronous orbit,[7] where they would have to shoot at targets over distances ten times longer than if they were at the lower orbits proposed by Pentagon analysts. Both types of assumptions tend to require unreasonably high laser powers, and hence vast supplies of fuel.

Proponents of beam weapons also make some reasonable-sounding points. President Reagan has advocated missile defense systems as a better strategy for protecting our country from nuclear attack than reliance on the threat of a massive nuclear counterattack. Senator Wallop has said that an antimissile laser battle station in space "threatens nothing except weapons of mass destruction."[8]

Yet proponents of crash development programs ignore the very real technological limitations and uncertainties present in all beam weapon concepts: chemical lasers, conventional short-wavelength lasers, free-electron lasers, particle beams, X-ray lasers, and microwaves. More money might speed the pace of development somewhat, but throwing billions of dollars at the problem will not put an orbiting shield of laser battle stations over our heads overnight. Without carefully planned experiments that are performed in a logical sequence, those billions of dollars could be wasted in exploring blind alleys. There are also serious logistic problems inherent in trying to vastly accelerate any large program, prominent among them is the need to train the engineers who will build the system and to develop the industrial base that will manufacture the hardware.

When will high-energy laser weapons work? The question is a complex one because the *laser* technology is the easiest part of the problem, something which is also true for most other types of beam weapons. The most severe difficulties come in getting the beam to the target and seeing that it does the needed amount of damage.

The United States and the Soviet Union both are probably close to having enough laser firepower on the ground to disable a satellite by blinding its sensors. In fact, both sides may already have that firepower. It is not at all clear that existing beam-direction and fire-control systems could hit a satellite on demand. Nor, for that matter, is it clear that such lasers could do anything militarily useful. If they have to wait until an enemy satellite obligingly floats by slowly, due overhead, they would be of little practical value.

BLOOM COUNTY **by Berke Breathed**

Beam weapons are easier to dream up than to build. All it takes is a fertile imagination, such as that of *Bloom County* cartoonist Berke Breathed. (Reprinted with permission.)

Nonetheless, the technology for building antisatellite laser weapons seems fairly close. Current satellites are easy targets, particularly spy satellites, because of the inherent vulnerability of their optical sensors to low levels of laser illumination. Because of the limited range of a ground-based laser, addition of an orbiting battle mirror or the use of an aircraft-mounted laser seems more valuable from a military standpoint. Orbiting antisatellite lasers would probably be much harder to build than those for use on the ground or in the air, although far easier than reaching the much higher powers needed for orbiting antimissile lasers. Putting an antisatellite laser into orbit would have another drawback: the actual deployment of an antisatellite weapon in orbit could be considered a provocative act by the other side.

The main limitation on antisatellite weapons may be their limited military usefulness. For better or worse, spy satellites serve a stabilizing function, reassuring each side that the other is not in process of launching an attack. Taking potshots at such satellites in peacetime, and especially during a period of international tension, doesn't make sense. The other side is far too likely to interpret those potshots as the first shots of a war and react accordingly. Once the fighting starts, antisatellite weapons could be valuable, particularly if the war is expected to last a while as in the unsettling scenario of a protracted nuclear war.

But care must be taken that these antisatellite laser capabilities not be too provocative, lest they trigger action by the other side that *causes* a war.

If battlefield lasers work, there could be two generations: a first aimed at optical and infrared sensors and a second aimed at causing physical damage to the target. The first type would be easier to build but would be of limited utility. Although many battlefield weapons rely on optical and infrared sensors for guidance, you can't be sure that the missile being fired at you has a guidance system you can knock out with a laser. However, such lasers should be fairly easy to build and, in fact, may be within reach now.

Lasers capable of causing physical damage to a target would be much more desirable on the battlefield. A weapon that blew up an antiaircraft missile would be far more reassuring to the pilot of the target plane than a weapon that blinded sensors but caused no obvious harm to the missile. The laser power needed has been demonstrated, although the technology needed has yet to be packaged in a form resembling the compact and rugged gear that would be needed on a battlefield. The big problem, however, is getting the beam to the target. It is far from clear that the problems presented by the atmosphere can be tamed well enough–and at a low enough cost–for many military uses.

Recent cuts in Navy and Air Force tactical laser weapon programs seem to indicate that beam-transmission and other problems have not been solved. Although much of the impetus for the cuts has come from Congress, the Pentagon has gone along without strenuous objection. Unofficial reports indicate that the Navy's MIRACL laser worked better than the equipment intended to get its beam to the target. The Airborne Laser Laboratory also had difficulties in delivering a lethal dose of laser energy. The Navy's program is now considered dead, while the Air Force effort is being wound down. Unless these programs are replaced, which seems unlikely, only the Army will have a tactical laser weapon program–and that a comparatively modest one which Congress cut back in fiscal 1983.

The cutbacks may make it hard for developers to even match the conservative predictions in the *Arms Control Impact Statements,* that high-power laser weapons would not be ready for battlefield use until the mid- to late 1990s.[9] Anti-sensor lasers might be ready earlier. However, interest in laser weapons seems to be shifting to antisatellite and antimissile applications, where the targets would be outside the beam-distorting atmosphere.

The prospects are unclear for charged particle beams for warship defense. If charged particle beams can make their way accurately through the atmosphere, if the beam can be steered precisely, and if compact, powerful systems can be developed, the technology might be on the scene in the late 1990s. Those first two "ifs" are big ones, and there is no assurance that the answers will come out right for beam weaponry. The same considerations apply to defense of point targets against ballistic missile attack.

If beam weapons work on the battlefield, something by no means certain, their main impact would probably be to speed the pace of battle. From a human standpoint every weapon now on the battlefield takes a perceptible amount of time to reach its target. Beam weapons would not. As far as the human eye could see, they would hit their targets almost instantaneously, although laser beams probably would take a few seconds for the "kill." Such weapons would probably provide even further impetus to speed the automation of the battlefield.

Beam Weapons for Missile Defense

The type of space-based laser defense against ballistic missiles envisioned by President Reagan, Senator Wallop, and others presents a quite different set of problems than battlefield lasers. If the laser is put into orbit, there would be no need to worry about the difficulty of getting a high-energy laser beam through the atmosphere–a problem that could make laser weapons impractical on the battlefield. Offsetting that advantage is the need to make any system that goes into space extremely compact and reliable–areas in which much work remains. One technical compromise between these problems is the idea of leaving the laser on the ground and using a set of orbiting battle mirrors to redirect the beam around the globe. This approach seems to make most sense for the free-electron laser because its output wavelength could be adjusted to match "windows" of good atmospheric transmission and because the electron accelerator it requires would be very hard to get off the ground. Nonetheless, any laser-based ballistic missile defense system would also face the very serious problem of finding, tracking, and hitting a target thousands of kilometers from the laser. This last problem could prove to be the most serious.

There is a real risk that efforts to develop laser systems for ballistic missile defense will not succeed. The Pentagon has said, "we will not be seeking to develop a single system which can intercept and flawlessly defend against all missiles and all attacks. Such a system may not be possible."[10] Better capabilities may be possible with a "layered" defense in which attacking missiles would have to run a gauntlet of defenses, which might include lasers and/or particle beams as well as other weapons systems. Yet even an imperfect defense system can deter an attack by making its outcome uncertain, as mentioned in Chapter 14. (Admittedly, a perfect system, capable of destroying all attackers, would be the ideal, but such perfection would be much more important in actually fighting a war than in deterring an attack.)

At a time when human needs around the world urgently call out for attention, developing a new generation of weaponry cannot be our first priority. Yet without some breakthroughs in international understanding, defense needs must claim some of our resources. I see enough promise in laser technology to feel that it would be foolish to walk away from its opportunities. The technology is not going to go away; it has come a long way in the past two decades, and progress is continuing around the world. We should keep on looking at the prospects for missile defense, remembering that any real payoff is likely to be a couple of decades away. At the same time I cannot see adequate reasons for a crash-priority, multibillion dollar program. The obstacles are large enough that even such a massive program would not deploy an effective laser weapon system by the end of the decade, and I am far from certain that the job could be finished by the end of the century. Such an effort would waste billions of dollars. Given the present level of technical uncertainty, I think it's premature to even try to assign a firm timetable.

There has been a growing debate over the idea of shifting the emphasis of space-based laser research from infrared chemical lasers to those emitting at shorter wavelengths. Although there are clearly advantages to shorter-wavelength lasers, I worry that a shift now might be premature. The Department of Defense has invested millions of dollars in the ALPHA chemical laser and the Large Optics Demonstration Experiment, and it would seem foolish to stop work on those programs as long as there was any prospect of getting useful results. Moreover, rapid shifts in direction of a long-term research program tend to be disruptive to the program itself, a problem shared by many other long-term government research programs that Congress funds on a year-to-year basis. And though chemical lasers clearly have their limitations, I am unaware of any that are fatal to the idea of using them in space. There are also problems at short wavelengths. Although smaller-diameter optics can be used at such wavelengths, the optics must be made with surfaces much more perfect than needed for infrared lasers. The lasers themselves are also not well developed. The chemical oxygen iodine laser described in Chapter 4 could offer important advantages over hydrogen fluoride, but so far high enough powers have not been demonstrated. I am far from convinced that practical weapon-scale excimer lasers can be built; so far powers fall far short of those attained with chemical lasers, and problems with laser physics and optical damage from the short-wavelength light could put an upper limit on output power. Free-electron lasers seem very promising in the longer term, but the technology is immature, and putting the needed electron accelerator into a satellite is likely to take a lot of work (thus the idea of sending the beam from a laser on the ground to a mirror in space). Short-wavelength lasers deserve continued attention, but not at the expense of abandoning chemical lasers.

The X-ray laser battle station is probably a weapon whose time should never come. As indicated in Chapter 6, the technology is extraordinarily difficult. The original report of the X-ray laser demonstration at the Lawrence Livermore National Laboratory generated widespread skepticism from outside observers.[11] One of the more polite descriptions of the weaponization concept was "premature," and one observer gave a blunt one-word assessment I would not repeat here. After two years of following the strange story of the X-ray laser, I remain unconvinced that X-ray laser weapons are feasible in the foreseeable future. Once again, the problem is that it's easier to build a powerful laser than to hit anything with it.

Some defense analysts seem enchanted by the notion of a "pop-up" weapon system in which the equipment needed in orbit–X-ray lasers or battle mirrors for ground-based lasers–would not be launched until they were needed. In theory, keeping this hardware on the ground or in submarines, on rockets ready to boost it into orbit, would reduce its vulnerability to enemy attack. The concept is new enough that it has gotten little critical attention, but my own analysis leaves me skeptical of its technical feasibility and worried about its strategic consequences. It is far from clear that a complex battle station can be put into orbit, stabilized adequately, and readied to zap a fleet of attacking missiles in the 30-min interval between the launch of a missile attack and the impact of the warheads at their targets. If the pop-up system has to be launched *before* warning of the actual start of an attack, that launch becomes a visible strategic maneuver. The other side can only view it as a sign of preparation for war, perhaps even as a sign of plans for a first strike. Unlike battle stations permanently stationed in orbit, pop-up systems do not seem adaptable to operation by international peacekeeping groups. The idea may deserve a hearing, but its capabilities seem to fit much better with the desire to plan a first strike than with a need to protect against one. That would only serve to increase instability.

There are other problems, too. The X-ray laser's reliance on a nuclear bomb would open up a veritable Pandora's box of issues. Existing treaties permit only small underground tests of nuclear explosives. Development of X-ray laser weapons would require other types of nuclear tests in the upper atmosphere or in space that are prohibited by the 1963 Nuclear Test Ban Treaty.[12] Putting any nuclear weapons into orbit, even the "small" ones needed to power X-ray lasers, is flatly prohibited by the 1967 Outer Space Treaty.[13] Actual deployment of antimissile lasers would also violate the 1972 ABM Treaty.[14] In short, a serious effort to develop X-ray laser weapons would probably require repudiation of a foundation of nuclear arms control–the test ban treaty–if they were to be tested under realistic conditions. Going ahead and putting such battle stations into orbit would violate two other major arms-control agreements. Such actions would probably get a hostile reaction from around the world, particularly from American allies in Europe who are very sensitive to the issue of nuclear armaments. Even though the X-ray laser might be nominally a "defensive" device, the country that first deployed one in space or in a pop-up system probably would be viewed as a prime promoter of the nuclear arms race. Given the real likelihood that the system would not provide an effective defense except *after* a first strike against the other side, the program seems a good one to pass up.

It is also far from certain that neutral particle beams could serve as the basis of space-based ballistic missile defense. The problem is generating a neutral beam focused tightly enough in angular direction and with a small enough spread in particle energy that it would be able to deliver a lethal dose of energy to a target a few thousand kilometers (or a couple of thousand miles) away. The prevalent idea is to generate a beam of negatively charged hydrogen ions and pass them through a gas to strip off the extra electrons. This inevitably involves collisions, which would tend to spread out the beam. Exactly how much spreading will be produced is one of the questions the Pentagon hopes to resolve in experiments planned at the Los Alamos National Laboratory, but even if the answer is encouraging much work remains to be done.

The Pentagon has yet to decide if microwave weaponry is anything more than a fuzzy idea. Microwaves could offer some attractive capabilities for disabling electronics, but the need for very large antennas would probably make high-power microwave systems designed to "cook" targets unwieldy on the battlefield.

The Soviet Beam Weapon Program

Is there really a beam weapon gap between the United States and Soviet Union? Although there does seem to be a beam weapon race, the difference between the two sides seems more apparent than real. I base that opinion on the following three reasons:

• First, the Soviet Union historically tends toward making risky demonstrations early in a development program. The United States, in contrast, has traditionally concentrated on thoroughly checking out systems to get rid of potential problems *before* a demonstration. The space race was an excellent example. The Soviet Union managed a first few spectacular shots, most notably Sputnik, but was eventually overtaken by the United States.[15] In keeping with this pattern, the Soviet Union might be the first to demonstrate (or try to demonstrate) a laser weapon, but that demonstration might not indicate a commanding lead–a possibility in keeping with comments of Pentagon officials mentioned earlier.

• The second reason is what I call the "Soviet submarine effect," the selective leaking of confidential information to the press. Years ago, defense-industry insiders noted that American press reports of sightings of Soviet submarines off the U.S. coast tended to soar whenever Congress was debating the Navy budget. The reports apparently came from Navy officials, who were trying to remind people about the importance of the Navy. Leaks can be a source of valuable information, but they can also present problems because by their nature they are almost impossible to verify. They may be accurate, they may be spurious, or they may be one sided. The latter may be the case when the source of the leak is a person who has just been on the losing side of a decision and who decides to appeal his case to the public through the press. *Aviation Week's* coverage of particle-beam weaponry began in just that way when retired Air Force General George Keegan went public with charges of a massive Soviet program. The magazine's coverage of X-ray lasers started with an unnamed source evidently unhappy because the Pentagon was unwilling to support the program.[16] Most observers would agree that while both articles carried a core of truth, they both also suffered from exaggeration.

• The third problem is the "threat inflation" that tends to occur during intelligence gathering. The information gathered during intelligence operations inherently tends to be imprecise. Technical details may be misunderstood, conversations may be only partly overheard, future goals may be confused with present realities, and the most interesting details may not be resolved by a spy satellite image. Intelligence analysts interpreting this information have to estimate a range of possibilities. Military planners often assume the "worst" case, which seems prudent when planning how to deal with possible threats.

However, all too often worst-case estimates are transformed into seemingly authoritative reports of new equipment about to roll off Soviet assembly lines, when in reality the Soviet Union may have built only a single experimental model that doesn't work very well.

One case in point is the warning that a large Soviet space booster under development "will have the capability to launch . . . even larger and more capable laser weapons," which appeared in the Pentagon's October 1981 brochure, *Soviet Military Power.*[17] In the middle of the following year *Aviation Week* reported the Soviets were working on a booster able to put 220,000 kg (240 tons) into orbit–twice the capacity of the Saturn 5 booster the United States developed for the Apollo program.[18] However, Barry J. Smernoff points out that the arrival of that booster had been expected since the late 1960s, but that as of early 1983 it had yet to get off the ground successfully.[19]

The same type of delays seems to plague plans to launch Soviet laser weapons reported in *Aviation Week.* An October 1981 report cited "hard information" on three new Soviet spacecraft, including one which appeared to be "a laser weapon packaged for space tests." The tests were expected to be performed in weeks or months, depending on the availability of boosters needed to put the large craft into orbit. American analysts reportedly expected the tests to be of antisatellite capabilities, "but do not rule out other tests."[20] Two years later, nothing further had been reported on the test plans. Perhaps the giant satellite is waiting for the big booster to be perfected.

Impact of Strategic Beam Weapons

How would the deployment of antimissile beam weapons affect the international balance of power? What risks would it raise? And in the light of these concerns, should we go ahead with President Reagan's plans to try to develop them? The answers are not always clear and depend on such factors as how much the missile defense system would cost.

The first thing to realize is that the existing Anti-Ballistic Missile (ABM) Treaty bans the actual deployment of a ballistic missile defense system. There might be efforts to bend or interpret the wording enough to let some battle stations slip through a loophole; after all, the word "development" has been interpreted to mean something else, as noted in Chapter 16. Yet by most current interpretations, even those of the Reagan administration's Arms Control and Disarmament Agency,[21] beam weapons intended for missile defense could not be deployed without violating the ABM Treaty. That's one point President Reagan glossed over in his March 23, 1983, speech pushing development of missile defenses.[22]

Europeans did not welcome President Reagan's proposal for ballistic missile defense with open arms. This cartoon, one of a series that normally covers electronics, was originally published in the British trade paper, *Electronics Times,* on April 14, 1983, just three weeks after Reagan's speech. It reflects a general uneasiness and fear that the Reagan Administration is not serious about exploring arms-control possibilities. (Courtesy of *Electronics Times.)*

It is also important to realize that treaties are not always considered sacred by the countries that have signed them, particularly after a change in government. Many of us might consider the ABM Treaty to have power equivalent to that of law. Yet faced with concern over the possible vulnerability of the MX missile system to destruction by a Soviet missile attack, the Reagan administration has seriously considered building an antimissile system to protect MX. The fiscal 1983 *Arms Control Impact Statements* concede that going ahead with that antimissile system "could require amending the ABM Treaty or withdrawing from it."[23] Although the Reagan administration apparently put the idea of defending MX aside while officials sought a viable home for the new missile, they could revive the antimissile proposal if they failed to find another way to assure MX's "survivability."

The strategic capabilities offered by a beam weapon missile defense system would be more tempting to the military mind than mere enhanced survivability for an intercontinental ballistic missile such as MX. Thus, they would provide even stronger temptation to break–or renegotiate–the ABM Treaty. From the standpoint of nations other than the United States and Soviet Union, renegotation would probably be much more reassuring than simple abrogation of the treaty, as it is allowed by the treaty and would indicate that the two major powers were still interested in arms control.

There is a legitimate concern that if one major power were far ahead of the other, it might try to seize a decisive advantage by orbiting a network of battle stations and using them to keep the other side from launching anything into space. That type of scenario would be dangerous because the underdog might see little alternative but to attack the battle stations during construction. However, it seems more likely that the United States and the Soviet Union would develop similar capabilities close enough in time that neither would ever have a decisive lead.[24] Also, it would take long enough to build large beam-weapon battle satellites that there would be plenty of time to talk before taking potshots. The situation could be dangerous, but it should be survivable.

The implications of X-ray laser deployment have been given distressingly little public attention. One reason may be that the pop-up X-ray laser concept is rather new. But I suspect another is that the idea has just grown, almost of its own accord, assembled from pieces of promising technology and the wish-lists of military planners, seemingly without much critical examination. That is unfortunate, because *if* the technology works, it seems likely that its reaction time would be too slow to stop a nuclear attack that was launched before it was "popped up." Thus, the defense would work only if the system was launched *before* the attack–an act that itself might precipitate an attack. Or it might be seen as part of a first strike capability, intended to wipe out only the fraction of enemy missiles which weathered the attack.

I find it very hard to put much credence in the most pessimistic of published scenarios: antimissile laser weapons obviously won't work, but that attempts to put them into space could nonetheless trigger World War III.[25] Certainly military organizations are capable of producing, and putting on the battlefield, weapons that don't work, but the flaws were not obvious to the people building the systems. In the real world military budgets are limited and can't be stretched to buy all the hardware on the "wish lists" compiled by fertile military minds. Research programs are not expected to generate lethal hardware immediately, but eventually something useful has to be produced, or advocates of other military programs will divert the money to their efforts. And if an obviously ineffective system *did* make its way into orbit, it is hard to imagine the cautious, multilayer military bureaucracies in the United States and Soviet Union doing something wildly provocative. They would more likely go through careful and cautious probing of capabilities to decide what to do about it.

Seeking the Best Approach

At this point, far from the reality of beam weapon battle stations, it is hard to sort out the scenarios and pick out a best course. Beam weapons might not work at all, though it seems far too early to reach such a conclusion. Antimissile systems might prove so expensive, or so limited in effectiveness, that the United States and Soviet Union would willingly accept limitations on their deployment, much as they did a decade ago in agreeing to the ABM Treaty itself. Although it would probably be impossible to verify restrictions on development of beam weapons, limits on deployment should be verifiable. Large satellites are easy to watch, and even pop-up systems could be monitored to the extent we can count submarine-launched ballistic missiles.

If antimissile beam weapons prove to be both affordable and effective, it would be desirable to negotiate treaties covering their deployment, which at the same time would lead to a phasing down of nuclear armaments. It might even be possible to create some sort of international peacekeeping organization to operate the defensive satellites, something like the concept proposed by Paul Nahin described in Chapter 14. Of course, safeguards would be necessary, perhaps even elaborate ones, to make sure that the peacekeeping force would serve to protect every country in the world, not just particular ones. It could be something as simple as making half the satellites American and half Soviet, which together would provide a worldwide defense, but separately would be unable to defend (or attack) adequately. Realizing such a goal will not be easy, but it would be far better than another round of arms race.

I would like to think that such a goal would be attainable by the time beam weapons capable of ballistic missile defense have been developed. I fear it is not attainable now. President Reagan's words are full of high-spirited ideals, as exemplified by his offer to share U.S.-developed missile defense technology with the Soviets.[26] Unfortunately, the reality of the Reagan administration's actions have fallen far short of those lofty ideals, with more effort devoted to rattling sabers than to making peace. Soviet leaders, too, have been strong on rhetoric and weak on substance. Instead of trying to negotiate seriously about arms control, the two countries seem to be putting most of their efforts into scoring propaganda points and finding excuses to avoid substantive talks. Thankfully, time remains to deal with the beam weapon issue, time that should bring new generations of leaders in both the United States and Soviet Union. We can only hope the new leaders will be up to the challenge.

If space-based beam weapons can provide missile defense, they would tip the strategic balance of power to the defense. Like any transition, this would bring some uncertainty and instability. Some danger would be inevitable, but there is also danger inherent in the present nuclear balance of terror. If the great powers can cope with the challenge of a transition to a strategic defense, the result could be a more stable world. The transition to a powerful defense might, as observers like Barry Smernoff hope, devalue nuclear arsenals enough that the United States and Soviet Union would be willing to dismantle theirs.

Major uncertainties would have to be overcome to reach that optimistic conclusion. Technological barriers would have to be overcome. The United States and Soviet Union would have to change their defense strategies. The two countries would have to negotiate changes in the existing ABM Treaty, without scrapping the whole framework of arms control. National leaders would have to raise their goals beyond short-sighted self-interest toward an ultimate demilitarization of the world that in the long term is the only thing which can provide security to all the people of this planet. We can only hope that such action is not impossible.

A defense-based strategic balance is not enough by itself. There is no assurance that it could be maintained as long as there is a continuing arms race. New technology alone cannot end the arms race. History warns us that there is no such thing as an "ultimate" weapon. In 1917, after seeing the airplane go to battle in World War I, a saddened Orville Wright wrote: "When my brother and I built the first man-carrying flying machine we thought that we were introducing into the world an invention which would make further wars practically impossible."[27] A generation later, the scientists who built the atomic bomb would voice the same futile hopes.

If beam weapons are merely another round in the arms race, they will be followed by a new generation of weapons with countermeasures designed to foil beam attack. Cruise-missile technology might be extended to longer ranges, as long as the low-flying missiles could keep out of the range of antimissile beams. More bombs would be built. Bigger beam weapons would be developed in response, and the cycle would go on.

The best we can hope is not that beam weapons will end the arms race, but that they will buy us the time we need to end it. The problem lies not in the technology, but in ourselves. It lies in the inability of the United States and Soviet Union to develop a mutual respect and trust for each other. We need to learn that we are not enemies but neighbors on a small planet with different ways of life. We need to learn to tolerate each other and to avoid trying to force our wills on others. Alas, those tasks seem far harder than building arsenals of beam weapons.

Given these realities, I see little alternative for now but to continue working on beam weapons. But at the same time we–both the United States and the Soviet Union–must strive as well to build the mutual understanding and respect that can lead to a peace far more meaningful than one imposed by the fear of each other's arsenals. If humankind can develop new technology but not better ways of understanding others, we will be only children, playing with ever more dangerous toys.

References

1. Arthur C. Clarke, *Profiles of the Future* (Bantam Books, New York, 1964), p. 14.
2. John Tierney, "Take the A-plane," *Science* 82, Jan/Feb 1982, pp. 46-55, gives an excellent account of the nuclear aircraft episode. The fundamental problem was that the program took on a life of its own, even without a discernable mission for such a plane.
3. The RCA holographic system is mentioned in: Brian J. Thompson, "Holography Technology," in Ernst Weber, Gordon K. Teal, and A. George Schillinger, eds, *Technology Forecast for 1980* (Van Nostrand Reinhold, New York, 1971), p. 168.
4. Ben Bova, guest of honor speech at Boskone XIV, a February 1977 science fiction convention conducted in Boston by the New England Science Fiction Association.
5. I suspect that the most widely read article critical of laser weapons is Kosta Tsipis, "Laser weapons," *Scientific American* 245 (6), 51-57 (December 1981); the article has been influential because of *Scientific American's* prestige. Because of this highly visible publication, reporters seeking a critical view of laser weaponry to balance their stories often turn to Tsipis.
6. The worst example that comes to mind is the assumption that a 1.4-m (4.6-ft) mirror would be used with a chemical laser; in Michael Callaham and Kosta Tsipis, *High-Energy Laser Weapons: A Technical Assessment* (Program in Science and Technology for International Security, Massachusetts Institute of Technology, Cambridge, Massachusetts, November 1980), p. 46. A 2.4-m (8-ft) mirror has already been built for NASA's space telescope, though it cannot handle high laser powers.
7. See, for example, Richard L. Garwin, "Ballistic missile defense (BMD) silos and space," popular version of paper presented at March 1982 meeting of the American Physical Society in Dallas. The idea of putting

a battle station in geosynchronous orbit is attractive because it would remain over one point on the globe, but the high cost of putting it in such a high orbit, and the Jong distances involved, are generally considered prohibitive.

8. Malcolm Wallop, "Opportunities and imperatives of ballistic missile defense," *Strategic Review,* Fall 1979, pp. 13-21.
9. Arms Control and Disarmament Agency, *Fiscal Year 1983 Arms Control Impact Statements* (U.S. Government Printing Office, Washington, D.C., March 1982), p. 300.
10. Donald L. Lamberson, plenary talk on Department of Defense Directed Energy Program, presented May 17, 1983 at Conference on Lasers & Electro-Optics, Baltimore; quote is from printed version supplied to reporters.
11. See, for example, Jeff Hecht, "The X-ray laser flap," *Laser Focus* **17** (5), 6 (May 1981), and Jeff Hecht, "X-ray laser controversy," *New Scientist* 92, 166 (October 15, 1981).
12. Text of the treaty appears in: Bhupendra Jasani, ed, *Outer Space-A New Dimension of the Arms Race* (Oelgeschlager, Gunn & Hain, Cambridge, Massachusetts, 1982), pp. 368-369.
13. *Ibid.,* pp. 370-374.
14. *Ibid.,* pp. 375-379; there are also separate statements of understanding.
15. For a history of the space race, see: Richard Hutton, *The Cosmic Chase* (Mentor, New York, 1981).
16. The particle-beam controversy opened with Clarence A. Robinson, Jr., "Soviets push for beam weapons," *Aviation Week & Space Technology,* May 2, 1977, and was followed by a series of follow-up pieces. The X-ray laser work was revealed in Clarence A. Robinson, Jr., "Advance made on high energy laser," *Aviation Week & Space Technology,* February 23, 1981, pp. 25-27, which was followed by a long silence, evidently because of the sensitive nature of the leak.
17. Department of Defense, *Soviet Military Power* (U.S. Government Printing Office, Washington, D.C., 1981), p. 79.
18. "Soviets outspending U.S. on space by $3-4 billion," *Aviation Week & Space Technology,* July 19, 1982, pp. 28-29.
19. Barry J. Smernoff, "The strategic value of space-based laser weapons," *Air University Review,* Mar/Apr 1982, pp. 2-17.
20. "Washington roundup," *Aviation Week & Space Technology,* October 5, 1981, p. 17.
21. Arms Control and Disarmament Agency, *op. cit.,* p. xv.
22. The text of President Reagan's talk appears in "President's speech on military spending and a new defense," *New York Times,* March 24, 1983, p. A20.
23. Arms Control and Disarmament Agency, *op. cit.,* p. viii.
24. Sometimes hysterical-sounding warnings that the United States is dangerously behind the Soviet Union in developing beam weapons have been made by a few advocates of a massive defense buildup, but some observers without a budgetary axe to grind hold that the U.S. is ahead or at least even, despite the supposedly greater Soviet effort. For example, Barry J. Smernoff *(op. cit.)* holds that the United States has a clear lead in space development that can be carried over to space-based lasers. A. M. Din of the University of Lausanne in Switzerland also believes that the United States will probably win out in development of beam weapons; see: A. M. Din, "The prospects for beam weapons," in Bhupendra Jasani, ed., *Outer Space-A New Dimension of the Arms Race* (Oelgeschlager, Gunn & Hain,

Cambridge, Massachusetts, 1982), pp. 229-239. As mentioned earlier, gloomy prophecies of potent Soviet beam weapons orbiting the globe seem to be nowhere near fulfillment.

25. See, for example, Michael Callaham and Kosta Tsipis, *op. cit.*
26. Benjamin Taylor, "US could offer ABM to Soviets, Reagan says," *Boston Globe,* March 30, 1983, p. 1 & 7.
27. Quoted in C. D. B. Bryan, *The National Air and Space Museum Vol. 1: Air* (Peacock Press/Bantam Books, New York, 1982), p. 90.

18.
A Brief Epilogue: The View from 2015: Downscaling Beam Weapons

The past 30 years have been a long strange trip, and we have learned a lot about beam weapons, technology, and defense policy.

The name "Star Wars" stuck to the Reagan Administration's program in the press, but formally it became the Strategic Defense Initiative or SDI. "Star Wars" became a politically incorrect term among defense contractors and military officials during the Reagan years. It grew to a massive multi-billion dollar program in the mid-1980s. SDI's fiscal 1986 budget included over $1 billion for lasers and optical technology, the vast bulk of it for high-energy laser weapons.[1]

Yet the future that Ronald Reagan's advisors envisioned in 1983 never arrived. No laser battle stations are in orbit, and no X-ray lasers are ready to pop up into space to block a Soviet nuclear attack. The only hardware produced so far by many billions of dollars spent on nuclear missile defense is a few dozen ground-based interceptors of questionable effectiveness, targeted at blocking small-scale attacks by "rogue states."

Many technologies important for defense have progressed dramatically. The Reagan era was a time of 1200-baud dial-up modems and computers running MS-DOS on floppy disks; today's smartphone has the computing power of late 1980s supercomputer, and my home has a 25-megabit data connection. Solid-state lasers and optical technology also have advanced dramatically. The 2014 Nobel Prize in Physics went to Isamu Akasaki, Hiroshi Amano, and Shuji Nakamura for developing efficient light-emitting diodes that were thought impossible in the mid-1980s. A new generation of solid-state lasers based on optical fibers generate kilowatt beams that slice metal sheets quickly and easily.

But major technologies crucial for SDI beam weapons did not work as advertised. The backbone of our space program, the space shuttle, peaked at nine missions per year in 1985, far below the rate needed to launch orbiting laser battle stations and supply them with massive amounts of chemical fuels. No chemical laser ever operated in space, or proved practical outside of a laboratory. Particle-beam weapons never got off the ground. The prospects of a bomb-driven X-ray laser scared the Soviets, but underground nuclear tests never yielded a workable version.

The contrast represents one central lesson about technology -- progress is uneven. Computing, communications and information technology in general have advanced dramatically since the 1960s. Spaceflight technology advanced dramatically from the 1950s through the Apollo Moon landings, but advanced very slowly for decades afterwards.

'What Happens If peace breaks out?'

Cold War tensions fueled the early growth of the Strategic Defense Initiative. Defense contractors saw SDI as a golden opportunity. After the first edition of this book came out, a local company invited me to talk about beam weapons. At lunch a vice president waxed enthusiastic about the prospects for company growth, made certain by decades of Cold War. He asked what I thought about the future.

"What happens if peace breaks out?" I replied.

It was a joke, prompted by my discomfort with how he was cheering on the continuation of the Cold War. Mikhail Gorbachev was just another member of the Soviet Politburo. Ronald Reagan was still an arch cold warrior. Yet in the end, the two and their immediate successors realized not only that the nuclear arms race was madness, but that they could do something to stop it. The Soviet Union fell in the process, and Russia and the United States downsized their nuclear arsenals and took them off high alert. The Cold War was over.

The American public was stunned to find that the economy of the former Soviet Union was a basket case. The Soviets did have a major program in beam weapons, but when Russia opened it up for inspection, it proved less formidable than had been expected. The giant laser at Sary Shagan, rated in the megawatt class in the *Soviet Military Power* brochure, could generate only about 20 kilowatts.

U. S. beam weapons programs had their own problems. Space-based chemical weapons began to look impractical. Issues included fuel supply and burning chemical fuels without disrupting the laser's aim at distant targets. The X-ray laser project slowly collapsed after test results proved to be misreported. Reporter William Broad blasted that as a billion-dollar deception in his book *Teller's War*.[2] But SDI just kept shifting its focus to new technologies, such as Lowell Wood's idea for a space-based fleet of "brilliant pebbles," computer-guided projectiles that would smash into nuclear missiles and destroy them in space.

Could SDI itself have been a bluff, a technological Potemkin village built to scare the Soviet leadership? That's the plausible conclusion of Nigel Hey's book *The Star Wars Enigma*.[3] I have heard tell that Livermore's cadre of X-ray laser developers claim credit for toppling the Soviet Union with their colossal bluff. But I have to wonder if it began that way. Viewed from afar, the quick shifts to new technologies was a sign that things were not working as well as hoped.

Perhaps it was more like a lengthy hand of draw poker, where the U.S. and Soviets started with high hopes but poor hands. After the draw, the U.S. had only a pair of deuces, but opted to bluff that it had a stronger hand. The Soviets, with an even weaker hand, perhaps jack high, folded.

But whatever the case, Reagan, Gorbachev and George H.W. Bush saw their chance to wind down the Cold War and managed that more deftly than most of us had thought possible.

The Airborne Laser

As the threat of a massive nuclear attack waned, the proliferation of nuclear weapons brought a new fear. "Rogue states" such as North Korea and Iran might threaten the United States and its allies with a smaller – but still devastating – nuclear strike. To counter that threat, the Air Force proposed building a fleet of Airborne Lasers, jumbo jets each fitted with a megawatt-class chemical oxygen-iodine laser (COIL). A COIL emits at 1.3 micrometers, a much shorter wavelength than deuterium-fluoride chemical lasers, allowing the use of smaller optics. The plan was for the planes to fly near the rogue states, prepared to fire their lasers at any nuclear missiles observed rising from the atmosphere during their vulnerable boost phase. The megawatt beam was expected to have enough power to destroy target missiles hundreds of kilometers away.

It seemed like a reasonable idea at the time. It's much easier to install a big laser in a jumbo jet than to put it into orbit, and the targets would be orders of magnitude fewer in number and a factor of ten closer. The Pentagon established the Airborne Laser program office in 1993, and issued a $1.1 billion contract in 1996 to Boeing, TRW and Lockheed Martin to build and test the first plane in the planned fleet.[4] The original plans called for the first firing of the laser to be in early 2003, but that slipped 20 months to November 2004. After that test, program manager Col. Ellen Pawlikowski said modifications were needed before the laser could meet its final milestone, shooting down a test missile in the air. The Missile Defense Agency estimated that would take an additional $1.39 billion.[5]

In the end, the Airborne Laser destroyed boosting missiles in the air for the first time in February 2010, more than five years later. The Missile Defense Agency wrote in a press release, "The revolutionary use of directed energy is very attractive for missile defense, with the potential to attack multiple targets at the speed of light, at a range of hundreds of kilometers, and at a low cost per intercept attempt compared to current technologies."[6] But by that point Secretary of Defense Robert Gates had already stopped plans to build a fleet of laser-equipped planes, telling Congress in May 2009 that the program "has significant affordability and technology problems, and the program's operational role is highly questionable."[7] The Airborne Laser was retired in 2012. Major problems included the logistics of handling the chemical fuels and problems arising from firing six massive laser modules, each the size of an SUV, from the plane.

Other Nuclear Missile Defense

In parallel with abandoning space-based lasers in favor of the Airborne Laser, the Pentagon also downsized its other nuclear missile defense projects. Brilliant Pebbles began to shrink during the last years of the Bush Administration. The new Clinton Administration further shrank it in March 1993 and renamed it "Advanced Interceptor Technology" before finally canceling development of space-based interceptors in December 1993.[8]

They also reorganized SDI into the Missile Defense Agency, which pressed forward with plans two types of ground-based interceptors:

• The Ground-Based Midcourse Defense is based on launching projectiles from the ground to hit long-range nuclear warheads after they have been launched into space but before they descend toward their target. They are actively guided kinetic-energy weapons that must hit the targets to disable them. Launched in December 2004, the program had spent over $40 billion through August 2012, with another $4.4 billion allocated in from fiscal 2013 to 2017, according to a General Accounting Office Study.[9] According to MDA, a total of 30 interceptors were deployed in California and Alaska as of the end of 2010, the most recent count listed on their web site.[10] A series of tests have produced mixed results, raising lingering questions about the system's effectiveness.

• The Terminal High Altitude Area Defense (THAAD) system uses movable batteries of hit-to-kill interceptors based on the ground to defend against short- to medium-range ballistic missiles. The most recent count of operational interceptors listed on the MDA web site is 50 interceptors as of 2012.[11] Although THAAD has also had its problems, the GAO report found "THAAD's major components are mature and its design is stable."[12]

Tactical High-Energy Lasers

In the mid-1990s, Israel and the United States teamed on a program called the Tactical High-Energy Laser (THEL) to test a deuterium-fluoride chemical laser emitting over 100 kilowatts for defense against rockets, artillery, and mortars at the White Sands Missile Range in New Mexico. The goal was to develop an effective defense against short-range weapons used by insurgents, a problem that had become acute in the middle east.

Several years of tests showed that such a high-energy laser could be remarkably effective against such weapons. But when field commanders were asked about the idea, they pointed to serious logistic problems. Chemical lasers are bulky, and require a continual supply of two special chemical fuels on the battlefield. Without ample supplies of both fuels, chemical lasers would be useless as weapons. Moreover, the laser fuels and the laser exhaust were hazardous. "A chemical laser on a battlefield is more of a hazard than the threat it is trying to mitigate," said John Boness of the aerospace contractor Textron Systems.[13] Field commanders asked for a different kind of laser, that ran on electricity which could be produced by generators burning diesel fuel, which is already on the battlefield, to fuel military vehicles and supply electric power.

That led the Pentagon's Joint Technology Office to launch a program called the Joint High Power Solid State Laser (JHPSSL) in December 2002. Solid-state lasers had come a long way since they were deemed unsuitable for laser weapons in the 1960s. Then they could convert only about one percent of input electrical power into laser light, and the waste heat could not be removed fast enough from the laser rod to fire a series of shots. But by the 1990s, the inefficient lamps that had been used to pump solid-state lasers were being replaced by semiconductor diode lasers, which could convert more than half of the input electrical power into laser light. The diode laser beams were not focused tightly enough to use as weapons, but they could excite atoms in thin slabs, thin disks, or even optical fibers that converted up to half the pump energy into a high-quality laser beam, and dissipated the waste heat readily.

In 2009, Northrop Grumman used diode pumping of thin slabs of laser material to generate a steady 100 kilowatt beam for five minutes, converting nearly 20% of the input electrical energy into laser light.[14] That met the goals of the JHPSSL project. Textron later reached the same power level with a different design. Meanwhile industrial materials-working lasers had been developed that generated more than 10 kilowatts by using diode lasers to excite atoms in the cores of optical fibers.

Those technologies are now being tested by the armed forces. In 2014, the Navy tested a laser weapon system called LaWS on board the USS Ponce in the Persian Gulf.[15] The system combines the outputs of six industrial fiber lasers to generate tens of kilowatts of light at about one micrometer in the infrared. Videos from the Office of Naval Research posted on YouTube show the system shooting down a drown and igniting explosives on a small boat. The Navy tested an array of a half-dozen industrial fiber lasers emitting a total of tens of kilowatts on the USS Ponce in the Persian Gulf. The Army, Air Force, and Marines have their own experiments in progress or under construction.

This new generation of laser weapons differs dramatically from those planned by the Reagan Administration. The current 100-kilowatt systems will deliver only 2% of the 5-megawatt output planned for an orbiting chemical laser battle station. Their targets will be a few kilometers away, not thousands of kilometers, and moving relatively slowly. The lasers will be on military trucks, ships or aircraft, not in a spacecraft. And the targets will be cheap weapons fired by insurgents, not sophisticated nuclear warheads equipped with elaborate countermeasures. So their task should be comparatively easy.

However, the new laser weapons still face serious technology issues. One is protecting the laser optics from the dirt and dust inevitably found on the battlefield. Dirt on focusing optics absorbs light, heating the lens or mirror to the point where the intense flux of laser power damages the surface. Military planners worry about how to minimize the chance of damage and how to repair it when it occurs without hauling the laser weapon back to a clean room. Another crucial issue is developing cooling systems able to withstand rugged field conditions; current test systems have relied on laboratory-grade equipment. Another unanswered question is whether they can be fielded at reasonable cost and in a reasonable time, and live up to the high expectations of developers.[16]

I wrote about the first half-century of laser weapons in 2009 for *Optics & Photonics News.*[17] All in all, high-energy laser technology has come a long way, but it still falls far short of what "Star Wars" sought to build three decades ago. Thankfully, today's arms race in beam weapons is not struggling to prevent nuclear apocalypse. But we still need to develop mutual understanding and respect.

References

1 Jeff Hecht, "Government Laser Spending: The Outlook," *Lasers & Applications 4* (6) pp. 65-70 (June 1985)

2 William Broad, *Teller's War: The Top-Secret Story Behind the Star Wars Deception* Simon & Schuster, New York, 1992)

3 Nigel Hey, *The Star Wars Enigma: Behind the Scenes of the Cold War Race for Missile Defense* (Potomac Books, 2006)

4 Fact Sheet: Airborne Laser (YAL-1A), United States Air Force, dated 27 February 2003

5 Jeff Hecht, "Long road ahead for Airborne Laser," *Optoelectronics Report 11* Dec 1, 2004 http://www.laserfocusworld.com/articles/oer/print/volume-11/issue-23/features/long-road-ahead-for-airborne-laser.html

6 "Airborne Laser Testbed Successful in lethal intercept experiment," Missile Defense Agency, Feb. 11, 2010 http://www.mda.mil/news/10news0002.html

7 Statement of Secretary of Defense Robert M. Gates, House Armed Services Committee, Wednesday, May 13, 2009 -- 10:00 A.M.

8 Doug Beason, "The Rise and Fall of Brilliant Pebbles," *The Journal of Social, Political, and Economic Studies 29* (2) 143-190 (Summer 2004)

9 General Accounting Office, *Defense Acquisitions: Assessments of Selected Weapon Programs Fiscal-Report GAO-13-294SP* March 2013, pp. 51-52

10 http://www.mda.mil/system/gmd.html checked June 6, 2015

11 http://www.mda.mil/system/thaad.html checked June 6, 2015

12 General Accounting Office, *Defense Acquisitions: Assessments of Selected Weapon Programs Fiscal-Report GAO-13-294SP* March 2013, p. 136

13 Jeff Hecht, "Ray guns get real," *IEEE Spectrum* July 2009

14 *Ibid*

15 David Smalley, "Historic Leap: Navy shipboard laser operates in Persian Gulf," Office of Naval Research, Press release Dec 10, 2014 http://www.onr.navy.mil/Media-Center/Press-Releases/2014/LaWS-shipboard-laser-uss-ponce.aspx

16 Jeff Hecht, "Photonic Frontiers: Rugged Battlefield Lasers. Ruggedizing high-energy lasers for the battlefield," *Laser Focus World 50* July 2014 http://www.laserfocusworld.com/articles/print/volume-50/issue-07/features/photonics-frontiers-rugged-battlefield-lasers-ruggedizing-high-energy-lasers-for-the-battlefield.html

17 Jeff Hecht, "Half a Century of Laser Weapons," *Optics & Photonics News 20* (2) 14-20 (Feb. 2009)

Index

CPSIA information can be obtained
at www.ICGtesting.com
Printed in the USA
BVOW06s0828071117
499685BV00032B/189/P